Environment and Livelihoods in Tropical Coastal Zones

Managing Agriculture–Fishery–Aquaculture Conflicts

Environment and Livelihoods in Tropical Coastal Zones

Managing Agriculture–Fishery–Aquaculture Conflicts

Edited by

Chu Thai Hoanh

International Water Management Institute (IWMI), Regional Office for South-east Asia, Penang, Malaysia

To Phuc Tuong

International Rice Research Institute (IRRI), Metro Manila, Philippines

John W. Gowing

University of Newcastle, Newcastle upon Tyne, United Kingdom

and

Bill Hardy

International Rice Research Institute (IRRI), Metro Manila, Philippines

In association with
the International Rice Research Institute (IRRI)
and
the International Water Management Institute (IWMI)

CABI is a trading name of CAB International

CABI Head Office
Nosworthy Way
Wallingford
Oxon OX10 8DE
UK

Tel: +44 (0)1491 832111
Fax: +44 (0)1491 833508
E-mail: cabi@cabi.org
Website: www.cabi.org

CABI North American Office
875 Massachusetts Avenue
7th Floor
Cambridge, MA 02139
USA

Tel: +1 617 395 4056
Fax: +1 617 354 6875
E-mail: cabi-nao@cabi.org

A catalogue record for this book is available from the British Library, London, UK.

A catalogue record for this book is available from the Library of Congress, Washington, DC.

Library of Congress Cataloging-in-Publication Data
Environment and livelihoods in tropical coastal zones : managing agriculture-fishery-aquaculture conflicts / edited by Chu Thai Hoanh ... [et al.].
p.cm. -- (Comprehensive assessment of water management in agriculture series ; 2)
Includes bibliographical references and index.
ISBN-13: 978-1-84593-107-0 (alk. paper)
ISBN-10: 1-84593-107-6 (alk. paper)
1. Aquaculture--Tropics--Congresses. 2. Agriculture--Tropics--Congresses. 3. Fisheries--Tropics--Congresses. 4. Coastal zone management--Tropics--Environmental aspects--Congresses. I. Chu, Thai Hoanh, 1949- II. Title. III. Series.

SH134.6.E58 2006
639.80913--dc22

2005033368

Typeset in 9/11 pt Palatino by Columns Design Ltd, Reading
Printed and bound in the UK by Biddles, Norfolk

Contents

Contributors vii

Preface xi

Series Foreword: Comprehensive Assessment of Water Management in Agriculture xiii

1. **Land and Water Management in Coastal Zones: Dealing with Agriculture–Aquaculture–Fishery Conflicts** 1
   J.W. Gowing, T.P. Tuong and C.T. Hoanh

2. **Adapting to Aquaculture in Vietnam: Securing Livelihoods in a Context of Change in Two Coastal Communities** 17
   C. Luttrell

3. **Livelihood Systems and Dynamics of Poverty in a Coastal Province of Vietnam** 30
   M. Hossain, T.T. Ut and M.L. Bose

4. **Social and Environmental Impact of Rapid Change in the Coastal Zone of Vietnam: an Assessment of Sustainability Issues** 48
   J.W. Gowing, T.P. Tuong, C.T. Hoanh and N.T. Khiem

5. **Brackish-water Shrimp Cultivation Threatens Permanent Damage to Coastal Agriculture in Bangladesh** 61
   Md. Rezaul Karim

6. **Coastal Water Resource Use for Higher Productivity: Participatory Research for Increasing Cropping Intensity in Bangladesh** 72
   M.K. Mondal, T.P. Tuong, S.P. Ritu, M.H.K. Choudhury, A.M. Chasi, P.K. Majumder, M.M. Islam and S.K. Adhikary

7. **Coastal Shrimp Farming in Thailand: Searching for Sustainability** 86
   B. Szuster

8. **Tracing the Outputs from Drained Acid Sulphate Flood Plains to Minimize Threats to Coastal Lakes** 99
   B.C.T. Macdonald, I. White, L. Heath, J. Smith, A.F. Keene, M. Tunks and A. Kinsela

9. **From Conflict to Industry – Regulated Best Practice Guidelines: a Case Study of Estuarine Flood Plain Management of the Tweed River, Eastern Australia** 107
I. White, M. Melville, B.C.T. Macdonald, R. Quirk, R. Hawken, M. Tunks, D. Buckley, R. Beattie, L. Heath and J. Williams

10. **Mangrove Dependency and the Livelihoods of Coastal Communities in Thailand** 126
E.B. Barbier

11. **Mangroves, People and Cockles: Impacts of the Shrimp-farming Industry on Mangrove Communities in Esmeraldas Province, Ecuador** 140
P. Ocampo-Thomason

12. **Interrelations among Mangroves, the Local Economy and Social Sustainability: a Review from a Case Study in North Brazil** 154
U. Saint-Paul

13. **Mangrove: Changes and Conflicts in Claimed Ownership, Uses and Purposes** 163
M.-C. Cormier-Salem

14. **Comparing Land-use Planning Approaches in the Coastal Mekong Delta of Vietnam** 177
N.H. Trung, L.Q. Tri, M.E.F. van Mensvoort and A.K. Bregt

15. **Applying the Resource Management Domain (RMD) Concept to Land and Water Use and Management in the Coastal Zone: Case Study of Bac Lieu Province, Vietnam** 193
S.P. Kam, N.V. Nhan, T.P. Tuong, C.T. Hoanh, V.T. Be Nam and A. Maunahan

16. **Developing a Consultative Bayesian Model for Integrated Management of Aquatic Resources: an Inland Coastal Zone Case Study** 206
E. Baran, T. Jantunen and P. Chheng

17. **Aquatic Food Production in the Coastal Zone: Data-based Perceptions on the Trade-off Between Mariculture and Fisheries Production of the Mahakam Delta and Estuary, East Kalimantan, Indonesia** 219
P.A.M. van Zwieten, A.S. Sidik, Noryadi, I. Suyatna and Abdunnur

18. **Managing Diverse Land Uses in Coastal Bangladesh: Institutional Approaches** 237
M. Rafiqul Islam

19. **Widening Coastal Managers' Perceptions of Stakeholders through Capacity Building** 249
M. Le Tissier and J.M. Hills

20. **Can Integrated Coastal Management Solve Agriculture–Fisheries–Aquaculture Conflicts at the Land–Water Interface? A Perspective from New Institutional Economics** 258
C. Brugere

21. **Responding to Coastal Poverty: Should We be Doing Things Differently or Doing Different Things?** 274
J. Campbell, E. Whittingham and P. Townsley

22. **Achieving Food and Environmental Security: Better River Basin Management for Healthy Coastal Zones** 293
S. Atapattu and D. Molden

Index 303

Contributors

Abdunnur, *Faculty of Fisheries and Marine Sciences, University of Mulawarman, Samarinda, East Kalimantan, Indonesia.*

S.K. Adhikary, *Khulna University, Khulna, Bangladesh.*

S. Atapattu, *International Water Management Institute (IWMI), PO Box 2075, Colombo, Sri Lanka.*

E. Baran, *WorldFish Center, PO Box 582, Phnom Penh, Cambodia.*

E.B. Barbier, *Department of Economics and Finance, University of Wyoming, Department 3985, 123 Ross Hall, 1000 E. University Avenue, Laramie, WY 82071-3985, USA.*

R. Beattie, *New South Wales Sugar Milling Co-operative Limited, 117 Pacific Highway, Broadwater, NSW 2472, Australia.*

V.T. Be Nam, *Integrated Resources Mapping Center (IRMC), Sub-National Institute for Agricultural Planning and Projection, 20 Vo Thi Sau Street, District 1, Ho Chi Minh City, Vietnam.*

M.L. Bose, *WorldFish Center (WFC), PO Box 500 GPO, 10670 Penang, Malaysia (formerly with the International Rice Research Institute, Philippines).*

A.K. Bregt, *Center for Geo-Information, Wageningen University, Droevendaalsesteeg 3, 6708 PB Wageningen, The Netherlands.*

C. Brugere, *Fishery Policy and Planning Division, Fisheries Department, Food and Agriculture Organization of the United Nations, Viale delle Terme di Caracalla, 00100 Rome, Italy.*

D. Buckley, *Tweed Shire Council, PO Box 816, Murwillumbah, NSW 2482, Australia.*

J. Campbell, *IMM Ltd., Innovation Centre, University of Exeter, Rennes Drive, Exeter EX4 4RN, United Kingdom.*

A.M. Chasi, *HEED-Bangladesh (Health, Education and Economic Development), Dhaka, Bangladesh.*

P. Chheng, *Inland Fisheries Research and Development Institute, PO Box 852, Phnom Penh, Cambodia.*

M.H.K. Choudhury, *Proshika Manobik Unnyan Kendra, Dhaka, Bangladesh.*

M.-C. Cormier-Salem, *UR169-IRD/MNHN, Département Hommes, Natures, Sociétés (HNS), CP26, 57 rue Cuvier 75231, Paris cedex 05, France.*

J.W. Gowing, *School of Agriculture, Food and Rural Development, University of Newcastle, Newcastle upon Tyne, NE1 7RU, United Kingdom.* E-mail: j.w.gowing@ncl.ac.uk

R. Hawken, *New South Wales Canegrowers, Condong Sugar Mill, Condong, NSW, Australia.*

L. Heath, *Centre for Resource and Environmental Studies, Australian National University, Canberra, ACT 0200, Australia.*

J.M. Hills, *Envision Partners LLP, University of Newcastle, Newcastle upon Tyne, NE1 7RU, United Kingdom.*

C.T. Hoanh, *International Water Management Institute (IWMI), Regional Office for South-east Asia, PO Box 500, GPO, 10670, Penang, Malaysia.*

M. Hossain, *Social Sciences Division, International Rice Research Institute (IRRI), DAPO Box 7777, Metro Manila, Philippines.*

M.M. Islam, *Khulna University, Khulna, Bangladesh.*

M. Rafiqul Islam, *Program Development Office for Integrated Coastal Zone Management, House 4/A, Road 22, Gulshan 1, Dhaka 1217, Bangladesh.*

T. Jantunen, *Environmental Consultant, Phnom Penh, Cambodia.*

S.P. Kam, *WorldFish Center (WFC), PO Box 500 GPO, 10670 Penang, Malaysia (formerly with the International Rice Research Institute, Philippines).*

Md. Rezaul Karim, *Urban and Rural Planning Discipline, Khulna University, Khulna, Bangladesh.*

A.F. Keene, *School of Environmental Science and Management, Southern Cross University, Lismore, NSW 2480, Australia.*

N.T. Khiem, *Faculty of Agricultural Economics, An Giang University, An Giang, Vietnam.*

A. Kinsela, *School of Biological, Earth and Environmental Sciences, University of New South Wales, Sydney, NSW 2052, Australia.*

M. Le Tissier, *Envision Partners LLP, University of Newcastle, Newcastle upon Tyne, NE1 7RU, United Kingdom.*

C. Luttrell, *Overseas Development Institute, 111 Westminster Bridge Road, London SE1 7JD, United Kingdom.* E-mail: c.luttrell@odi.org.uk

B.C.T. Macdonald, *Centre for Resource and Environmental Studies, Australian National University, Canberra, ACT 0200, Australia.*

P.K. Majumder, *Department of Agricultural Extension, Khulna, Bangladesh.*

A. Maunahan, *International Rice Research Institute (IRRI), DAPO Box 7777, Metro Manila, Philippines.*

M. Melville, *School of Biological, Earth and Environmental Sciences, University of New South Wales, NSW 2057, Australia.*

D. Molden, *International Water Management Institute (IWMI), PO Box 2075, Colombo, Sri Lanka.*

M.K. Mondal, *Bangladesh Rice Research Institute, Gazipur, Bangladesh.*

N.V. Nhan, *Integrated Resources Mapping Center (IRMC), Sub-National Institute for Agricultural Projection and Planning, 20 Vo Thi Sau Street, District 1, Ho Chi Minh City, Vietnam.*

Noryadi, *Faculty of Fisheries and Marine Sciences, University of Mulawarman, Samarinda, East Kalimantan, Indonesia.*

P. Ocampo-Thomason, *School of Geography, Politics and Sociology, University of Newcastle, Newcastle upon Tyne, NE1 7RU, United Kingdom.*

R. Quirk, *New South Wales Canegrowers, Condong Sugar Mill, Condong, NSW, Australia.*

S.P. Ritu, *Bangladesh Rice Research Institute, Gazipur, Bangladesh.*

U. Saint-Paul, *Centre for Tropical Marine Ecology (ZMT), Fahrenheitstr. 6, 28359 Bremen, Germany.*

A.S. Sidik, *Faculty of Fisheries and Marine Sciences, University of Mulawarman, Samarinda, East Kalimantan, Indonesia.*

J. Smith, *School of Biological, Earth and Environmental Sciences, University of New South Wales, Sydney, NSW 2052, Australia.*

I. Suyatna, *Faculty of Fisheries and Marine Sciences, University of Mulawarman, Samarinda, East Kalimantan, Indonesia.*

B. Szuster, *Department of Geography, University of Hawai'i at Manoa, 445 Social Sciences Building, 2424 Maile Way, Honolulu, HI 96822, USA.*

P. Townsley, *IMM Ltd., Innovation Centre, University of Exeter, Rennes Drive, Exeter EX4 4RN, United Kingdom.*

L.Q. Tri, *College of Agriculture, Can Tho University, 3/2 Street, Can Tho City, Vietnam.*

N.H. Trung, *College of Technology, Can Tho University, 3/2 Street, Can Tho City, Vietnam.*

M. Tunks, *Tweed Shire Council, PO Box 816, Murwillumbah, NSW 2484, Australia.*

T.P. Tuong, *Crop, Soil, and Water Sciences Division, International Rice Research Institute (IRRI), DAPO Box 7777, Metro Manila, Philippines.*

T.T. Ut, *Nong Lam University, Thu Duc, Ho Chi Minh City, Vietnam.*

M.E.F. van Mensvoort, *Laboratory of Soil Science and Geology, Wageningen University, Duivendaal 10, 6701AR Wageningen, The Netherlands.*

P.A.M. van Zwieten, *Aquaculture and Fisheries Group, Wageningen University, P.O. Box 338, 6800 AH Wageningen, The Netherlands.*

I. White, *Centre for Resource and Environmental Studies, Australian National University, Canberra, ACT 0200, Australia.*

E. Whittingham, *IMM Ltd., Innovation Centre, University of Exeter, Rennes Drive, Exeter EX4 4RN, United Kingdom.*

J. Williams, *New South Wales Department of Primary Industries, 1243 Bruxner Highway, Wollongbar, NSW 2477, Australia.*

Preface

The coastal zone is a big place: some 40% of the world's population lives within 100 km of the sea and this zone is under increasing pressure. Sustainable development and management of coastal zone resources are vitally important to human well-being, to national economies and to the ecosystems on which we depend.

In simple spatial terms, the coastal zone is the interface between the land and the ocean. It comprises inshore waters below low-tide level, intertidal areas and tracts of land above high-tide level. It is an area of transition where terrestrial and marine environments interact, characterized by a complex web of interactions among people, resources and ecosystems. This is a functional aspect of the definition rather than a simple spatial relationship, which is critical to our understanding of how it should be managed.

The coastal zone environment that is of particular interest to us is represented by river deltas, mangrove swamps, salt marshes and estuaries where the land–water interface is gradual, extensive and seasonally varying. It has the following characteristics:

- The aquatic environment is subject to seasonally varying salinity.
- The terrestrial environment is vulnerable to both tidal and riverine flooding.
- The natural resource base supports aquaculture, agriculture and fisheries.

The off-shore limit may be arbitrarily defined according to legal and administrative considerations, but the inland boundary requires more careful consideration because of the hydrological linkage between the coastal zone and inland river basins. There are examples of many coastal zones being affected by a reduced flow of fresh water and sediment as a result of dams, barrages and water diversions (e.g. Indus, Nile, Volta) that occur very long distances upstream.

The focus of the book is around the challenges people face in managing crops, aquaculture, fisheries and related ecosystems in inland areas of coastal zones in the tropics. A priority issue that emerges from the case studies presented here is the impact of change on poor people whose livelihoods depend upon open-access resources. Any development decision that aims at enhancing production from aquaculture and/or agriculture is likely to adversely affect access to and the productivity of these resources. Conflicts arise between different stakeholders and in this book we discuss the nature of these conflicts and identify what is known and not known about how to manage them. The book will therefore help planners, resource managers and donors to make better-informed investment decisions in connection with development of the coastal zone.

The chapters in this book were selected from papers presented at the International Conference on Environment and Livelihoods in Coastal Zones: Managing Agriculture–Fishery–Aquaculture Conflicts, organized in Bac Lieu, Vietnam, on 1–3 March 2005. We would like to express our thanks to the Comprehensive Assessment of Water Management in Agriculture, the Challenge Program on Water and Food, the WorldFish Center, the International Rice Research Institute (IRRI), the International Water Management Institute (IWMI), the many donors who support these institutes and programmes, the People's Committee of Bac Lieu Province, Vietnam, and Can Tho University, Vietnam, for their assistance in sponsoring and organizing the conference. Grateful acknowledgement is extended also to the many anonymous reviewers who provided invaluable assistance in the process of editing the papers into the versions that appear in this book.

The Editors

Series Foreword
Comprehensive Assessment of Water Management in Agriculture

There is broad consensus on the need to improve water management and to invest in water for food as these are critical to meeting the Millennium Development Goals (MDGs). The role of water in food and livelihood security is a major issue of concern in the context of persistent poverty and continued environmental degradation. Although there is considerable knowledge on the issue of water management, an overarching picture on the water–food–livelihoods–environment nexus is missing, leaving uncertainties about management and investment decisions that will meet both food and environmental security objectives.

The Comprehensive Assessment of Water Management in Agriculture (CA) is an innovative, multi-institute process aimed at identifying existing knowledge and stimulating thought on ways of managing water resources to continue meeting the needs of both humans and ecosystems. The CA critically evaluates the benefits, costs and impacts of the past 50 years of water development and challenges to water management currently facing communities. It assesses innovative solutions and explores consequences of potential investment and management decisions. The CA is designed as a learning process, engaging networks of stakeholders in producing knowledge synthesis and methodologies. The main output of the CA is an assessment report that aims to guide investment and management decisions in the near future, considering their impact over the next 50 years in order to enhance food and environmental security in supporting the achievement of the MDGs. This assessment report is backed by CA research and knowledge-sharing activities.

The primary assessment research findings are presented in a series of books that will form the scientific basis for the Comprehensive Assessment of Water Management in Agriculture. The books will cover a range of vital topics in the areas of water, agriculture, food security and ecosystems – the entire spectrum of developing and managing water in agriculture, from fully irrigated to fully rainfed lands. They are about people and society, why they decide to adopt certain practices and not others and, in particular, how water management can help poor people. They are about ecosystems – how agriculture affects ecosystems, the goods and services ecosystems provide for food security and how water can be managed to meet both food and environmental security objectives. This is the second book in the series, the first being *Water Productivity in Agriculture: Limits and Opportunities for Improvement.*

Effectively managing water to meet food and environmental objectives will require the concerted action of individuals from across several professions and disciplines – farmers, fishers, water managers, economists, hydrologists, irrigation specialists, agronomists and social scientists. The material presented in this book represents an effort to bring a diverse group of people together to present a truly cross-disciplinary perspective on water, food and environmental issues within the coastal zone. The complete set of books should be invaluable for resource managers, researchers and field implementers. These books will provide source material from which policy statements, practical manuals and educational and training material can be prepared.

The CA is performed by a coalition of partners that includes 11 Future Harvest agricultural research centres supported by the Consultative Group on International Agricultural Research (CGIAR), the Food and Agriculture Organization of the United Nations (FAO) and partners from some 80 research and development institutes globally. Co-sponsors of the assessment, institutes that are interested in the results and help frame the assessment, are the Ramsar Convention, the Convention on Biological Diversity, FAO and the CGIAR.

For production of this book, financial support from the governments of The Netherlands and Switzerland for the Comprehensive Assessment is appreciated.

David Molden
Series Editor
International Water Management Institute
Sri Lanka

1 Land and Water Management in Coastal Zones: Dealing with Agriculture–Aquaculture–Fishery Conflicts

J.W. Gowing,[1] T.P. Tuong[2] and C.T. Hoanh[3]
[1]*School of Agriculture, Food & Rural Development, University of Newcastle, Newcastle upon Tyne, United Kingdom, e-mail: j.w.gowing@ncl.ac.uk*
[2]*International Rice Research Institute, Metro Manila, Philippines*
[3]*International Water Management Institute, Regional Office for South-east Asia, Penang, Malaysia*

Abstract

The coastal environment has undergone rapid change in recent times. Change in the state of the environment is multifaceted, but a key concern is the way that natural habitats – principally mangrove forests and salt marshes – have been extensively cleared and converted to shrimp farming and other uses. The expansion of shrimp farming has also encroached onto agricultural lands. Coastal shrimp farming has been practised for a very long time in some countries as part of the traditional livelihood system, but recent strong demand in global markets, together with technological advances, has fuelled rapid expansion. These rapid, and generally unplanned changes, have provoked conflicts among the three dominant resource-dependent livelihoods in the inland coastal zone: agriculture, shrimp farming and fishing.

The coastal zone is characterized by ambiguities of resource ownership and a complex web of interactions among people, resources and ecosystems. Conflicts exist between the drive for short-term financial gain and the desire for long-term sustainable development. Conflicts exist between the priorities of people who derive their livelihoods from aquaculture and those who depend upon agriculture. Conflicts exist between the needs of people who may gain from intensification of land use for agriculture and/or aquaculture and other people (e.g. fisherfolk), whose livelihoods may be adversely affected by environmental impacts. This synthesis chapter presents a discussion of trends, problems and approaches to managing change in the inland coastal zone. We identify key messages from previous research and development experience and consider the supporting evidence for these messages.

Introduction

Coastal zones are home to 40% of the world's population and support much of the world's food production and industrial, transportation and recreation needs, while also delivering vitally important ecosystem services. The coastal environment is under pressure and has undergone rapid change in recent times. The scale of the stresses imposed on this environment poses a threat to the resilience of both natural and human systems. Driving forces are demographic, economic, institutional and technological. These build up

 Environment and Livelihoods in Tropical Coastal Zones (eds C.T. Hoanh, T.P. Tuong, J.W. Gowing and B. Hardy)

environmental pressure through land-use change, intensification of resource exploitation, urbanization, industrial development, tourism and recreational demand. Changes occurring in the state of the environment include altered nutrient, sediment and water fluxes; degradation of habitats and loss of biodiversity; and pollution of soils, groundwater and surface water. These in turn affect human welfare through their effects on productivity, health and amenity.

One of the key issues is land-use change, in particular the rapid growth of shrimp aquaculture. Change in the state of the environment is multifaceted, but a key concern is the way that natural habitats – principally mangrove forests and salt marshes – have been extensively cleared and converted to shrimp farming and other uses. The reduction in area occupied by mangrove forest is well documented and has provoked widespread concerns over environmental and social impacts. However, it is important to recognize that recent expansion of shrimp farming has also encroached onto agricultural lands (Karim, Chapter 5, this volume; Szuster, Chapter 7, this volume). This is the second key issue in the coastal zone, but it is important to see both in the context of the range of conflicting demands of the different stakeholders who live within and depend upon the resource base in this environment.

In this synthesis chapter, we discuss trends, conditions, responses and scenarios for the coastal zone. A problem analysis follows in which we examine the main environmental and social impacts of change. We then identify key messages from previous research and development experience and provide supporting evidence for these messages. Crucial to the achievement of a more sustainable approach is the adoption of appropriate evidence-based policy for which we identify knowledge gaps.

Land-use Change in Coastal Zones

Shrimp-farming trends

Extensive shrimp farming has been practised for a very long time in some countries as part of the traditional livelihood system, but recent strong demand in global markets together with technological advances has fuelled rapid expansion and intensification. The annual percentage rate of growth at 17% between 1970 and 2000 was considerably higher than that of other food production sectors. However, the double-digit growth rates of 23% in the 1970s and 25% in the 1980s slowed to 7% in the 1990s (FAO, 2003). To some commentators (e.g. Fegan, 1999) this is seen as a success story, having developed from a cottage industry based upon a backyard production system to a global industry in little more than 30 years; to others (e.g. EJF, 2003), this rapid growth is associated with serious negative environmental and social impacts.

In 2000, brackish-water aquaculture comprised 4.6% by weight of total aquaculture production but 15.7% by farm-gate value. Since 1970, shrimp farming has emerged as a major source of foreign earnings and an important source of income and employment. Estimates for the main shrimp-producing countries put the total employment generated by shrimp farming at around 2 million people. The top 25 producer countries are listed in Table 1.1. Of these, the top ten account for 90% of world production.

Shrimp-farming methods are classified according to the level of technology adopted, stocking density and yield. Terminology varies between sources and countries, but we can recognize the following types:

Extensive: traditional methods rely on natural recruitment of shrimp postlarvae from wild sources and natural productivity of the ecosystem; built-in intertidal areas with water exchange by tidal action; pond size typically >10 ha; trap and hold wild shrimp at a density of 1–3/m^2; yield typically less than 200 kg/ha/year.

Semi-intensive: the first stage of development usually involves some stocking of shrimp postlarvae from a hatchery; natural productivity may be enhanced by fertilizers and occasionally some use of feeds; pond size 2–10 ha; water exchange usually provided mainly by tidal action, supplemented by low-lift axial-flow pumps; stocking density of 3–10/m^2; yield typically 1000–2500 kg/ha/year.

Table 1.1. The top 25 producers of farmed shrimp in 2000 by weight and value (from FAO, 2003).

No.	Country	Production (mt)	Production (000 US$)
1	Thailand	299,700	2,125,384
2	China	217,994	1,307,964
3	Indonesia	138,023	847,429
4	India	52,771	393,938
5	Vietnam	69,433	319,392
6	Ecuador	50,110	300,660
7	Philippines	41,811	271,385
8	Bangladesh	58,183	199,901
9	Mexico	33,480	194,184
10	Brazil	25,000	175,000
11	Malaysia	15,895	124,577
12	Colombia	11,390	91,120
13	Sri Lanka	6,970	78,342
14	Taiwan	7,237	60,483
15	Honduras	8,500	59,500
16	Venezuela	8,200	34,030
17	Australia	2,799	27,557
18	Madagascar	4,800	24,000
19	Nicaragua	5,411	17,423
20	USA	2,163	14,513
21	Belize	2,648	12,710
22	New Caledonia	1,723	12,061
23	Costa Rica	1,350	11,475
24	Panama	1,212	6,399
25	Peru	512	3,741

mt = metric tonnes

Intensive: progression to advanced production systems relies on artificial stocking at high density (>10/m^2) in small ponds (1–2 ha) with heavy feeding rate; involves mechanical aeration; sometimes incorporates water recirculation and/or treatment; generally above high-tide level to allow drainage and drying of pond bottom between crops; yield of over 7500 kg/ha/year possible with multiple cropping, but 5000 kg/ha/year is typical.

Mangrove trends

Mangrove forests occupy intertidal areas along tropical and subtropical coasts, especially where large river systems deposit alluvial sediment and salinity is moderated by high freshwater discharge. They therefore represent a dominant natural ecosystem of tropical and some temperate coastal zones. The global extent of mangroves has been estimated at 181,077 km^2 (Spalding *et al.*, 1997), but this represents a much reduced area compared with their undisturbed extent. Historically, they have been exploited for forest products or converted to various alternative forms of land use (salt pans, aquaculture ponds, agriculture, urbanization and industrial development). In some countries, the reduction exceeds 50% (Table 1.2).

Shrimp farming has no doubt contributed to the overall loss of mangroves, as documented in some countries, but it is by no means the only factor. For example, mangrove loss in Thailand over the last two decades has been exacerbated by the expansion of shrimp farming (see Szuster, Chapter 7, this volume). In Chakoria, Bangladesh, shrimp areas expanded from 10,000 ha to 75,000 ha between 1967 and 1988, at the expense of a decrease in mangrove area from

Table 1.2. Estimated loss, in % of the original forest cover, of mangrove area in selected countries (from WRI, 2000).

	Loss (%)
Asia	
Brunei	20
Indonesia	55
Malaysia	74
Myanmar	75
Pakistan	78
Philippines	67
Thailand	84
Vietnam	37
Africa	
Angola	50
Côte d'Ivoire	60
Gabon	50
Guinea Bissau	70
Tanzania	60
Latin America	
Costa Rica	0
El Salvador	8
Guatemala	31
Mexico	30
Panama	67
Peru	25

70,000 ha to 7,000 ha (see Islam, Chapter 18, this volume). Many commentators (e.g. Gujja and Finger-Stich, 1996; Janssen and Padilla, 1999; Nickerson, 1999; EJF, 2003) have attributed much of the global loss to the expansion of shrimp farming, but the evidence for this assertion is not always strong because the unavailability or unreliability of data prevents the assessment of the true extent of the link worldwide. In-depth studies of the history of mangrove exploitation are few (see, for example, Walters, 2003).

The relationship between the decline of mangrove and expansion of shrimp farming is examined by Lewis *et al.* (2003), who showed that in Thailand half of the mangrove area present in 1960 had been lost before the shrimp boom in the 1980s. Elsewhere in South-east Asia, conversion for salt production, urbanization and agriculture has removed large areas. Within southern Asia, mangroves in India and Bangladesh have been heavily exploited for timber, fuelwood and other forest products for centuries, and population pressure has led to serious forest degradation. These countries also show evidence that before the expansion of shrimp farming large tracts of mangrove had already been converted to rice farming. In Africa, it is apparent from Table 1.2 that the extent of mangrove destruction is comparable with that in Asia even though coastal aquaculture in general, and shrimp farming in particular, is not widespread. A comprehensive survey of 5000 shrimp farms in Asia (ADB/NACA, 1997) showed that less than 20% of the area occupied by intensive and semi-intensive farms was former mangrove. Most of the shrimp farms on former mangrove land were extensive ponds. The estimated total of 400,000 ha of ponds on former mangrove land represents only 5% of the mangrove resource. In an analysis conducted for the World Wide Fund for Nature, Clay (1996) concluded that '... the extent of mangrove destruction worldwide resulting from shrimp farming is only a tiny fraction of the total lost to date ...'. Nevertheless, conflicts do exist between shrimp farmers and mangrove resource users.

Conversion of rice farms

Coastal rice lands in tropical regions often suffer from saline intrusion that prevents crop production in the dry season. This is a natural phenomenon in deltaic and estuarine environments because of seasonally varying freshwater input; however, it may be aggravated by upstream river basin management (see Atapattu and Molden, Chapter 22, this volume). Agricultural lands in the brackish-water zone generally have lower productivity than those in the freshwater zone. As demand for shrimp increased, many farmers found that shrimp farming could bring them higher income than agriculture and they converted their rice fields into shrimp ponds. It is difficult to estimate the extent to which brackish-water shrimp farming has encroached on to agricultural land worldwide, but available data from a few countries illustrate the significance of the trend.

A large number of rice farmers in central Thailand converted irrigated paddy fields into shrimp ponds during the latter half of the 1990s (Szuster *et al.*, 2003). Surveys conducted by the Thai Department of Land Development and the Department of Fisheries suggested that shrimp farms operating within freshwater areas could have accounted for as much as 40% (or approximately 100,000 t) of Thailand's total farmed shrimp output in 1998 (Limsuwan and Chanratchakool, 1998). In Bangladesh, because of commercial interests of shrimp, many coastal polders constructed since the 1960s to protect agricultural land from inundation of salt water were turned into large shrimp culture *ghers* (ponds) during the 1990s. Shrimp area expanded from 51,812 ha in 1983 to 137,996 ha in 1994 and to 141,353 ha in 2002 (DoF, 1995, 2003). In the coastal zone of the Mekong River Delta of Vietnam, rice area decreased from 970,000 ha in 2000 to 800,000 ha in 2002, whereas shrimp area increased from 230,000 ha to 390,000 ha in the same period (MNRE, 2002, unpublished).

Problem Analysis

Coastal zones support three distinct types of resource-dependent livelihood: agriculture, shrimp farming and fishing (and/or extraction of other common-property resources). There are many cautionary tales about environmental and social problems arising from the impacts of land-use change (see Table 1.3). In order to promote evidence-based policy, a worthwhile problem analysis depends upon recognizing and evaluating interactions (and trade-offs) among agriculture, aquaculture and fisheries in this environment. While the focus here is necessarily on negative impacts of change, this is not inevitable and the aim is to identify possibilities for co-existence and win-win scenarios for future resource use.

Bailey and Pomeroy (1996) argue that the complexity and high natural productivity of the environment provide many niches for these different activities, but sustainable development depends upon their co-existence rather than specialization. Lewis *et al.* (2003) similarly argue that the promotion of coastal aquaculture in an environmentally (and socially) responsible manner requires 'adopting the principles of co-existence of mangroves and aquaculture'. They note that 'implying co-existence is possible and documenting its actual occurrence are two different things'.

Environmental impacts

Aquaculture's effects on mangrove resources include cutting trees and clearing land, hydrological changes due to the construction of canals and roads, and the spread of disease to wild shrimp (Lewis *et al.*, 2003). Other environmental impacts related to the development of extensive shrimp culture in mangrove forests include coastal erosion, saline intrusion into agricultural lands, decrease in shrimp postlarvae and mud crab, increased malarial incidence in coastal areas and acidification of soils and waters (Boyd and Clay, 1998; GESAMP, 1991; Paez-Osuna, 2001). Several authors have pointed out the irony that mangrove destruction itself is sometimes the main reason for the unsustainability of shrimp farming because of erosion, loss of natural productivity, water acidity and contamination. As a consequence, some extensive shrimp farm developments have been abandoned (Dierberg and Kiattisimkul 1996) and environmental activists have criticized such occurrences, calling them 'slash-and-burn' exploitation or 'swidden aquaculture'.

Table 1.3. Common environmental and human impacts of aquaculture development in the coastal zone.

Environmental problems	Human problems
Destruction of mangrove, wetlands and other sensitive aquatic habitats	Restricted access to common-property resources
Water pollution resulting from pond effluents, excessive use of bio-active chemicals in aquaculture ponds, excessive use of pesticides and fertilizers in agriculture	Loss of land because of indebtedness or coercion
Salinization of land and water by drainage and seepage from ponds	Reduced employment opportunities for landless people
Acidification arising from development of acid sulphate soils	Loss of subsistence fishery
Spread of aquatic animal diseases to native populations	Increased vulnerability as a result of less diverse sources of income
Negative effects on biodiversity caused by exploitation of wild shrimp larvae/ brood-stock and destruction of habitat	Health and social impacts arising from degraded domestic water supply
Negative impact on vegetation cover and terrestrial livestock	Higher economic values but increased inequity and social unrest

Where shrimp ponds have expanded into rice farms, salinization of soil and water is a major concern (Szuster and Flaherty, 2002). This is particularly detrimental when shrimp farms encroach on to the originally freshwater area as in Thailand, where rice farmers realized that the potentially high profits derived from shrimp production could easily offset the costs associated with trucking salt water to their land. Seepage and percolation from shrimp ponds can salinize adjacent rice fields and the long-term build-up of salt threatens the sustainability of agriculture on neighbouring farms. This is not the case where shrimp farms are located in the brackish-water zone, where salinity intrusion is a natural seasonal phenomenon, such as in the Mekong Delta of Vietnam (Phong *et al.*, 2003).

The adoption of shrimp–rice production systems in the brackish-water zone may also lead to the encroachment of shrimp ponds onto homestead lands. Karim (Chapter 5, this volume) reports that in Bangladesh before 1975 most shrimp farms (>80%) were located further than 500 m from homestead areas, but, in 1999, 46% of the shrimp ponds were found within 10 m of homestead land. Fruit trees and many plant species have gradually decreased because of salinity and the shrinking of homestead areas. Similarly, grazing land and its vegetation cover also declined.

One of the key concerns is the effect of the periodic discharge of shrimp-pond water that contains high concentrations of suspended solids, nutrients and bio-active chemicals. This occurs as natural drainage after heavy rain, when ponds are emptied at the end of the season, and when water is exchanged during the season. The discharge of high loads of nutrients and suspended solids has the potential to have adverse effects on the receiving waters, including the stimulation of algal blooms and the creation of anoxic conditions (Naylor *et al.*, 1998). Graslund and Bengtsson (2001) provide a comprehensive review of chemicals used in shrimp farming and the potentially adverse effects of discharges into the environment, but in general their impact on coastal waters is poorly documented in rigorous scientific studies. The characteristics of shrimp-farm discharges are qualitatively different from agricultural and urban effluents. The resulting discharge water has high concentrations of inorganic particles, phytoplankton, particulate and dissolved organic compounds, and ammonium derived from feeds. Burford *et al.* (2003) linked ecological processes in intensive shrimp ponds with impacts downstream, but it should be noted that extensive and low-level semi-intensive shrimp farms do not cause appreciable chemical discharge pollution.

Just as shrimp ponds may pollute their environment and provoke downstream problems, they may also suffer from poor water quality due to upstream users. There is much debate about this as an issue affecting decisions about the appropriate intensity of shrimp-farm development in a given area (i.e. carrying capacity), but the wider issue of land use within the river basin merits consideration. A case study from Honduras (Dewalt *et al.*, 1996) is illustrative in that it examines the dispute between shrimp producers and other people from coastal communities (farmers and fishermen), which has led to serious confrontations. Their study demonstrates the importance of a wider perspective in that growing population and increased intensity of farming, especially in upstream hillside communities, is seen to have contributed to increased sediment and pesticide loads in the coastal environment. Environmental change in the coastal zone may be the result of actions farther up the river basin.

The development of acid sulphate soils (ASS) for both aquaculture and agriculture also merits consideration. Such soils are associated with inland coastal zones (salt marshes, mangrove forests and other estuarine wetlands) and when oxidized these pyrite (FeS_2)-rich sediments generate sulphuric acid. On-site impacts affect shrimp ponds (Sammut, 1999) and agricultural fields (Minh *et al.*, 1998) because of low pH and high concentrations of aluminium and iron. Cultural practices developed to reclaim these soils depend upon liming and leaching of toxic substances. Leaching results in the transfer of acidity to the surroundings, leading to severe acidification of the local aquatic environment (Minh *et al.*, 1997). Acid can also be exported farther into tidal creeks and estuarine waters, where mass mortalities of fish have been recorded, and there is evidence of chronic long-term effects on coastal habitats (Sammut *et al.*, 1996; Wilson *et al.*, 1999; see also Macdonald *et al.*, Chapter 8, this volume).

Human impacts

Socio-economic impacts associated with change in resource-dependent livelihoods have often been underestimated or oversimplified. This can perhaps be explained in part by failures to recognize off-site effects, which occur when the boundaries of the system under consideration are not properly defined (Phillips *et al.*, 2001). The nature and extent of impacts inevitably differ between countries, but general lessons do emerge.

In Bangladesh, shrimp farming has become a major export industry, but concern has grown about negative socio-economic impacts (Deb, 1998). Shrimp farming itself is less labour-intensive than rice cultivation (Deb (1998) estimates a 75% reduction), thus giving rise to concern for impact on poor people whose livelihoods depend on selling labour. However, the overall labour requirement of the shrimp industry is higher than that of rice production because of the level of employment in ancillary activities. In 1990, total on-farm and off-farm labour requirements were 22.6 million person-days, and the corresponding figure for 2005 was projected to be 60 million. It is logical to assume that the shrimp industry should absorb the surplus rural labour force in coastal areas, but in reality benefits to local people are less because many shrimp producers prefer hiring labour from outside. Social tension arises because of this and also because of coercive methods (seizure and intimidation) adopted by investors wishing to gain access to land for conversion into shrimp ponds (see also Karim, Chapter 5, this volume).

The social impact of shrimp farming in India was assessed by Patil and Krishnan (1997), who surveyed 26 coastal villages in Andhra Pradesh. Respondents were asked to rank the degree of severity of specific factors arising in shrimp development areas. For fishing communities, blocked access to the beach and saline well water were scored as the most severe problems. Salinization of land and shortages of fodder and fuelwood were the main problems identified in farming communities, together with the problem of saline well water. Similar issues have been identified in Bangladesh.

The social unrest arising from the socio-economic impacts reported in Bangladesh and India is less evident in Vietnam, but similar underlying problems have been reported (EJF, 2003; Hoanh *et al.*, 2003). Results from a study involving ecosystem changes from agriculture to aquaculture in Quang Ninh Province of Vietnam (Adger *et al.*, 2000) demonstrate that conversion of part of a mangrove forest for agriculture and aquaculture affects property rights and imposes additional stress on local livelihoods. There is evidence of increased inequality since poorer people are more dependent on common-property resources that are degraded or made less accessible. It appears that common-property management of the remaining mangrove and fishing areas is also undermined by the changes in property rights and inequitable benefits derived from enclosure and conversion (see also Luttrell, Chapter 2, this volume, and Ocampo-Thomason, Chapter 11, this volume).

In spite of widely reported environmental and social impacts, the potential for substantial profits attracts both local farmers and outside entrepreneurs, and shrimp farming continues to expand and to dominate the debate on land use in the coastal zone. The debate has tended to polarize between those who emphasize the economic benefits and those who emphasize negative impacts. Planning for sustainable development requires consideration of both perspectives and trade-offs between them (GESAMP, 2001). The starting point should be the recognition that private and social benefits often diverge. Be *et al.* (1999) outlined this conflict in the context of an analysis of alternative land uses (shrimp monocrop, rice monocrop and rice–shrimp) for the Mekong Delta in

Vietnam. They identified the critical policy issue that farmers have not received appropriate signals about the cost of externalities associated with private investment decisions. This point has been echoed by many commentators on the shrimp–mangrove conflict (e.g. Janssen and Padilla, 1999; Huitric *et al.*, 2002). It is sometimes argued (e.g. Fegan, 1999) that this has resulted from the 'gold-rush mentality' associated with the early stages of an immature industry. The challenge therefore is to move quickly to put in place the measures necessary to develop a sustainable industry.

Approaches to Managing Change

Regulating farm operations

Governments have responded mainly with specific regulations relating to shrimp-farm operation (such as effluent limits, design standards, best management practices and codes of conduct). Many tropical nations (e.g. Belize, Brazil, Ecuador, India, Mexico, Thailand and Venezuela) have made aquaculture effluent regulations, which are designed to prevent effluents from causing negative impacts on receiving waters. These farm-level measures have often been ineffective. Some non-government organizations have also proposed effluent standards for aquaculture. Among them, the Global Aquaculture Alliance (GAA) has suggested that members adopt environmentally responsible culture methods to comply with effluent standards. These standards consist of initial, rather lenient, limits, and stricter target limits with which the members should comply within 5 years (Boyd and Gautier, 2000).

A large number of producer associations, governmental fishery agencies, international development organizations, environmental non-government organizations and others have formulated codes of conduct for aquaculture (Boyd *et al.*, 2002). A code of conduct in its most basic form is a set of guiding principles consisting of broad statements about how management and other operational activities should be conducted. Most aquaculture codes reference the Code of Conduct for Responsible Fisheries presented by the Food and Agriculture Organization (FAO) of the United Nations (FAO, 1995, 1997) and the general principles of the codes usually reflect those of the FAO code. Most codes do not have any legal authority, and adoption is usually voluntary.

Given that tens of thousands of small farms are operated by individuals with relatively little technical knowledge, it is virtually impossible to effectively regulate aquaculture effluents by applying traditional water quality standards. An alternative is to require the application of specific practices called best management practices (BMPs). A BMP is the best available and practical means of preventing a particular environmental impact while still allowing production to be economically efficient. The best inducement is when adoption of BMPs clearly increases profit. Thus, BMPs should be related back to farm economic performance. For example, suppose that the BMPs are to lower stocking rates and use better feed management. The lower stocking rates and lower feed inputs will result in better water quality, less stress, faster growth, better feed conversion ratios and less waste produced. This scenario will also increase efficiency and profits. Another example is the storage of rainfall in ponds to avoid overflow. Less overflow means that less water will need to be pumped into ponds to maintain water levels. A reduction in pump operation will reduce costs and increase profits.

The main disadvantages of relying on codes of practice are summarized below:

- Adoption is voluntary, so some producers may not follow codes of conduct despite promotional efforts.
- Producers who adopt a code of conduct may selectively adopt BMPs and avoid those that are expensive or difficult to implement.
- There are many obstacles to effective self-evaluation and third-party verification.
- Small producers may lack technical knowledge for using BMPs, and education and training will be difficult and expensive.

- Implementation of programmes could be slow and result in substantial costs to farmers.
- Effectiveness of BMPs in codes of conduct is assumed, but monitoring is needed to verify this assumption.
- Unless all stakeholders are involved in preparing codes of conduct, the BMPs may not address significant socio-economic issues.

Because of the necessity of preserving mangroves and recycling aquaculture wastes, some researchers have proposed integrated shrimp–mangrove systems (Robertson and Phillips 1995; Dierberg and Kiattisimkul, 1996). Expected benefits of integrated systems include enhancement of coastal fisheries, minimization of contamination of the coastal environment and provision of a higher-quality water supply for shrimp farming. Integrated mangrove–shrimp farming systems have the advantage of combining mangrove conservation with the high-income potential of aquaculture (Macintosh, 1998). One approach is to transform current extensive shrimp farming into 'silvo-fishery' systems (Macintosh, 1998). The Indonesian *tambak* is a traditional form of integrated system in which extensive aquaculture is sustained by mangrove productivity (Hambrey, 1996; Macintosh, 1998). Binh *et al.* (1997) demonstrated that integrated mangrove–shrimp farms (mangrove covering 30–50% of the pond area) in Vietnam have higher economic returns than farms where mangrove had been cleared. Johnston *et al.* (1999) investigated yields of shrimp and wood from mixed systems in Vietnam, but raised concerns over their sustainability.

Another way of integrating shrimp ponds and mangrove areas is to discharge pond effluents into a mangrove wetland, which is used as a biofilter to remove suspended solids, lower BOD[1] and absorb nutrients in order to limit the risk of eutrophication of the adjacent waters (Twilley, 1992; Robertson and Phillips, 1995; Rivera-Monroy *et al.*, 1999). How mangrove forests work as sinks for phosphorus and nitrogen is poorly understood, but Corredor and Morell (1994) reported their effectiveness in removing nutrients from effluents. However, it is not possible to make any general recommendation about an appropriate ratio of mangrove–shrimp pond area while the nutrient assimilation capacity of different kinds of sediments and plants remains unknown (Gautier, 2002).

Integrated coastal zone management

Although some problems can be addressed at the farm level, many problems require strategic intervention at a wider landscape or basin scale and call for collective action (see also Szuster, Chapter 7, this volume). In many ways, what we have is a classic example of why integrated coastal zone management (ICZM) is needed:

- Coastal aquaculture commonly straddles the boundary between land and sea.
- Resource (land, water) ownership or rights allocation, and related administration, is often complex or ambiguous in prime aquaculture locations.
- Aquaculture may be seriously affected by water quality and habitat degradation caused by other activities.
- Aquaculture itself may affect environmental quality and the interests of other users through conversion of natural habitat; through pollution of recipient waters with nutrients, organic substances and potentially toxic (hazardous) chemicals; and through the spread of disease.
- Poorly planned aquaculture may result in negative feedback and self-pollution.

Unfortunately, there are few clear examples of the successful integration of aquaculture into comprehensive ICZM. It is arguable that this is because there have been very few genuinely integrated initiatives, where aquaculture has been assessed alongside the full range of existing or potential activities in the coastal zone using consistent and rational assessment criteria, agreed upon across a

[1] BOD, biochemical oxygen demand.

range of interests and agencies. To do this thoroughly takes time, however, and this poses a dilemma in many developing-country situations where aquaculture is developing very rapidly. The case of Ecuador, where population pressure, industrial development and shrimp farming have had significant negative impacts on estuarine resources throughout a period in which a long-term ICZM project was under way, is particularly notable (GESAMP, 2001). Also, shrimp farming has developed uncontrollably in Sri Lanka, with adverse environmental consequences and self-pollution, despite a strong ICZM awareness and a variety of initiatives in place (Nichols, 1999).

Based on a review of experience, GESAMP (2001) concluded that comprehensive ICZM may be effective as a starting point where coastal aquaculture is in the early stages of development, where institutions for resource management are flexible or undeveloped and where appropriate legal and institutional frameworks are in place or can be developed rapidly. The available scientific and technical capacity is often a constraint and there is a need to develop appropriate planning tools (see Trung *et al.* and Baran *et al.*, Chapters 14 and 16, respectively, this volume, for examples). However, technical competence does not guarantee success, since institutional inertia may mean that planning authorities do not respond quickly to rapidly changing circumstances (Hoanh *et al.*, 2003). Also, well-laid plans are often undermined by the strength of economic and political interests. Evidence from Thailand (GESAMP, 2001) suggests that more locally focused initiatives (e.g. relating to an estuary or lagoon system) may offer the most practical starting point, since they retain the benefits of integration but at a smaller scale.

An integrated strategy for sustainable development might include

- zones with development and environmental objectives specifically related to aquaculture and other compatible activities, and
- allocation of environmental capacity, in terms of waste production/emission limits, for aquaculture and other activities within these zones.

Zoning (an *allocation of space)* implies bringing together the criteria for locating aquaculture and other activities in order to define broad zones suitable for different activities or mixes of activities. Geographic information systems (GIS) are particularly well suited to facilitating this task (see Kam *et al.*, Chapter 15, this volume). Zoning may be used either as a source of information for potential developers (for example, by identifying those areas most suited to a particular activity) or as a planning and regulating tool, in which different zones are identified and characterized as meeting certain objectives. Zoning of land (and water) for certain types of aquaculture development may

- help to control environmental deterioration at the farm level,
- reduce adverse social and environmental interactions,
- serve as a focus for estimates of environmental capacity and
- serve as a framework for providing or improving infrastructure to small-scale farmers.

The strength of zoning lies in its simplicity, its clarity and its potential in terms of streamlining procedures (see Islam, Chapter 18, this volume). For example, once a zone is established and objectives defined, then developments that meet the objectives and general conditions for the zone may need no further assessment (such as an environmental impact assessment). What is allowed and what is not allowed is clear, and developers can plan accordingly. Any monitoring required can be applied to the whole zone rather than to individual farms. Its weakness lies in its rigidity, and farmers must adapt to the situation within the zone. No zone is perfect, land/water capability assessment may have been inadequate, boundaries are frequently arbitrary and conditions may change. Flexibility and farmer choice are limited by the zone criteria. On the other hand, the task of catering to highly diverse needs is also quite difficult, and can break down. There may be small pockets of land or water of high potential for aquaculture that were not recognized in the resource-assessment process. Exclusion of these lands from an aquaculture zone could prevent appropriate development, subject it to inappropriate regulation or restrict access of poor people to opportunities for aquaculture development. Furthermore, zoning may actually be undesirable for encouraging a concentration of aquaculture because of the associated environmental and social impacts.

Environmental capacity measures the resilience of the natural environment in the face of impact from human activities. Some assessment of environmental capacity is desirable and is of particular relevance to the problem of cumulative effects. It has been argued (GESAMP, 2001) that environmental capacity must be assessed, even if only at the most elementary level, if sustainable development is to have any practical meaning. Environmental capacity (otherwise referred to as assimilative capacity) is 'a property of the environment and its ability to accommodate a particular activity or rate of an activity ... without unacceptable impact' and must be measured against some established standard of environmental quality. In the case of aquaculture, it will be applied to a specified area (e.g. a bay, lagoon or estuary) and might be interpreted as

- the rate at which nutrients can be added without triggering eutrophication,
- the rate of organic flux to the benthos without major disruption to natural benthic processes and
- the rate of dissolved oxygen depletion that can be accommodated without causing mortality of the indigenous biota.

A set of planning interventions in the form of incentives and constraints (planning regulations) will be required to implement the strategy and ensure that objectives are met, standards are not breached and environmental capacity is not exceeded. These might apply to

- location of aquaculture development,
- waste emissions,
- the quantity or quality of inputs used (e.g. food, chemicals) and
- design, technology and management practices.

Given the nature of coastal aquaculture as a mainly small-scale activity, the implementation of recommendations may be difficult for farmers, and the enforcement of regulations difficult for authorities (see, for example, Murthy, 1997). This may be made more effective if responsibility for design, implementation and enforcement is located at the proper administrative level, and full use is made of self-management and self-enforcement capacity by industry and farmers' associations (see White *et al.*, Chapter 9, this volume).

Incentives, on the other hand, do not suffer from problems of evasion and non-compliance, and in some cases can be used to stimulate innovation leading to more environmentally friendly technologies. The use of economic instruments to influence both siting and operation holds considerable promise. Although some positive incentives may be costly, it should be possible to pay for them with negative incentives (e.g. taxes on undesirable locations, activities, technologies). However, incentives may need to be underpinned or reinforced through complementary regulation.

Environmental impact assessment (EIA) is a standard planning tool for evaluating the potential consequences of development decisions, and has been used widely in coastal management. Strategic environmental assessment (SEA) is a relatively recent tool that has been developed to evaluate the environmental effects of policies, plans, programmes and other strategic actions. The likely environmental and social impacts of a range of technologies or development options in different locations can be compared, and planning interventions to minimize environmental impact can be devised. Alongside EIA and SEA, properly informed planning requires consideration of the impact of development decisions on the livelihoods of people who depend upon the natural resource base. Luttrell, Ocampo-Thomason, Saint-Paul and Campbell *et al.* (Chapters 2, 11, 12 and 21, respectively, this volume) provide insights from experiences in different countries. They demonstrate the dependence of poor people on open-access resources as, for example, in mangrove forests. The failure to detect and respond to this adverse impact may be due in part to a lack of capacity among decision-makers to engage with and understand the perceptions of stakeholders (Le Tissier and Hills, Chapter 19, this volume). The explanation may also be due in part to the inherent difficulty of detecting impacts and attributing changes to causes (van Zwieten *et al.*, Chapter 17, this volume).

Conclusions

The coastal zone is home to 40% of the world's population and supports much of the world's food production, while also delivering important ecosystem services. But it is under increasing pressure that threatens the resilience of both natural and human systems. The problem is multifaceted, but contributions to this publication have focused in particular on land-use change within the tropical coastal zone.

Among the diverse environments that make up the coastal zone, the land-use issue considered here is most pertinent to river deltas and estuaries, which are characterized by a gradual and seasonally varying land-water interface. The various contributors to this publication have presented evidence from different countries, but in each case we can recognize common features:

1. The aquatic environment is subject to seasonally varying salinity.
2. The terrestrial environment is vulnerable to both tidal and riverine flooding.
3. The natural resource base supports agriculture, aquaculture and fisheries.

The critical land-use issue has been shown to be the expansion and intensification of brackish-water shrimp production. This activity takes place in ponds that may have been developed by clearing natural habitats (principally mangrove forest and salt marsh) or by converting agricultural land (notably rice farms). Extensive shrimp farming has been a part of the traditional livelihood system, but recent strong demand in global markets, together with technological advances, has provided the impetus for rapid and generally unplanned change.

The potential for quick profits from shrimp production attracts both local farmers and outside entrepreneurs, but questions are raised about short-term risk and long-term sustainability. The debate has tended to polarize between those who see increased productivity of land and water resources and others who emphasize the negative impacts. Policymakers, planners and others concerned with environmental protection face a real dilemma in making development decisions. The widely reported problems can be attributed in part to the 'gold-rush mentality' associated with the early stages of an immature industry. This has been exacerbated by institutional weaknesses that have allowed unplanned and unregulated development, leading to environmental stresses that in turn affect human welfare. It is often only at this late stage that policy responses are triggered. The question that this publication has sought to answer is: Can we achieve socially and environmentally sustainable development?

The complexity and high natural productivity of the environment lead us to believe that co-existence of alternative natural resource-based livelihoods is the key to sustainable development. However, implying that co-existence is desirable and making it happen are two different things. Zoning can be seen as an essential element of planned development, but no zone is perfect and livelihood choice will inevitably be limited within any zone. Seasonal zoning may provide the best compromise as, for example, in a rice–shrimp rotation system with alternate freshwater and brackish-water conditions. The spatial scale at which zoning occurs also merits careful consideration, with an assumption in favour of smaller units allowing more flexibility. Creation of a buffer zone around homesteads will be necessary to prevent close encroachment of shrimp ponds and the resulting salinity problems that affect the daily living environment of farmers.

One advantage of more localized zoning is that adoption of a participatory approach becomes more feasible. This should ensure better-informed decisions and greater likelihood of compliance. There is a need to develop appropriate planning tools and, in particular, tools for proper assessment of both environmental capacity and the value of ecosystem services. Improved knowledge should in time allow for the establishment of economic instruments to incentivize appropriate use of the natural resource base (e.g. through resource-use charges or environmental capacity charging). However, in the short term, control will depend upon establishing a regulatory framework and imposing penalties for any infringement.

Regulation is required both to control land use within any zone and to exercise control over the nature of production activities. However, given that there are many thousands of small-scale producers, it will be very difficult to effectively regulate on-farm activities. Promotion of best management practices and codes of conduct should be seen as a priority for all concerned institutions. Investment is needed in capacity building within local extension services and in creating effective farmers' organizations to empower community participation in natural resource management. A key issue here is the control over intensification and, in particular, the intensity of shrimp production. The progression from extensive to intensive systems brings trade-offs between economic benefit on the one hand and environmental and social impact on the other. Many cases have shown clearly that intensive shrimp farming is not sustainable.

A priority issue that emerges from the case studies presented in this publication is the impact of change on poor people. There is evidence of an increasing gap between the rich and the poor. Spending of public resources on coastal zone infrastructure (e.g. tidal sluices and polders) has been shown to deliver economic benefit while still causing relative poverty to increase. The livelihoods of poor people depend upon open-access resources, which include, but are not limited to, fisheries. Any development decision that aims to enhance production from aquaculture and/or agriculture is likely to impact adversely on access to and productivity of these resources. Planners and decision-makers should recognize this conflict and ensure that they have adequate information on the importance and value of open-access resources.

In spite of frequent calls for integrated water resource management, the coastal zone is generally considered in isolation from the river basins to which it is linked. The coastal zone sits at the tail-end of river systems and suffers the impact of upstream river basin development. Changed flow regimes, sediment yields and pollution loads all add to the direct local pressure on the coastal zone. Arguably, the health of the coastal zone can be seen as an indicator of river basin health. Improved river basin management will seek to increase water productivity and manage multiple uses while delivering essential environmental flows. Institutional barriers between those concerned with river basin management and those responsible for coastal zone management must not be allowed to threaten the sustainable development of this vital resource base and the livelihoods of those who depend on it.

References

ADB/NACA (1997) *Aquaculture sustainability and the environment*. Asian Development Bank and Network of Aquaculture Centres in Asia-Pacific, Bangkok.

Adger, W.N., Kelly, P.M., Ninh, N.H. and Thanh, N.C. (2000) Property rights, institutions and resource management: coastal resources under the transition. In: Adger, W.N., Kelly, P.M., Ninh, N.H., Thanh, N.C. (eds) *Living with Environmental Change: Social Vulnerability, Adaptation and Resilience in Vietnam*. Routledge, London.

Bailey, C. and Pomeroy, C. (1996) Resource dependency and development options in coastal South-east Asia. *Society and Natural Resources* 9(1), 191–199.

Be, T.T., Dung, L.C. and Brennan, D. (1999) Environmental costs of shrimp culture in the rice-growing regions of the Mekong Delta. *Aquaculture Economics & Management* 3(1), 31–42.

Binh, C.T., Phillips, M.J. and Demaine, H. (1997) Integrated shrimp-mangrove systems in the Mekong delta of Vietnam. *Aquaculture Research* 28, 599–610.

Boyd, C.E. and Clay, J.W. (1998) Shrimp aquaculture and the environment. *Scientific American* 278(6), 43–49.

Boyd, C.E., Hargreaves, J.A. and Clay, J.W. (2002) *Codes of practice and conduct for marine shrimp aquaculture.* Report prepared under the World Bank, NACA, WWF and FAO Consortium Program on Shrimp Farming and the Environment: 31.

Boyd, C.E. and Gautier, D. (2000) Effluent composition and water quality standards. *Global Aquaculture Advocate* 3(5), 61–66.

Burford, M.A., Costanzo, S.D., Dennison, W.C., Jackson, C.J., Jones, A.B., McKinnon, A.D., Preston, N.P. and Trott, L.A. (2003) A synthesis of dominant ecological processes in intensive shrimp ponds and adjacent coastal environments in NE Australia. *Marine Pollution Bulletin* 46, 1456–1469.

Clay, J.W. (1996) *Market potentials for redressing the environmental impact of wild captured and pond produced shrimp.* World Wildlife Fund, Washington, D.C.

Corredor, J.E. and Morell, M.J. (1994) Nitrate depuration of secondary sewage effluents in mangrove sediments. *Estuaries* 17, 295–300.

Deb, A.K. (1998) Fake blue revolution: environmental and socio-economic impacts of shrimp culture in the coastal areas of Bangladesh. *Ocean and Coastal Management* 41, 63–88.

Dewalt, B.R., Vergne, P. and Hardin, M. (1996) Shrimp aquaculture development and the environment: people, mangroves and fisheries on the Gulf of Fonseca, Honduras. *World Development* 24(7), 1193–1208.

Dierberg, F.E. and Kiattisimkul, W. (1996) Issues, impacts and implications of shrimp aquaculture in Thailand. *Environmental Management* 20(5), 649–666.

DoF (Department of Fisheries) (1995) *Shrimp resources statistics.* Central Shrimp Cell, Department of Fisheries, Dhaka, Bangladesh.

DoF (Department of Fisheries) (2003) *Fishery Statistical Yearbook of Bangladesh: 2002–2003.* Department of Fisheries, Dhaka, Bangladesh, 41 pp.

EJF (2003) *Risky Business: Vietnamese Shrimp Aquaculture – Impacts and Improvements.* Environmental Justice Foundation, London.

FAO (1995) *Code of Conduct for Responsible Fisheries.* Food and Agriculture Organization of the United Nations, Rome.

FAO (1997) *Towards sustainable shrimp aquaculture development: implementing the FAO Code of Conduct for Responsible Fisheries.* Food and Agriculture Organization of the United Nations, Rome.

FAO (2003) *Review of the State of World Aquaculture.* Food and Agriculture Organization, Rome, 95 pp.

Fegan, D.F. (1999) Research issues in sustainable coastal shrimp farming: a private sector view. In: Smith, P.T. (ed.) *Towards Sustainable Shrimp Culture in Thailand and the Region.* Proceedings of a workshop held at Hat Yai, Songkhla, Thailand. Australian Centre for International Agricultural Research, Canberra, Australia.

Gautier, D. (2002) *The Integration of Mangrove and Shrimp Farming: a Case Study on the Caribbean Coast of Colombia.* Report prepared under the World Bank, NACA, WWF, and FAO Consortium Program on Shrimp Farming and the Environment: 26.

GESAMP (1991) *Reducing Environmental Impacts of Coastal Aquaculture.* IMO/FAO/Unesco/WMO/WHO/IAEA/UN/UNEP Joint Group of Experts on the Scientific Aspects of Marine Pollution, Rome.

GESAMP (2001) *Planning and Management for Sustainable Coastal Aquaculture Development.* IMO/FAO/Unesco/WMO/WHO/IAEA/UN/UNEP Joint Group of Experts on the Scientific Aspects of Marine Pollution, Rome.

Graslund, S. and Bengtsson, B.-E. (2001) Chemicals and biological products used in South-east Asian shrimp farming, and their potential impact on the environment – a review. *The Science of the Total Environment* 280, 93–131.

Gujja, B. and Finger-Stich, A. (1996) What price prawn? Shrimp aquaculture's impact in Asia. *Environment* 38(7), 12–39.

Hambrey, J.B. (1996) Comparative economics of land use options in mangrove. *Aquaculture Asia* 1(2), 10–15.

Hoanh, C.T., Tuong, T.P., Gallop, K.M., Gowing, J.W., Kam, S.P., Khiem, N.T. and Phong, N.D. (2003) Livelihood impacts of water policy change: evidence from a coastal area of the Mekong River delta. *Water Policy* 5, 475–488.

Huitric, M., Folke, C. and Kautsky, N. (2002) Development and government policies of the shrimp farming industry in Thailand in relation to mangrove ecosystems. *Ecological Economics* 40, 441–455.

Janssen, R. and Padilla, J.E. (1999) Preservation or conversion? Valuation and evaluation of a mangrove forest in the Philippines. *Environmental & Resource Economics* 14(3), 297–331.

Johnston, D., Clough, B., Xuan, T.T. and Phillips, M.J. (1999) Mixed shrimp-mangrove forestry farming systems in Ca Mau Province, Vietnam. *Aquaculture Asia* 4, 6–12.

Lewis, R.R., Phillips, M.J., Clough, B. and Macintosh, D.J. (2003) *Thematic Review on Coastal Wetland Habitats and Shrimp Aquaculture.* Report prepared under the World Bank, NACA, WWF and FAO Consortium Program on Shrimp Farming and the Environment: 81.

Limsuwan, C.S. and Chanratchakool, P. (1998) *Closed Recycle System for Sustainable Black Tiger Shrimp Culture in Freshwater Areas.* Proceedings of the Fifth Asian Fisheries Forum, 11–14 November 1998, Chiang Mai, Thailand.

Macintosh, D. (1998) Mangroves and coastal aquaculture sustainability. In: *Aquaculture Sustainability and the Environment. Report on a Regional Study and Workshop.* Asian Development Bank/Network of Aquaculture Centres in Asia, Bangkok, Thailand, pp. 242–251.

Minh, L.Q., Tuong, T.P., van Mensvoort, M.E. and Bouma, J. (1997) Contamination of surface water as affected by land use in acid sulphate soils in the Mekong River Delta, Vietnam. *Agriculture Ecosystems & Environment* 61(1), 19–27.

Minh, L.Q., Tuong, T.P., van Mensvoort, M.E. and Bouma, J. (1998) Soil and water table management effects on aluminum dynamics in an acid sulphate soil in Vietnam. *Agriculture Ecosystems & Environment* 68(3), 255–262.

Murthy, H.S. (1997) Impact of the Supreme Court judgement on shrimp culture in India. *INFOFISH International* 3(97), 30–34.

Naylor, R.L., Goldburg, R.J., Mooney, H., Beveridge, M., Clay, J., Folke, C., Kautsky, N., Lubchenco, J., Primavera, J. and Williams, M. (1998) Ecology: nature's subsidies to shrimp and salmon farming. *Science* 282(5390), 883–884.

Nichols, K. (1999) Coming to terms with 'integrated coastal management': problems of meaning and method in a new arena of resource regulation. *Professional Geographer* 51(3), 388–399.

Nickerson, D.J. (1999) Trade-offs of mangrove area development in the Philippines. *Ecological Economics* 28, 279–298.

Paez-Osuna, F. (2001) The environmental impact of shrimp aquaculture: causes, effects, and mitigating alternatives. *Environmental Management* 28(1), 131–140.

Patil, P.G. and Krishnan, M. (1997) The Kandaleru shrimp farming industry and its impact on the rural economy. *Agricultural Economics Research Review* 10(2), 293–308.

Phillips, M.J., Boyd, C. and Edwards, P. (2001) *Systems Approach to Aquaculture Management.* Aquaculture in the Third Millennium, NACA/FAO, Bangkok, Thailand.

Phong, N.D., My, V., Nang, N.D., Tuong, T.P., Phuoc, T.N. and Trung, N.H. (2003) Salinity dynamics and its implication on cropping patterns and rice performance in shrimp-rice system in My Xuyen and Gia Rai. In: Preston, N. and Clayton, H. (eds) *Rice-shrimp Farming in the Mekong Delta: Biological and Socioeconomic Issues.* ACIAR Technical Reports No. 52e, Canberra, Australia, pp. 70–88.

Rivera-Monroy, V.H., Torres, L., Bahamon, N., Newmark, F. and Twilley, R. (1999) The potential use of mangrove forests as nitrogen sinks of shrimp aquaculture pond effluents: the role of denitrification. *Journal of the World Aquaculture Society* 30(1), 12–25.

Robertson, A.I. and Phillips, M.J. (1995) Mangroves as filters of shrimp pond effluents: predictions and biogeochemical research needs. *Hydrobiologia* 295, 311–321.

Sammut, J. (1999) Amelioration and management of shrimp ponds in acid sulfate soils. In: Smith, P.T. (ed.) *Towards Sustainable Shrimp Culture in Thailand and the Region.* Proceedings of a workshop held at Hat Yai, Songkhla, Thailand. Australian Centre for International Agricultural Research, Canberra, Australia.

Sammut, J., White, I. and Melville, M.D. (1996) Acidification of an estuarine tributary in eastern Australia due to drainage of acid sulfate soils. *Marine and Freshwater Research* 47(5), 669–684.

Spalding, M., Blasco, F. and Field, C. (1997) *World Mangrove Atlas.* The International Society for Mangrove Ecosystems, Okinawa, Japan.

Szuster, B. and Flaherty, M. (2002) Cumulative environmental effects of low salinity shrimp farming in Thailand. *Impact Assessment and Project Appraisal* 20(2), 189–200.

Szuster, B., Molle, F., Flaherty, M. and Srijantr, T. (2003) Socio-economic and environmental implications of inland shrimp farming in the Chao Phraya Delta. In: Molle, F. and Srijantr, T. (eds) *Perspectives on*

Social and Agricultural Change in the Chao Phraya Delta. White Lotus Press, Bangkok, Thailand, pp. 177–194.

Twilley, R. (1992) Impacts of shrimp mariculture practices on the ecology of coastal ecosystems in Ecuador. In: Olsen, O. and Arriaga, L. (eds) *Establishing a Sustainable Shrimp Mariculture Industry in Ecuador*. The University of Rhode Island, Narragansett, Rhode Island, pp. 91–120.

Walters, B.B. (2003) People and mangroves in the Philippines: fifty years of coastal environmental change. *Environmental Conservation* 30(3), 293–303.

Wilson, B.P., White, I. and Melville, M.D. (1999) Floodplain hydrology, acid discharge and change in water quality associated with a drained acid sulfate soil. *Marine and Freshwater Research* 50(2), 149–157.

WRI (2000) *A Guide to World Resources 2000–2001: People and Ecosystems: The Fraying Web of Life*. World Resources Institute, Washington, D.C.

2 Adapting to Aquaculture in Vietnam: Securing Livelihoods in a Context of Change in Two Coastal Communities

C. Luttrell
Overseas Development Institute, London, United Kingdom,
e-mail: c.luttrell@odi.org.uk

Abstract

This chapter examines the effects of aquaculture development on the livelihoods of households in two historically and geographically distinct coastal communities in north and south Vietnam. It is shown that the importance of open-access resources for livelihoods increases in line with the poverty and vulnerability of the social group. This increase has occurred at the same time as a decrease in the availability of open-access products because of the privatization of resources associated with aquaculture. Many open-access resources are accessed illegally or with unofficial access rights and this has implications for livelihood security, since open-access resources have uncertain rights and legislative status, a status that is open to change and frequently subject to privatization. The high capital investment required and the loans necessary to raise that capital are also increasing the vulnerability of the wealthier households involved in aquaculture.

Introduction

Vietnam's entry into the shrimp market lagged behind other Asian countries such as Thailand. However, the favourable agro-climatic conditions, particularly in the south, the opening of the economy and the spread of shrimp disease in other shrimp-producing Asian countries all led to a spectacular entry into the market (Flaherty *et al.*, 1999). Traditional coastal aquaculture has been a part of the livelihood structure in Asia for hundreds of years (Ling, 1977). However, in a climate of new technologies and increasing global demand, traditional extensive systems supplying local markets have rapidly changed to resource-intensive, high-production systems catering to the global market (Barraclough and Finger-Stich, 1996; Flaherty *et al.*, 1999). Shrimp became a high development priority in Vietnam as an important means of earning foreign exchange (Vo Van Trac and Pham Thuoc, 1995) and a commodity that has a high comparative advantage. Indeed, foreign earnings from aquaculture are high and are increasing annually (Vietnam News, 1999; Vietnam News Agency, 1999; Asia Pulse, 2001).

This chapter examines the effects of aquaculture development on the livelihoods of households in two historically and geographically distinct communities in coastal

 Environment and Livelihoods in Tropical Coastal Zones (eds C.T. Hoanh, T.P. Tuong, J.W. Gowing and B. Hardy)

areas in north and south Vietnam. In the two case study areas, there has been official support for aquaculture at the district level. Preferential taxation and financial support are readily available to traders and exporters of aquatic products in the form of credit insurance funds, development support funds and an export insurance fund (Vietnam News Agency, 2000). Most of the capital resources for the 'shrimp boom' came from government-supported banks (Dang Kim Son, 1998), with the Vietnam Bank for Agriculture (Agribank) in the Mekong Delta providing medium-term 3-year loans for shrimp-pond building and short-term 1-year loans for inputs, up to a maximum of 70% of the total costs (Euroconsult, 1996).

Research Context

The southern research site, Nam Hai Commune, is located in Thinh Binh District,[1] Ca Mau Province (the southernmost province of the country). The northern research site, Da Rang Commune,[1] is an island, located in Xuan Giao District, Quang Ninh Province (the northernmost province of the country), 100 km from the Chinese border (Fig. 2.1). These case studies provide two examples of very different ecological and social systems, particularly in terms of mangrove ecology, geography and social history (Luttrell, 2000). Such different contexts, as well as varying history, tenurial systems, community cohesion and social structures, have had very different impacts on the process of adaptation to change in these case study areas (Luttrell, 2001, unpublished PhD thesis).

In Da Rang Commune, the mangrove forest and associated mud flats have, until recently, provided open-access resources. Gatherers, including those from other communes, have been free to carry out low-intensity exploitation of marine and mud-flat products. Here, the spread of shrimp farming occurred later than in the south and with more caution. However, since the 1990s, large amounts of this open-access area have been in effect privatized by the granting of private contracts to individuals, many of whom were from different provinces, for the construction of shrimp ponds for extensive and semi-intensive shrimp farming. This has drastically reduced the area available to the local community for the collection of mangrove products and is said to have lowered the quality of the environment, resulting in a further reduction in these products (Luttrell, 2000).

The area of Nam Hai Commune is divided into the forest enterprise and the area of private land, both of which have very different property regimes. The forest enterprise area is managed by a state-owned company, which is directly under the jurisdiction of the province. Some of the area has been contracted out to households on 20-year leases, which gives the right to construct shrimp ponds on a proportion of their area but not to use the trees without authorization. The 'protection' area of the forest enterprise is managed as forest by the enterprise staff, and local people are forbidden to carry out any kind of activity, although it is common for the staff to grant unofficial gathering permits. The rest of the commune area is private household land, which has been completely converted to shrimp ponds carrying out semi-intensive aquaculture. There are no official areas of common or open-access land, and forest enterprise land is consequently used illegally by many to collect mangrove products.

The proximity of Da Rang Commune to the Chinese border and the opening of the border in 1990 have resulted in the recent emergence of the mangrove areas as the source of a significant amount of household income from the trading of similar mangrove products for every household in the commune. The high prices for such products in the Chinese market have caused a shift in the perception and valuation of the mangrove and mud-flat areas, and this has led to a debate about the possibility of the privatization of such areas. In Nam Hai Commune, because of the lack of a market in the vicinity, such products do not have a high

[1] The district and commune names mentioned in this chapter have been changed.

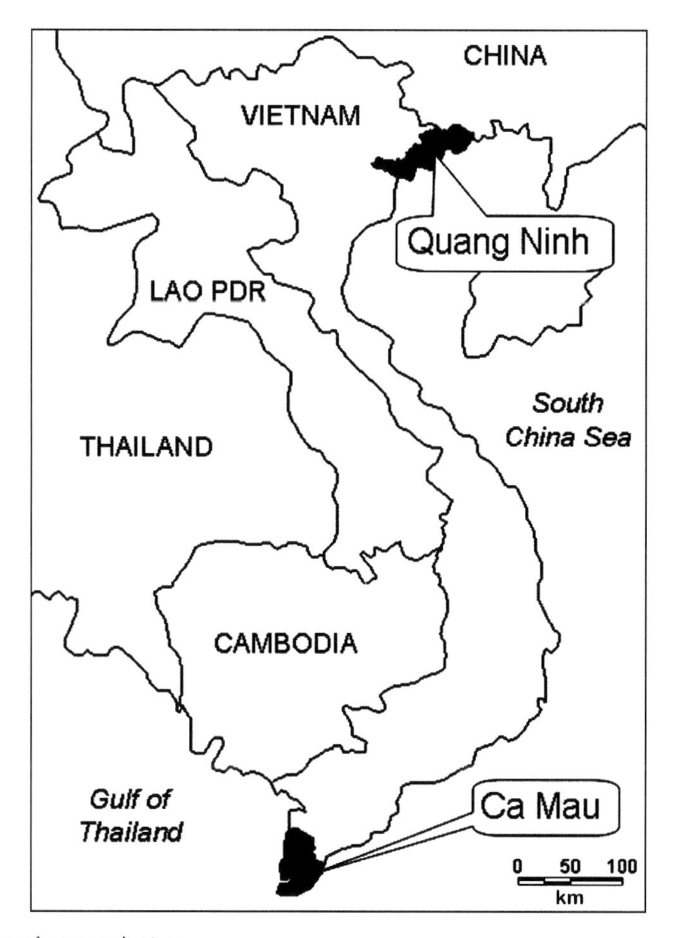

Fig. 2.1. Location of case study sites.

value and their collection is limited to landless households with no alternative income.

Research Method

Data on livelihoods and institutional change were collected during two research periods in each of the two case studies within a period of 16 months between May 1998 and August 2000. Semi-structured interviews with officials, key informants and individual permanent and temporary commune members were carried out. The research and analysis methodology primarily used qualitative techniques and the analysis employed an inductive approach drawing on principles of grounded theory (Glaser and Strauss, 1967). However, each interview also included the collection of some quantitative data.

Sampling was initially stratified by hamlet to allow recognition of the disparities between the hamlets. Administratively, Nam Hai Commune is divided into two sections, the forest enterprise land and the area of private land under the jurisdiction of the commune, and within each hamlet the sample was further differentiated according to this land ownership. Following this, respondents were initially identified by using the technique of 'snowball sampling'. This technique is a non-random method of data collection whereby interviewees are asked to nominate further informants (Eland-Goossense *et al.*, 1997; Faugier and Sargeant, 1997). In the choice of respondents, attempts were made to maintain a balance between broad categories of household based on wealth, occupation, gender, age, location of residence and ethnicity or origin. The sampling emphasis constantly evolved through the research process. In the first stage of the research, the selection of the respondents sought to capture diversity. During the second stage of data collection, an attempt was made to return to as many households from the first sample as possible, and, in addition, to focus on specific issues identified by the first stage of the analysis. Sampling was brought to an end when it was considered that a 'data saturation'[2] point had been reached in assimilating information and that little further could be added to the understanding (Table 2.1).

For each of the sampled households, interviews were carried out with one or more of the household members and in some cases two interviews with the same household took place, one in each of the field-work periods. For the other households, adaptations to change in the recent past have been elicited through recall data. This permits conclusions as to the type of access on which the livelihood portfolio is based, how this access has shifted in the face of change and what this shift has meant for livelihood vulnerability. The households have been classified into 'social groups' based on the criteria described in Table 2.2. At both of the case study sites, shrimp farming is currently dominated by semi-intensive systems, which involve the stocking of the ponds with fry. In Nam Hai Commune, all households in the NH-FOREST and NH-LAND groups are involved in shrimp farming, the majority of which are stocking with tiger shrimp. In Da Rang Commune, a distinction has been made in the categorization between those households stocking with tiger shrimp (DR-SHRIMP) and those with other kinds of shrimp (DR-POND).

Experience of Shrimp Farming and Failure

The southern case study: Nam Hai Commune

Aquaculture production and fisheries play a major role in the economy of Thinh Binh District. Although the conversion from rice to shrimp farming was initially illegal, the district later gave its support to it. Official government-supported loans are granted to all shrimp farmers, and tax reductions were introduced in times of poor harvest. Despite the official encouragement of shrimp farming, inadequate attention has been given to the technical details necessary for effective shrimp production and environmental protection. In most cases, productivity in Nam Hai Commune was high, at 100–200 kg/ha for the first 3 years of shrimp production, but fell sharply after this. This is a situation reported throughout the Mekong Delta (Euroconsult, 1996). Many households now regret the conversion to shrimp and they prefer to convert back to rice, which would at least guarantee them a subsistence livelihood source. Others recall how rice yields were good before the conversion to shrimp. Conversion back to rice, however, could be carried out only with external assistance or loans, since it would require substantial earth-moving and dyke-building activity and would result in the absence of any returns for at least one year. Requests for assistance are therefore being put to the District. Reconversion would also require agreement of

Table 2.1. Sample numbers in the two case studies.

Number	Da Rang Commune	Nam Hai Commune
Of individuals interviewed	115	179
Of households interviewed	88	129
Of households revisited	17	31

[2] In qualitative research, there are no published guidelines or tests of adequacy for estimating the sample size required to reach saturation equivalent to those formulas used in quantitative research. Rather, in qualitative research, the signals of saturation are determined by the investigator and by evaluating the adequacy and the comprehensiveness of the results (Gubrium, 1995).

Table 2.2. Categorization of 'social groups' as used in the analysis.

Criteria	Social group	Number of respondents
Da Rang Commune		
Tiger shrimp growers	DR-SHRIMP	11
Trading households (those not in social group DR-SHRIMP)	DR-TRADE	4
Pond owners (those not in social groups DR-SHRIMP or DR-TRADE)	DR-POND	21
Landowners (those not in social groups DR-SHRIMP, DR-TRADE or DR-POND)	DR-AGRIC	61
Households with only fallow land	DR-FALLOW	3
Ethnic minorities	DR-MINORITY	15
Landless households	DR-NOLAND	2
Nam Hai Commune		
Households allocated forest enterprise land	NH-FOREST	53
Households owning commune land	NH-LAND	50
Households renting or farming agricultural land/shrimp pond without land certificate	NH-RENT	8
Landless households	NH-LANDLESS	46
Households living in the fishing settlement of Lang Ca or owning fixed bed nets on the inland rivers	NH-FISH	22

all landowners in the area because the building of dykes to prevent the intrusion of saline water would have to be a communal effort.

Interviews carried out in January 1999 with the Thinh Binh District People's Committee showed district officials to be supportive of shrimp farming. The Vice-President of the District People's Committee (1999 interview) emphasized that efforts had been made to help farmers intensify their aquacultural systems, but he was unenthusiastic about farmers' recent requests for assistance to convert their land back to rice. However, recommendations made by the Department of Agricultural Development (DARD) of the district in interviews carried out 6 months later contradicted those of the People's Committee. Contrary to the District People's Committee policy, DARD recommended the conversion of shrimp ponds back to rice cultivation.

The northern case study: Da Rang Commune

Since 1993, the Xuan Giao District People's Committee has leased the mangrove area of Da Rang Commune to private entrepreneurs for 10 to 20 years for the payment of a one-off fee. Nguyen Thanh Manh and Phan Anh Thi Dao (1998) reported a stable yield from the shrimp ponds of 250–300 kg/ha/year. Other authors, however, record low aquacultural productivity, with only 20–30% of ponds achieving an annual shrimp output of 70–100 kg/ha (Nguyen Duc Cu, 1998).

This fall in productivity has been attributed to many factors, including the collapse of dykes; loss of natural habitat for larvae of shrimp, crabs and fish (Nguyen Duc Cu, 1998); pollution; dynamite; and electric fishing, which also decreases the levels of natural fry (Da Rang Commune, 1998). By 1997, there were seven experimental tiger shrimp ponds in the commune stocked with an average of 4000 seeds (Nguyen Thanh Manh and Phan Anh Thi Dao, 1998), but by 1998 only one household had succeeded in cultivating tiger shrimp. During 1999, a few of the richest households were growing tiger shrimp from locally adapted seed that was available for sale in the commune.

Changes in the Structure of Livelihoods

Northern case study: Da Rang Commune

Social groups DR-SHRIMP and DR-TRADE consist of households that trade in aquatic products or cultivate tiger shrimp. In all cases where tiger shrimp are grown, this activity has been adopted recently. It does not, however, represent the dominant livelihood source for all of these households, and the livelihood sources of social groups DR-SHRIMP and DR-TRADE are the most diversified of all the groups in the commune, being characterized by capital-intensive activities such as trading and renting out of machinery.

The livelihood profiles of other land-owning groups in the commune (social groups DR-POND, DR-AGRIC and FALLOW) vary according to their labour availability. Those households that do not have surplus labour are less likely to engage in open-access collecting. For those who do engage in collecting, their livelihood profile is dominated by the income from open-access products, mostly mud-worms and clams. For others, rice growing and the cultivation of natural shrimp are important. Although it may not be the dominant livelihood source, rice remains an important safety net for the majority of these households (Fig. 2.2).

For the ethnic minorities and the landless group (social groups DR-MINORITY and NOLAND), the collection of open-access products clearly dominates their livelihood structures, but rice and hiring out of labour are also important elements (Fig. 2.3).

Southern case study: Nam Hai Commune

Land-owning households on commune land (social group NH-LAND) have livelihood profiles dominated by income from their own shrimp ponds, and in some cases from the trading of aquatic products and the hiring out of labour to other shrimp ponds (Fig. 2.4). Open-access products and those collected illegally from the forest enterprise are often supplementary livelihood sources, and these sources increase in importance in times of adversity. The second round of interviews, in combination with recall data, reveals a dichotomy between those households that have suffered a decline in their shrimp harvest and those whose shrimp harvests have improved since the last interview. Those households whose shrimp have declined since the initial high harvests have shifted their main income towards activities such as migrant labour, which is encouraged by the falling demand for labour in the area, and diversified their subsidiary livelihood sources. Open-access products make up a larger proportion of these livelihood sources. Where shrimp harvests have increased, the households have benefited from the increased demand for hired labour and for the trading of products.

The livelihood profiles of the landless groups in Nam Hai Commune (social groups NH-RENT and NH-LANDLESS) are dominated by a reliance on open-access resources, a large proportion of which are collected illegally in the forest enterprise and from private ponds (Fig. 2.5). Hiring out of labour for shrimp ponds and open-access activities such as fishing are also important, but the demand for hired labour relies heavily on the success of the shrimp harvests in the area. Unlike the richer groups, all in this group identified the extreme storm Typhoon Linda as having had a serious effect on their livelihood vulnerability and as an additional 'shock' to which they had to adapt. The second round of interviews, in combination with the recall data collected in the inter-

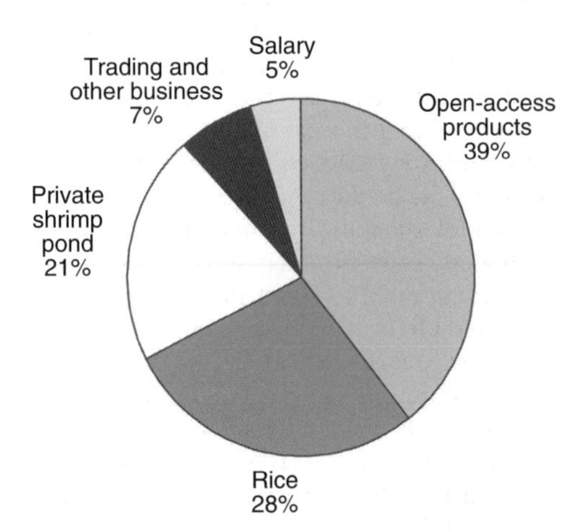

Fig. 2.2. The dominant livelihood sources (those ranked first by respondents) of land-owning households (social groups DR-POND, DR-AGRIC and FALLOW) in Da Rang Commune. The term 'private' shrimp pond is used to refer to shrimp from ponds owned by that group rather than illegal access to others' shrimp ponds (from household interviews, Da Rang Commune, 1998–1999 (n = 43)).

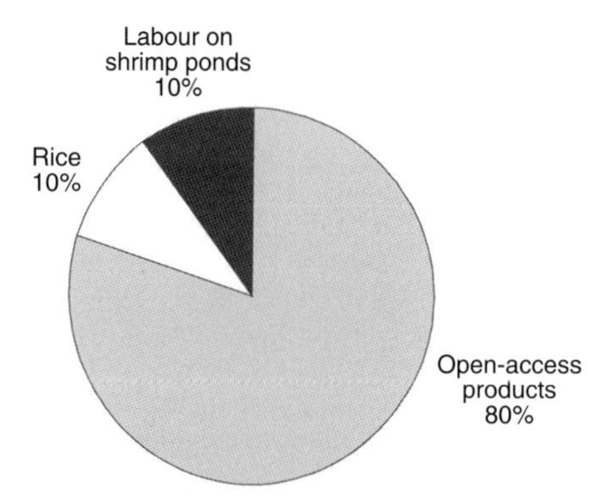

Fig. 2.3. The dominant livelihood sources of ethnic minorities and landless households (social groups DR-MINORITY and DR-NOLAND) in Da Rang Commune (from household interviews, 1998–1999 (n = 10)).

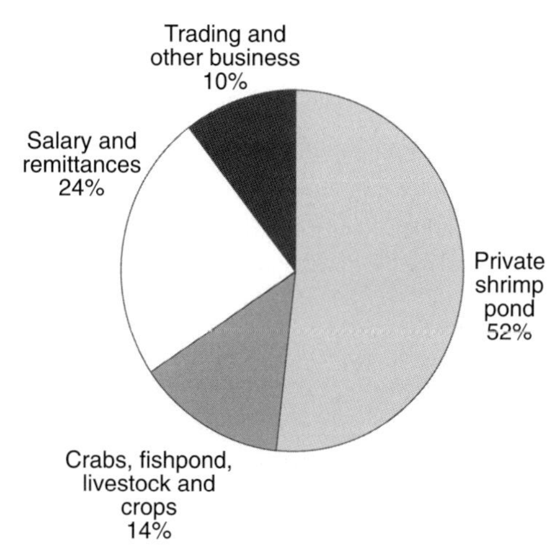

Fig. 2.4. The dominant livelihood sources of land-owning households (social group NH-LAND) in Nam Hai Commune (from household interviews, Nam Hai Commune, 1998–1999 (n = 29)).

views, reveals an increased importance of open-access and illegally accessed products for livelihoods, together with a marked diversification of livelihood sources.

The livelihood profiles of households with leases in the forest enterprise (social group NH-FOREST) are dominated by income from shrimp farming. A minority of this group, however, is more reliant on the hiring out of its labour to other shrimp-pond owners or on trading of aquatic products. Subsidiary livelihoods in this group are diverse and, for the most part, are based on open-access or illegal products collected from the forest enterprise. Some households maintain their links with land, shrimp ponds or jobs elsewhere and, for them, the land in the forest enterprise is a small part of a set of spatially diversified livelihood sources. The second round of interviews and the recall data showed that reliance on livelihoods from shrimp ponds in the forest enterprise had decreased because of the failure of local shrimp farming. There had been a shift in livelihood strategies away from the hiring out of labour and from open-access fishing towards income from shrimp ponds in other areas.

Impact on Resource Access: Implications for Livelihood Vulnerability

Box 2.1 summarizes the shifts in livelihood profiles for the various social groups in the two different communes. However, it must be acknowledged that the outcomes and shifts in activity are rarely the result of just one set of factors. The introduction of shrimp farming in Da Rang Commune occurred at a time close to the opening of trade across the Chinese border, with the accompanying expansion in export markets

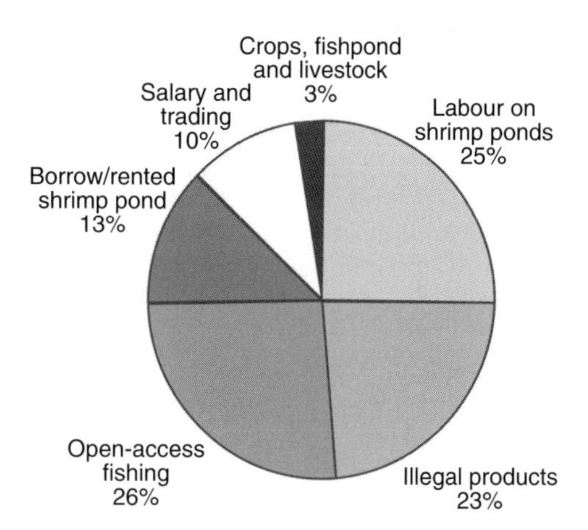

Fig. 2.5. The dominant livelihood sources of landless and tenant households (social groups NH-RENT and NH-LANDLESS) in Nam Hai Commune (from household interviews, Nam Hai Commune, 1998–1999 (n = 39)).

Box 2.1. General patterns of change in livelihood profiles of different social groups through the periods of expansion and subsequent failure of shrimp farming.[a]

Social group	Ranked livelihoods before introduction of shrimp farming	Ranked livelihoods during shrimp boom[b]	Ranked livelihoods after decline in shrimp and open-access products
Da Rang Commune			
	Pre-1992	1992	1995
DR-TRADE[c]	Insignificant as the market had not developed	Aquatic product trading	Aquatic product trading Other trading Other private business
My mother says '*Phi thuong bat phu*' ('if you do not trade, you will never be rich'). My family is getting involved in handicrafts, agricultural machinery and other trading as well (aquatic product trader, Da Rang Commune, 1998: interview no. 273b).			
DR-POND	Agriculture	Shrimp Open-access products	Shrimp Open-access products Small trading
DR-AGRIC	Agriculture	Open-access products Agriculture	Open-access products Agriculture
'If we just lived on paddy, we would die because the rice we grow is not enough for our family. I just need half a day collecting to get enough money from the mud flat to buy rice. In the morning we go collecting and by afternoon we have money' (farmer, Da Rang Commune, 1999: interview no. 239b).			
Nam Hai Commune			
	1978	1985	1992
NH-FOREST	Not available	Shrimp	Shrimp Open-access products Trading
NH-LAND	Agriculture	Shrimp	Open-access products Shrimp Labour Livestock
'Here people just work for hire; even from the shrimp pond owners we get nothing and we also have to work for hire ... it is just like having no land' (shrimp farmer, Nam Hai Commune, 1999: interview no. 131).			
NH-LANDLESS	Agricultural labour	Labour on shrimp ponds	Open-access products Labour

[a] The table was derived from a general summary based on qualitative and quantitative information from the households studied.
[b] At the northern study site, this also coincided with the opening of trade across the Chinese border.
[c] The categorization is based on 1999 data, so DR-TRADE is households involved in trading in the year 1999.

for open-access products. There has been a general shift in livelihood patterns and vulnerability. The move from a dependence on agriculture to shrimp farming has been superseded more recently by further changes in livelihoods accompanying the failure of the shrimp harvest. The relative importance of open-access products has increased for all social groups, particularly the poorest. In the case of wealthier social groups, this has been one part of a more diversified livelihood profile.

Social differentiation and upheaval

Successful shrimp farming brings immediate profits and people were encouraged to take this livelihood option on account of low income from rice and limited employment opportunities elsewhere. The subsequent 'gold rush mentality' (Flaherty *et al.*, 1999) has resulted in the poor construction of ponds and the phasing out of basic food crops, to be replaced by this higher-value export crop. At both research sites, the construction of shrimp ponds has had a marked social effect. Shrimp farming has further increased the disparity between those who can and those who cannot afford to continue to invest in this livelihood activity.

In Nam Hai Commune, the conversion to shrimp farming resulted in huge shifts in access rights because it represented a massive capital investment. Poorer households, which had no access to this level of capital, were forced to sell land to other households in the area or to new entrepreneurial households from other areas. In Nam Hai Commune, in particular, the poor who cannot afford the capital investment in shrimp ponds have been displaced by the inward migration of people with capital. This displacement is now leading to richer households moving out because of the low returns from shrimp farming. The so-called 'gold rush' of shrimp farming is over, ponds are being sold off more cheaply and households from a lower income group are moving in. The price of land has reportedly decreased to 25–35% of its value over 5 years.

Profit accruing to outsiders and indebtedness

In Da Rang Commune, it is mainly investors from outside the commune who have benefited from shrimp farming and, having made large profits, moved elsewhere when they realized that the harvests were failing. At first, only two local households had a share in the ponds, but, after 2 years, when returns began to decrease, the owners from outside the commune began to sell the ponds to locals (Nguyen Thanh Manh and Phan Anh Thi Dao, 1998). For the most part, those who own the shrimp ponds now face very low harvests and huge debts that cannot be repaid. In Da Rang Commune, labourers who are hired locally tend to be the relatives and friends of the pond owners, so that the ethnic immigrants, who are most in need of work, are at a disadvantage. Most outsider pond owners in Da Rang do not use much local labour, so benefit to the local residents from shrimp farming is limited.

At both research sites, those households involved in shrimp farming made huge initial profits; this was particularly the case with those able to make large investments in terms of constructing sluice gates and dredging ponds. Subsequently, many of the shrimp-farming households suffered losses, entering a downward spiral of debt, and were often forced to sell their land.

Of the 42 shrimp farmers interviewed in Nam Hai Commune, 74% had taken out loans to carry out shrimp farming and all these respondents are unable to repay the loans, with only a few able to make interest payments. The average loan, from the farmers interviewed who had taken out official loans, was 5.62 million VND. However, this is probably an understatement deriving from the sensitive nature of the topic. Furthermore, interest rates are high, ranging from 0.5% to 3.2% per month from the banks to 15% from private moneylenders. The picture is similar in Da Rang Commune: 165 households were operating ponds between 1993 and 1995 but by 1998 there were only 75 such households (Nguyen Thanh Manh and Phan Anh Thi Dao, 1998). Many suffered losses, are in debt and have been forced to sell their shares. Indeed, many of the landless households interviewed had once been involved in shrimp farming.

The results of this study do, however, show that, in the case of aquaculture, there is not a clear relationship between poverty and vulnerability as other authors note (Davies and Hossain, 1997; Ellis, 1998; Carney, 1999). In the case of the abnormal storms associated with Typhoon Linda, which hit the southern coast of Vietnam in 1997, it was the households that had invested in shrimp farming that appeared to have suffered the most because, having lost their investment, they were pushed deep into debt. Poor households that had not taken out loans were relatively unaffected economically.

Dramatic changes in the livelihood profiles

The death of natural shrimp has increasingly forced those who have invested in the shrimp-pond infrastructure to invest in tiger shrimp, thus increasing even further the costs and risks involved. For others who cannot afford this, the failure of the shrimp to provide the expected windfall has led to households being forced to look for alternative livelihoods. Some 25% of the landless households interviewed had previously been engaged in shrimp farming, and others affected included those who previously farmed but lost their land to the spread of aquaculture.

The loss of land induces households to carry out activities such as hiring out of labour (but it has become increasingly harder to find such work as the shrimp continue to die and investment decreases), pig breeding, wine making and illegal collection of natural products from the forest enterprise or private ponds. There has been a clear trend among farmers who have retained their ponds towards activities such as crab rearing, *Nipa* (a type of palm tree) growing and the planting of fruit trees and some other crops, as well as the collection of fish and mangrove products. Richer households are attempting to diversify their livelihood away from a reliance on natural resources by developing activities such as trading and small businesses (see NH-FOREST and NH-LAND, Box 2.1).

Enclosure but increased reliance on open-access resources

The conversion of rice land to shrimp ponds in Nam Hai Commune during the late 1980s resulted in few changes to the property rights over the land itself, because shrimp ponds were built on agricultural land over which households had official private land rights. However, the enclosure of the water, and of the fish/shrimp stocks that the ponds contained, did represent a privatization of these resources *in those areas*.

At both research sites, the mangrove and mud-flat areas, which are used for shrimp ponds, previously provided livelihoods for those who depended on open-access resources. There is a clearly perceived link between the construction of shrimp ponds and the deterioration of the quantity and quality of open-access products. In spite of the hostility in Da Rang Commune against shrimp farming, local residents were not able to present a united front against the pond owners from outside because some members of the community (usually the more influential members) wanted to profit from the ponds themselves. Many members of the community have invested in shrimp farming in a number of different ways and therefore they have much to lose from its cessation. In addition, fishing contracts have been introduced by the district[3] and, although contracts are not necessary for local residents to fish, contracts are needed for more intensive fishing techniques. This has resulted in the loss to the local community of open-access resources and added benefit to the district from the renting out of contracts. Therefore, élites are able to stake personal claims over open-access resources used by local residents and this results in the situation, as described by Bailey (1985), in which a complex ecosystem supporting multiple use is transformed into a greatly simplified system that becomes the private property of a small number of individuals.

Despite the increased enclosure of open-access resources and the decrease in quality owing to environmental deterioration, open-access resources have become an increasingly important part of the livelihoods of

[3] Fisheries in the district are regulated under the Law of Fisheries Protection, Decree 40 and Decision 34. The district is responsible for the contracts and the tax level is decided on agreement between the district and the commune (Director of Thinh Binh District (DARD), 1999: interview).

many households (see Box 2.1: NH-LANDLESS). There are two main reasons for this: first, those livelihood activities based on privately owned resources, such as rice and shrimp farming, have failed to expand successfully and, second, there has been a shift towards types of private land use that exclude the poorer groups of the community. Rice agriculture is an activity based on private resources, it requires relatively cheap inputs and it can therefore be carried out by a variety of social groups. At both case study sites, however, rice agriculture has failed to produce sufficient profits to compete with shrimp farming and, in Nam Hai Commune, this has resulted in the total conversion of rice land, thus excluding the poorer social groups from any private livelihood activity. Following initial high profits, shrimp farming itself has failed to produce significant profits and has, in fact, been the ruin of many a household economy. The conversion of failed shrimp ponds back into mud flats or mangrove forest with productivity levels similar to levels before conversion is problematic. Once again, a livelihood source based on a private resource has failed, thus pushing an increasingly wider group of people into a reliance upon open-access resources.

At both research sites, open-access resources represent the mainstay of the livelihoods of the landless and poorest groups (Figs 2.3 and 2.5). Increasingly, landlessness has led to more people depending on alternative livelihoods, such as hiring out of labour and collecting open-access resources. In Nam Hai Commune, hiring out of labour is the preferred activity of the landless households owing to the low availability and quality of wild products, but, in times of adversity when no other livelihood source is available, open-access products provide their main livelihood source. In Da Rang Commune, even households engaged in high-investment activities such as the cultivation of tiger shrimp or trading will also collect some open-access products to supplement their income. In Nam Hai Commune, the collection of open-access products by land-owning groups is less frequent, but these resources do represent an important subsidiary livelihood source that increases in importance during times of poor shrimp harvest.

Conclusions and Policy Implications

Many of those households at the research sites that could afford the investment required for the conversion to shrimp benefited from initially high profits. Conversely, the 'enclosure' of open-access resources and the inability of many households to afford the conversion resulted in serious social upheaval and a shift in livelihood profiles. The subsequent failure of shrimp farming has led to the collapse of the household economies of many shrimp farmers and to a high level of debt. This is a trend seen more recently in other areas of Vietnam such as the north-central coastal area (Luttrell *et al.*, 2004). Above all, the expansion of shrimp farming has produced a sharp differentiation between those social groups that have benefited from the conversion and those social groups that have suffered both directly and indirectly.

These results show that, despite the increasing trend toward the privatization of resources, which the introduction of aquaculture has brought about, it is open-access resources that are making an increasingly important contribution to livelihood profiles of the poorest households. The importance of open-access resources for livelihoods increases in line with the poverty and vulnerability of the social group. This has negative implications for livelihood vulnerability, since livelihood sources that are dependent on open-access resources are vulnerable because of their uncertain rights and legislative status, a status that is open to change and increasingly subject to privatization in these areas.

The way in which such an opportunity for growth has the potential to result in significant inequality (and environmental damage) emphasizes the need to make benefits more accessible to the poorest groups. It also questions the justification for state subsidies supporting infrastructure or credit for such activities such as shrimp farming.

Aquaculture is not an accessible livelihood option for the poorest because of the required levels of capital, technology, infrastructure and land. Labour opportunities in aquaculture are limited outside the family. Far more seriously, aquaculture can reduce other options by decreasing access by the poorest to open-access resources.

References

Asia Pulse (2001) Vietnam's shrimp farms a key part of fisheries five-year plan. *Asia Pulse*, 20 February 2001.

Bailey, C. (1985) The Blue Revolution: impact of technological innovation on Third World fisheries. *Rural Sociologist* 5, 259–266.

Barraclough, S. and Finger-Stich, A. (1996) *Some Ecological and Social Implications of Commercial Shrimp Farming in Asia.* United Nations Research Institute for Social Development, Geneva.

Carney, D. (1999) *Approaches to Sustainable Livelihoods for the Rural Poor.* ODI Poverty Briefing Paper 2: London, UK, January 1999.

Dang Kim Son (1998) Development of agricultural production systems in the Mekong Delta. In: Vo Tong Xuan and Matsui, S. (eds) *Development of Farming Systems in the Mekong Delta of Vietnam.* JIRCAS, CTU and CLRRI, Ho Chi Minh City, Vietnam.

Da Rang Commune[4] (1998) Summary for 1998 and Action Plan for 1999. Da Rang Commune People's Committee, Vietnam.

Davies, S. and Hossain, N. (1997) *Livelihood Adaptation, Public Action and Civil Society: a Review of the Literature.* Institute of Development Studies Working Paper 57. University of Sussex, Brighton, UK.

Eland-Goossense, M.A., Van de Goor, L.A.M., Vollemans, E.C., Hendriks, V.M. and Garretsen, H.F.L. (1997) Snowball sampling applied to opiate addicts outside the treatment system. *Addiction Research* 5(4), 14–23.

Ellis, F. (1998) Household strategies and rural livelihood diversification. *Journal of Development Studies* 35(1), 1–38.

Euroconsult (1996) *Coastal Wetlands Protection and Development, South Mekong Delta, Project Preparation.* Final Report, May 1996.

Faugier, J. and Sargeant, M. (1997) Sampling hard to reach populations. *Journal of Advanced Nursing* 26(4), 8–28.

Flaherty, M., Vendergeest, P. and Miller, P. (1999) Rice paddy or shrimp pond: tough decisions in rural Thailand. *World Development* 27(12), 2045–2060.

Glaser, B.G. and Strauss, A.L. (1967) *The Discovery of Grounded Theory.* Aldine, Chicago, Illinois.

Gubrium, J.F. (1995) Taking stock. *Qualitative Health Research* 5(3), 267–269.

Ling, S. (1977) *Aquaculture in South-east Asia: A Historical Overview.* University of Washington Press, Seattle, Washington.

Luttrell, C. (2000) Institutional change and natural resource use in coastal Vietnam. *Geojournal Special Issue: Communities, Environments and Post-Communist Transition: Case Studies* 55, 529–540.

Luttrell, C., Hoang Van Son, Ha Luong Thuan, Ngo Lan, Cao Tien Viet, Vu Dien Xiem and Dau Thi Le Hieu (2004) *Sustainable Livelihood Opportunities and Resource Management in Coastline Communes Facing Special Difficulties.* Partnership to Assist the Poorest Commune Report, Ministry of Planning and Investment, Hanoi, Vietnam.

Nguyen Duc Cu (1998) Critical environmental threats to the tidal wetlands ecosystems in the northeast coastal zone of Vietnam. In: Phan Nguyen Hong, Nguyen Hoang Tri and Quan Quynh Dao (eds) *Management and Conservation of Coastal Biodiversity in Vietnam.* Proceedings of the CRES/MacArthur Foundation Workshop, Ha Long City, Vietnam, 24–25 December.

Nguyen Thanh Manh and Phan Anh Thi Dao (1998) Preliminary results of research on the socio-economic situation of Da Rang Commune, Xuan Giao District, Quang Ninh. In: Phan Nguyen Hong, Nguyen Hoang Tri and Quan Quynh Dao (eds) *Management and Conservation of Coastal Biodiversity in Vietnam.* Proceedings of the CRES/MacArthur Foundation Workshop, Ha Long City, Vietnam, 24–25 December.

Vietnam News (1999) Ca Mau Province earns $94 million from exports. *Vietnam News*, 8 October 1999.

[4] All district and commune names mentioned in this paper have been changed.

Vietnam News Agency (1999) Import-export operations over past eight months. *Vietnam News Agency*, 1 September 1999.

Vietnam News Agency (2000) Vietnam to maintain rice acreage for food security. *Vietnam News Agency*, 27 June 2000.

Vo Van Trac and Pham Thuoc (1995) *Shrimp and Carp Aquaculture Sustainability and the Environment*. Vietnam study report, Research Institute of Marine Products, Hai Phong and Asian Development Bank RETA 5534.

3 Livelihood Systems and Dynamics of Poverty in a Coastal Province of Vietnam

M. Hossain,[1] T.T. Ut[2] and M.L. Bose[3]

[1] *Social Sciences Division, International Rice Research Institute, Metro Manila, Philippines, e-mail: m.hossain@cgiar.org*

[2] *Nong Lam University, Ho Chi Minh City, Vietnam*

[3] *WorldFish Center, Penang, Malaysia (formerly with the International Rice Research Institute, Philippines)*

Abstract

It is well recognized in the development literature that movement in and out of poverty is caused by several demographic, economic and natural factors that are within and outside the control of the household. This chapter uses primary data collected from a sample survey to understand the major factors behind changes in economic conditions in a coastal province in South Vietnam. The survey was conducted in 2001 in five purposively selected villages to: (i) understand the livelihood systems in the Vietnam coastal area; and (ii) analyse the impact of government intervention (construction of embankments and sluices to prevent saltwater intrusion) for water management on rural livelihoods. A comparative analysis of costs and returns of intensified rice farming, rice–shrimp farming and semi-intensive shrimp farming is conducted to study the effect of the change from the brackish-water to the freshwater system on the productivity of land, the most important asset possessed by rural households. An income determination function is estimated to analyse the effect of the endowment of various capital items on household income. Farmers were asked to report changes in economic conditions over the last 10 years and the reasons behind the changes. These qualitative data were related to the endowment of capital at the household level to analyse the factors contributing to the poverty dynamics in the region.

Introduction

Since the reunification in 1975, the government of Vietnam has sought to improve the livelihood of the rural population mostly through agricultural development focusing on technological progress in rice farming. The *doi moi* policy (renovation) from the late 1980s, with a market-oriented economy, had a positive impact on agricultural production, especially rice production. As a result, Vietnam turned from being a net importer to a major exporter of rice in the world market. The Mekong River Delta (MRD) has been the major source of the growth in rice production and exportable surplus over the last two decades. However, because of saline intrusion, the growth in rice production in the coastal parts of the delta lagged behind that of other regions. To cope with this problem, the government began investing in the early 1990s in the construction of a series of

 Environment and Livelihoods in Tropical Coastal Zones (eds C.T. Hoanh, T.P. Tuong, J.W. Gowing and B. Hardy)

embankments and sluices along the coast to prevent saline water incursion, with the objective of intensifying rice production and improving farmers' livelihoods (Hoanh *et al.*, 2003).

This study assesses the effects of government intervention on the livelihoods of the people in Bac Lieu Province, a coastal province in the MRD. We want to test the proposition that, although increasing rice-cropping intensification through control of saline water intrusion could lead to improvements in livelihoods of the people dependent on rice farming, it might adversely affect the livelihood strategies of the people who rely on fisheries and shrimp farming using brackish water, an important resource in the coastal area.

Conceptual Framework and Methodology

Sustainable livelihood framework

The Sustainable Livelihood Framework (SLF) (Chambers and Conway, 1992; Carney 1998; Dearden *et al.*, 2002) was used as the framework for the study. Livelihood constitutes the capacities, assets (including both material and social resources) and activities needed for a means of living. Livelihoods are sustainable when they are resilient in the face of external shocks and stresses, are not dependent on external support, maintain the long-term productivity of natural resources and do not compromise the livelihood options of others.

The SLF starts with five categories of *capital* (assets) and an understanding of how people use them as a means for livelihood strategies. These assets are: (i) human capital, for example, good health, skills and knowledge; (ii) natural capital such as land, water and biotic resources provided by nature; (iii) physical capital accumulated by the people themselves; (iv) financial capital comprising liquid resources such as cash in hand, savings, jewellery and credit; and (v) social capital such as networks and relationships with people and organizations, and personal qualities. The *livelihood strategy* depends on the household's endowment of these various assets.

The productivity of capital and livelihood outcomes are largely determined by organizations, institutions, policies and infrastructure that are beyond the control of the households. In the SLF analysis, these are called *transforming structures*. The sluice gates and the canal system developed by the government to control intrusion of saline water in the coastal area are examples of a transforming structure. The market forces and institutions that determine the prices of products and inputs are also part of the transforming structure, and they also determine the productivity of capital and returns to enterprises and ultimately have an impact on the *livelihood outcomes. Vulnerability context* frames the external environment in which people operate. These are shocks, trends and seasonal fluctuations, such as natural disasters and price fluctuations, that influence people's livelihoods and their belongings, but are beyond their control. An important livelihood strategy is to accumulate enough assets to develop resilience to these external shocks.

Poverty dynamics

The pathways for moving out of or into poverty that we call 'poverty dynamics' can be analysed using the sustainable livelihood framework. Many earlier researchers (such as Mellor, 1986; Lanjouw and Stern, 1993; Fields *et al.*, 2003) identified the major forces for the enrichment of livelihood. They include: (i) technological progress in agriculture; (ii) accumulation of physical and human capital; (iii) occupational mobility from farm to non-farm activities; and (iv) rural–urban migration.

There are also forces that drag households into poverty. The most important is the increasing population resulting from demographic factors. As children grow up and form their own households, the endowment of natural and physical capital declines (Hayami and Ruttan, 1985). The other factors causing downward mobility are death and disability of earning members, natural hazards and market-induced adverse movement in terms of trade for agriculture.

To assess the importance of various factors behind poverty dynamics, panel data for each household through longitudinal surveys are needed. But we do not have such data. Therefore, we asked respondents these questions: 'What has changed in the economic conditions of the household over the last 10 years?' and 'What are the factors behind the change?' We also made a quantitative analysis of the effect of the change from the shrimp-based to rice-based system on the people in different zones with data on costs and returns in rice and shrimp farming.

Sample selection

The study covers Bac Lieu Province, an area covering 254,000 ha, of which about 200,000 ha are cultivated. It supported a population of 764,000 in 2001. To increase agricultural production, particularly rice, from 1994 to 2000 the government constructed a series of sluices progressively from the east to the west to prevent saltwater incursion (Fig. 3.1). As a result, the canal water progressively became fresh from the east to the west, as indicated by the retreat of the isohalines in Fig. 3.1.

Based on the soil types and the duration since canal water has become fresh, the project area can be classified into three zones.

- *Zone 1* is found in the east of the project site (east of the 1998-isohaline, Fig. 3.1), is characterized by alluvial soil types, with water and environment changed from a brackish-water ecology to a freshwater ecology before 1998. Hereafter, this zone will be called an 'early intervention zone' (EIZ).
- *Zone 2* lies in the middle section of the study area, in between the 1998- and 2000-isohaline. Large areas of this zone have acid sulphate soils. Water and environment changed to a freshwater ecology after 1998 and before 2000. This zone will be referred to as a 'recent intervention zone' (RIZ).
- *Zone* 3 lies to the west and north of the 2000-isohaline. This area is not affected much by the closure of sluices because saline water can flow throughout the area from the East Sea when the sluice system is closed, and therefore is called a 'marginal intervention zone' (MIZ).

To assess the impact of the government intervention on peoples' livelihoods at the study site, we first conducted a participatory rural appraisal (PRA) in order to draw a general picture. Next, we conducted a household survey to gather relevant data on the operation of the household economy and to understand livelihood strategies.

The information obtained at a particular point in time can be used to depict the changes in the livelihood system over time induced by the salinity control intervention. The EIZ represents the situation of stability after the transition from the brackish-water system to the freshwater system is complete. The MIZ represents the situation in the coastal area in the absence (or with little indirect influence) of government intervention. The RIZ represents the situation in transition from the brackish-water system to the freshwater system.

Using this typology and the results from the PRA, we purposively selected five villages: two from the EIZ, two from the RIZ and one from the MIZ. All households in the selected villages were interviewed with a structured questionnaire to collect data on different aspects of the operation of the household economy. The name of the villages and the sample households covered by the survey can be seen in Table 3.1 and locations are shown in Fig. 3.1.

Results

Changes in production systems

The following information on the three zones was obtained from key informants through a PRA.

Early intervention zone (EIZ)

According to the key informants, the salinity in the area started to decrease in 1994, and in 1997 it was about 50% of the level of the pre-

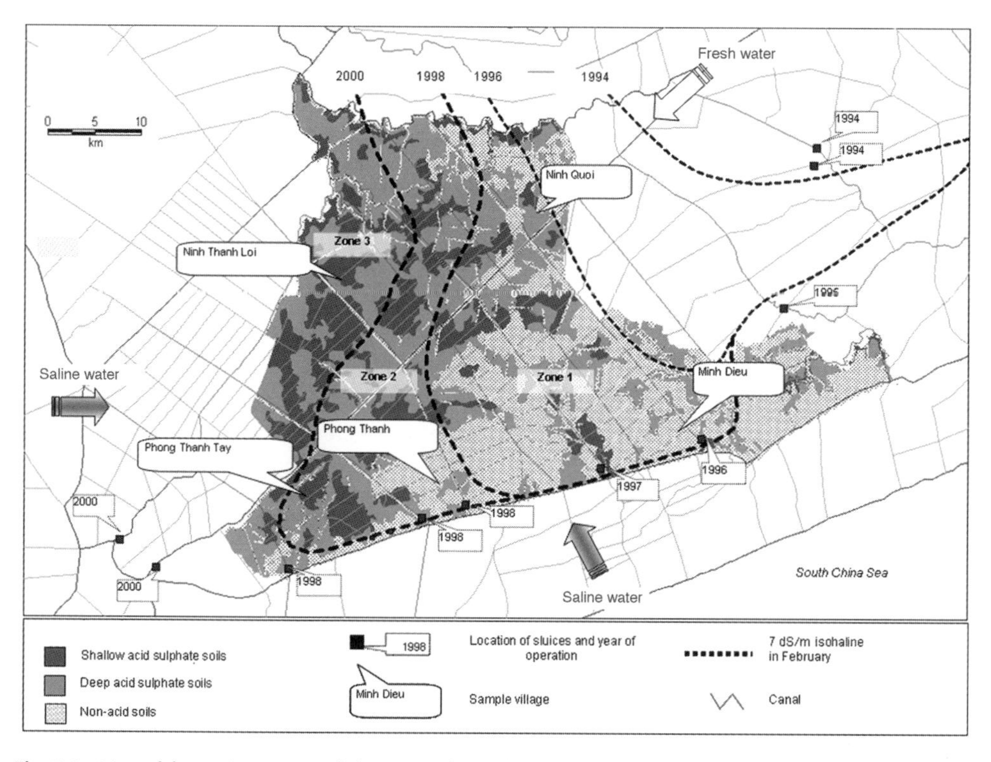

Fig. 3.1. Map of the project area with location of sampled villages and zones. Zone 1, early intervention; Zone 2, recent intervention; Zone 3, marginal intervention or little effects of the interventions.

vious years, and since then fresh water has been found in all canals almost all year round.

Before 1990, swamp land occupied a large area, with abundant supplies of natural shrimp and fish. Most of this swamp land was converted to rice land from 1996 to 1998. With the increased availability of fresh water, rice cropping in the village shifted from one crop of traditional varieties in the 1980s to double cropping by the adoption of short-duration rice varieties. In 1999, farmers tried further intensification by growing three rice crops. But, as the yield for each season dropped, farmers reverted to double cropping of rice. On the relatively high land, they started growing vegetables, maize, sugarcane and fruit trees as the area became flood-free. The natural catch of shrimp and fish decreased substantially, leading to a decline in employment and income from fisheries, particularly for poor households. However, intensification of rice farming generated opportunities for hired work for the landless and marginal farmers. Labourers from the different upper provinces migrated to compete with local residents for this work, so many poor farmers had to move to the provinces nearby to seek work after the second rice crop. Raising freshwater fish in ponds that were no longer flooded became a common source of supplementary income among the economically better-off households.

Recent intervention zone (RIZ)

Before 1996, farmers used to cultivate one traditional saline-tolerant rice crop in the rainy season and shrimp in the dry season. Poor households that did not own land worked as hired workers in shrimp fields, cultivated rice and caught natural shrimp and fish. After 1996, as canal water became progressively fresh, the production system changed. In 1998, 10% of the area under shrimp was converted to raising modern, high-yielding rice varieties. The area allocated to rice increased

Table 3.1. Distribution of samples of households in the study area.

Name of village	Zone	Type of village	No. of households	Average size of households	Percentage of poor households
Ninh Quoi	1	Early intervention	185	4.96	53
Minh Dieu	1	Early intervention	162	5.29	43
Phong Thanh	2	Recent intervention	197	5.34	55
Phong Thanh Tay	2	Recent intervention	213	4.99	60
Ninh Thanh Loi	3	Marginal (little effect from) intervention	154	4.86	39
Total			911	5.09	51

to 70% in 1999 and to 100% in 2000. Although fresh water was found in the canals, acid sulphate soil caused very low rice yields in spite of the use of high-yielding rice varieties. Farmers saw very low returns to rice production, and some even lost money. The intervention affected the livelihood of all categories of households in this area, but large farmers were particularly hard hit. Some farmers started to raise tiger shrimp but few succeeded in getting a good crop. Since the harvest of natural fish had been the largest share of income of the poor groups, their livelihood was particularly negatively affected by the loss of this source of income due to the enclosure of the sluices. The poor started migrating to other places to trade their labour.

Marginal intervention zone (MIZ)

Farmers cannot grow rice in the dry season because of the saline soil and water. After one or two shrimp crops in the dry season, farmers rotated with one salinity-tolerant traditional rice crop in the rainy season. This rice–shrimp system proves to have sustainable productivity. Lime is used in shrimp culture to ameliorate soil acidity; it has positive effects not only on shrimp cultivation but also on rice production in the next season. This resulted in an increase in rice yield from 2 t/ha in 1998 to 3–4 t/ha in 2000.

Livelihood systems

Endowment of capital

Human capital was the predominant asset in the area. An average household consisted of 5.1 persons. The proportions of males and females were equal. The economically active age group (15–67) constituted 67% of the population. About 92% of the males and 87% of the females in this age group participated in economic activities. The average number of workers per household was 3.3; 85% of them were engaged in agriculture (Table 3.2). The quality of human capital in terms of educational attainment of the workers was low. The average years of schooling for the household head (manager) were 4.4 and for the average worker 4.6. For the household head, 12% had had no formal schooling, 40% had attended primary school only and just 5.5% had graduated from secondary schools. Nearly 14% of the households are managed by women.

The area was well endowed with natural capital by Vietnamese standards. The average size of land owned by the household was 1.56 ha, compared with 1.15 ha for the Mekong River Delta and 0.72 ha for Vietnam as a whole. Of this land, 0.11 ha was used for homestead and another 0.11 ha as orchards and homestead gardens, and 1.33 ha as cultivated land (Table 3.2). Very little land was used for raising fish or shrimp, except in the village from the MIZ. The land area under shrimp farming in that village was 1.99 ha.

Households that operated no land constituted 16% of all households. The number varied from 6% to 22% among the five villages. The proportion of the non-farm households was higher in the MIZ. Thus, a sizable proportion of the households was dependent on only human capital for their livelihood. Smallholdings with a size of up to 1.0 ha constituted 37% of the households, while

Table 3.2. Endowment of capital per household, by zone,[a] 2001 (from sample household survey, IRRI).

Capital item	EIZ (*n* = 347)	RIZ (*n* = 410)	MIZ (*n* = 154)	All areas (*n* = 911)
Land owned (ha)	1.34	1.60	1.88	1.56
Land cultivated (ha)	1.12	1.41	1.60	1.33
Shrimp/fish pond (ha)	0.00	0.04	1.41	0.26
Acidic land (ha)	0.56	1.28	1.41	1.03
Irrigated land (ha)	1.07	1.20	1.58	1.22
Land rented in (ha)	0.01	0.02	0.17	0.04
Agricultural workers (units)	2.91	2.75	2.64	2.79
Non-agricultural workers (units)	0.31	0.43	1.05	0.49
Education of head (years of schooling)	4.72	4.48	3.66	4.43
Education of average worker (years of schooling)	4.85	4.64	4.01	4.59
Agricultural capital (US$)	282	174	112	205
Non-agricultural capital (US$)	261	278	210	260
Loan received (US$)	227	246	133	220

[a] EIZ, early intervention zone, RIZ, recent intervention zone, MIZ, marginal intervention zone.

farms of over 4.0 ha that could be considered as large constituted only 3.2% of the households. Some 22% of the households operated holdings of more than 2.0 ha and controlled over 52% of the total land. The incidence of tenancy cultivation was rare except in the village in the MIZ, where 10% of the cultivated land was tenant-operated (Table 3.2). Owing to the concentration of the canals, the area had good control over irrigation. Nearly 92% of the land was irrigated, of which 85% was irrigated by pumps and the remaining 15% by a gravity system.

The accumulation of physical capital was still at a low level. About 45% of the total value of non-land fixed assets was used for agriculture (Table 3.2). The amount of money borrowed from different sources was US$220 per household. Nearly 85% of the loan was obtained from institutional sources.

The level of social capital can be measured by membership of the households in different organizations and networks. Only 11% of the households reported membership in the Farmers' Union, 4.2% in the Veterans' Club and 2.5% in the Women's Union. Members from six out of the 911 households were actively involved with the Communist party.

The variation in capital endowment in the three zones can be noted in Table 3.2. The endowment of land was higher while the level of education of the workers was lower in the MIZ than in the other zones. For other capital items, there was only a marginal difference.

Among the assets, the physical capital was most unequally distributed, with the bottom 40% of the households owning only 7% of the capital, whereas the top 10% had a share of 42% (Table 3.3). The least unequally distributed capital was the level of skills of the workers. The distribution of land was also fairly unequal. The income originating from the improvement in productivity of land and physical capital is thus expected to be unevenly distributed among the households in the area.

Livelihood strategy

Households use their capital endowments in different ways to make a living. Table 3.4 provides information on the occupational distribution of the workers at the study sites. In the survey, the respondents were asked to report a maximum of two occupations, the principal (in which the major portion of time was allocated) and the subsidiary for each household member. The responses indicate that about 17% of the workers had a secondary occupation. The dominant source of livelihood was farming (Table 3.4). Other

Table 3.3. The degree of inequality in the distribution of assets, 2001 (from sample household survey, IRRI).

Rank of the household on the scale of asset ownership	Cumulative share of land (%)	Cumulative share of physical capital (%)	Cumulative share of human capital/schooling (%)
Bottom 40%	10.6	6.6	23.2
Middle 40%	40.8	34.0	44.2
Next 10% (ninth decile)	17.9	17.1	14.5
Top 10%	30.7	42.3	18.1
Gini coefficient	0.47	0.58	0.25

significant sources of livelihood were selling of labour to other farms and organizing handicraft production within the homestead (cottage industry), mostly making mats (with nipa leaves) and fishing nets. Non-farm agricultural activities, such as fishing, livestock raising or forestry, are taken up mostly as subsidiary occupations. Only 1.3% of the workers reported them as the principal occupation. Similarly, many more workers had paid agricultural labour as a subsidiary occupation. Very few workers were engaged in trade, business or services. The data show high dependence on land as the source of livelihood. Women were engaged in farming almost to the same extent as men. But some gender difference in the occupational structure can be noted in Table 3.4.

Table 3.5 reveals some differences in livelihood strategy among the three zones. A larger proportion of workers was dependent on farming in the EIZ than in the MIZ, indicating a positive effect of rice intensification on employment generation. But, the effect on the generation of employment for hired workers was marginal. Engagement in handicraft production was limited mostly to the MIZ. The higher engagement in handicrafts in this zone is presumably due to the availability of raw materials in the saline areas (nipa leaves, mangrove forests, etc.) and also to the demand for products in fishing, fish processing and storage, and fish trade. A substantially larger proportion of workers reported other agriculture (livestock raising, aquaculture, etc.) as a subsidiary occupation in the EIZ. With salinity control came a greater opportunity for earning income through vegetables, horticultural crops and livestock raising.

Table 3.4. Sources of employment by % sex of workers, 2001 (from sample household survey, IRRI).

	Male workers (*n* = 1419)		Female workers (*n* = 1348)		Total (*n* = 2767)	
Occupation	Primary	Primary and secondary	Primary	Primary and secondary	Primary	Primary and secondary
Agriculture	91.7	105.4[a]	86.7	96.5	89.2	101.1[a]
Farming	80.2	85.5	77.5	81.0	78.9	83.3
Other agriculture	1.5	4.3	1.2	5.3	1.3	4.9
Agricultural wage labour	10.0	15.6	8.0	10.1	9.0	12.9
Non-agriculture	8.3	14.2	13.3	18.3	10.8	16.3
Handicrafts	3.5	7.3	7.6	10.4	5.6	8.8
Trade and business	1.0	2.2	3.7	5.5	2.3	3.8
Services	3.3	4.2	1.9	2.3	2.6	3.3
Non-agricultural labour	0.5	0.8	0.1	0.1	0.3	0.4
Total	100.0	119.6[a]	100.0	114.8[a]	100.0	117.4[a]

[a] The total exceeds 100 because workers are counted twice if they reported a secondary occupation.

Table 3.5. Sources of employment of workers by zone, [a] 2001 (from sample household survey, IRRI).

Sources of employment	EIZ		RIZ		MIZ	
	Primary occupation	Primary and secondary occupation	Primary occupation	Primary and secondary occupation (% of all total)	Primary occupation	Primary and secondary occupation
Agriculture	92.6	109.7	89.8	98.4	80.4	88.6
Farming	83.4	88.7	77.9	81.1	71.4	77.1
Other agriculture	2.0	9.1	0.7	2.5	1.5	1.5
Agricultural wage labour	7.2	11.9	11.2	14.8	7.5	10.0
Non-agriculture	7.4	10.7	10.2	14.2	19.6	34.1
Handicrafts	2.4	3.2	4.1	5.7	16.4	29.2
Trade and business	2.1	3.5	2.7	4.3	1.9	3.2
Services	2.8	3.6	2.8	3.6	1.3	1.7
Non-agricultural labour	0.1	0.4	0.6	0.6	Nil	Nil
Total	100.0	120.4	100.0	112.6	100.0	122.7

[a] EIZ, early intervention zone, RIZ, recent intervention zone, MIZ, marginal intervention zone.

The data on the sources of household income show minor involvement of households in many economic activities (Table 3.6). A majority of the households reported some income from cultivation of other crops, livestock raising and forestry, but the amount of income earned from these activities was less than US$100 per annum. In terms of the contribution to household income, the major activities in order of importance were shrimp farming (43%), rice farming (22%), agricultural wage labour (10%) and non-agricultural labour, including handicrafts (6%).

Livelihood outcomes

The average income earned by a household was estimated at US$1032 during the year. With the household consisting of 5.1 members, per capita income was estimated at US$203 (Table 3.6). About 85% of the income came from agriculture (including aquaculture). However, there was a large variation in income in the three zones (Table 3.7). Per capita income was US$635 in the MIZ, owing to the large income contributed by shrimp farming and fisheries. Per capita income in the EIZ was less than a quarter of that in the MIZ. Although availability of fresh water facilitated intensification of rice farming, the loss in income from shrimp farming and fisheries far outweighed the increase in income from rice farming.

The average per capita income was only US$99 for the RIZ. The transition from the brackish-water to the freshwater regime indeed had a much more substantial negative impact on livelihood. Although the average size of cultivated land was higher in this zone than in the EIZ (see Table 3.2), the income from rice farming was less than a fifth, which pushed down the average household income in this zone. Agricultural workers selling labour to rice farmers earned more than the rice farmers themselves. Low rice yield could be attributed to the acid sulphate soil.

The income was also fairly unequally distributed, with a Gini coefficient of total income of 0.56 (Table 3.8). This high degree of income inequality is mainly due to the highly unequal distribution of income from shrimp farming (Gini coefficient = 0.73). The income from rice cultivation was more equally distributed (Gini coefficient = 0.48). The income inequality analysis at the village level shows relatively less inequality within a village. The Gini coefficient in the two villages in the EIZ was 0.46 and 0.37, whereas the estimate for the village in the MIZ was 0.63. The income situation in the two villages under transition (in the RIZ) was not normal during the year of the survey because of substantial income losses from rice cultivation. Actually, it is the larger land-owning households in these villages that suffered more. As a result, the income distribution was in fact better than in normal circumstances. The concentration coefficient of income in this zone was 0.38 and 0.44. These numbers at the village level suggest that the very high concentration of income for the study area as a whole is partly due to the high inequality of income across villages.

Since the level of income is low, and it is highly unequally distributed, the incidence of poverty is expected to be very high in the study area. We have measured different indices of poverty using the method suggested by Foster *et al.* (1984). The poverty-level income is taken as US$127 per capita per year, adjusting the norm used by the Vietnam government by changes in the cost of living index. It is substantially lower than the dollar-a-day norm that the World Bank uses for comparing poverty across countries (World Bank, 2001). The estimated numbers are reported in Table 3.9. For all five villages together, the number of households living below poverty was estimated at 44% for the year of the survey. The numbers varied from 23% for the MIZ to 37% in the EIZ, indicating that the intervention had a negative impact on poverty reduction. The poverty incidence was very high for the villages in the RIZ because of the drastic fall in income from the water management intervention. It can be noted that, although the income inequality was very high in the shrimp-producing village in the MIZ, the poverty situation was lower than in the villages that practised intensified rice farming with fresh

Table 3.6. The structure of household income, all zones, 2001 (from authors' estimate from household survey data).

Sources of income	Households reporting income from the source (%)	Average income from the source for those who earned from the source (US$)	Average income from the source for all households (US$/annum)	Share of the source in total household income (%)
Agriculture	96.6	905	874	84.7
Rice farming	59.2	391	231	22.4
Farming other crops	51.0	58	29	2.8
Fisheries/shrimp	28.5	1546	441	42.8
Livestock	58.5	73	42	4.1
Forestry	51.6	46	24	2.3
Agricultural labour	35.7	298	106	10.3
Non-agriculture	36.2	434	158	15.3
Trade and business	6.6	630	42	4.1
Services	6.3	355	22	2.1
Remittance	7.5	193	14	1.4
Non-agricultural labour[a]	17.0	382	65	6.3
Rents and others	3.8	387	15	1.5
Total income	100.0	1032	1032	100.0
Family size	–	–	5.09	–
Per capita income	–	–	203	–

[a]Includes transport operators, construction workers and workers engaged in cottage industry.

Table 3.7. The structure of household income, specific zones,[a] 2001 (from sample household survey, IRRI).

	Households reporting income from the source (%)			Annual income from the source (US$)		
Sources of income	EIZ	RIZ	MIZ	EIZ	RIZ	MIZ
Agriculture	98.8	94.6	96.8	646	326	2847
Rice farming	79.8	43.6	53.9	415	88	198
Other crop farming	68.0	42.7	35.1	33	33	11
Fisheries/shrimp farming	30.8	10.5	71.4	21	31	2481
Livestock farming	64.0	58.6	46.1	65	32	21
Forestry	55.0	44.4	63.0	22	15	53
Agricultural wage labour	36.3	41.2	19.5	91	128	83
Non-agriculture	26.1	37.3	45.5	90	184	239
Trade and business	4.9	7.8	7.1	15	69	28
Services	8.1	5.4	4.5	21	21	28
Remittance	8.6	8.5	1.9	12	20	5
Non-agricultural labour	8.4	17.3	35.7	24	57	178
Rents and others	4.0	4.9	Nil	18	17	–
All sources	100.0	100.0	100.0	736	510	3085
Family size	–	–	–	5.12	5	5
Per capita income	–	–	–	144	99	635

[a] EIZ, early intervention zone, RIZ, recent intervention zone, MIZ, marginal intervention zone.

Table 3.8. Degree of inequality in income distribution (from authors' estimate from household survey data).

Rank of household on per capita income scale	Cumulative share of total income (%)	Income from shrimp/fish	Income from rice
Bottom 40%	9.7	5.1	10.8
Middle 40%	29.2	17.0	38.7
Ninth decile	15.8	12.1	22.9
Top 10%	45.1	65.7	27.6
Gini coefficient	0.563	0.734	0.476

water. Thus, the change from shrimp farming to rice farming not only reduces the income of the better-off sections of society but also contributes to an increase in poverty.

What do the people themselves think about poverty in a locality? Interest is growing in the participatory measurement of poverty (Narayan, 2000; World Bank, 2001). We asked the respondents to report whether they would consider their households as 'extremely poor', 'moderately poor' or 'non-poor'. The findings on this self-perception of poverty are reported in Table 3.10. The qualitative data should reflect the poverty situation in a normal year, rather than for the year of the survey as measured by our objective analysis of the quantitative data. For the study area as a whole, 51% of the households considered themselves as 'poor', with 14% as 'extremely poor'. The proportion of poor households was reported to be lower in the shrimp-producing village (MIZ) than in the villages in the EIZ. These findings are similar to those from the objective analysis reported earlier.

Table 3.11 shows self-perception of poverty by major socio-economic groups. The incidence of poverty was substantially higher in the households managed by women than in the male-headed households. Extreme poverty is concentrated in households operating less than 0.4 ha of land,

Table 3.9. Measures of poverty by zone, 2001[a] (from authors' estimate from household survey data).

Measure of poverty	EIZ[b]	RIZ	MIZ	All areas
Head count index (%)	36.6	57.8	22.7	43.8
Poverty gap index (%)	17.2	31.0	9.4	22.1
Squared poverty gap index (%)	10.5	21.2	5.2	14.4

[a] The poverty line income for 2001 was measured at US$126.
[b] EIZ, early intervention zone, RIZ, recent intervention zone, MIZ, marginal intervention zone.

Table 3.10. Incidence of poverty: self-perception of households (from authors' estimate from household survey data).

	Percentage of households considering themselves as			
Type of zone/village	Extremely poor	Moderately poor	Non-poor	All households
Early intervention				
Ninh Quoi	15.1	38.4	46.5	100
Minh Dieu	14.2	28.4	57.4	100
Recent intervention				
Phong Thanh	12.7	44.2	43.1	100
Phong Thanh Tay	16.0	43.2	40.8	100
Marginal intervention				
Ninh Thanh Loi	14.3	24.7	61.0	100
All villages	14.4	36.8	48.8	100

those managed by the less educated and those that have agricultural wage labour as the means of livelihood. The information suggests that, at the present stage of development of the rural economy, ownership of land and a better quality of human capital are important factors behind livelihood improvement.

Pathways out of and into poverty

Water management interventions

Government investment in the construction of embankments and sluices for maintaining fresh water for rice cultivation could be considered an element of a 'transforming structure', an external force to promote agricultural intensification and diversification. Since natural capital (land and water) and human capital (workers) are the dominant assets, such an intervention is important to mark the first step towards enrichment by improving the productivity of land and labour.

Table 3.12 presents information on seasonal yields and annual productivity of the rice and shrimp system. The village Ninh Quoi, which produces three rice crops a year, had a total rice production of 11.8 t/ha/year valued at US$1381 and net returns to the household-owned resources of US$761. Growing three rice crops is too taxing, however, for maintaining soil fertility. In Minh Dieu, the other village in the EIZ, farmers grew two rice crops a season and had net returns to family resources of US$606 per year. Farmers of Ninh Thanh Loi village in the MIZ grew one rice crop using traditional varieties and had net returns of US$243 from rice. But, in addition, the farmers harvested shrimp and fish in the other two seasons from the same land and got a yield of 289 kg/ha/year, equivalent to 16.1 t of rice. The net returns from shrimp/fish cultivation per year were estimated at US$1337 per ha. The village got total returns of US$1580 per ha per annum from the rice–shrimp system. Thus, compared with the most intensive rice system that the government intervention has induced, the pre-existing rice–shrimp system

Table 3.11. Profile of the poor by socio-economic group (from sample household survey, IRRI).

Socio-economic group	Percentage of households in group	Percentage of households considering themselves as		
		Extremely poor	Moderately poor	Non-poor
Landholding (ha)				
Nil	15.8	47.2	43.8	9.0
<0.4	12.4	22.1	54.9	23.0
0.4–1.0	24.3	13.1	49.3	37.6
1.0–2.0	25.6	3.4	32.2	64.4
2.0–4.0	18.7	1.2	14.0	84.8
>4.0	3.2	0.0	3.4	96.6
Education of household head				
No schooling	12.2	32.4	39.6	27.9
Attended primary school	40.3	16.6	40.1	43.3
Attended secondary school	42.0	8.6	34.5	56.9
High school graduate	5.5	4.0	22.0	74.0
Source of livelihood				
Farming	76.1	8.9	35.2	55.8
Agricultural labour	7.0	55.6	39.7	4.8
Non-agriculture	16.9	22.6	41.9	35.5
Gender of the head				
Male	85.8	12.1	36.1	51.8
Female	14.2	28.7	40.3	31.0

in the coastal area provided almost twice as much income.

It should be mentioned here that shrimp prices were more favourable than rice prices in the year of the survey. One kg of shrimp was equivalent to 55 kg of rice (US$6.53 versus US$0.12 per kg). Both rice and shrimp prices fluctuate from year to year, but the shrimp market is more volatile. Even if the relative price of shrimp had declined by 50% from the level of 2001, the rice–shrimp system would remain more profitable than the intensive rice production system.

The other consideration for poverty reduction is generation of employment for resource-poor households. We estimate that the rice–shrimp system generates employment of 57 days per ha per year (35 days in two shrimp seasons and 22 days in one rice season), almost equivalent to the days of employment generated in three rice crops in the intensified rice system. Since shrimp is almost entirely marketed and is a perishable crop, it generates additional employment in processing, storage, transport and trade, which is higher than in postharvest and marketing operations of rice.

To conclude, the investment in water management intervention did not contribute to improvement in agricultural productivity and employment generation, a step on the road to livelihood enrichment. We have noted in the previous section that the off-farm and non-farm economic activities were still relatively unimportant because of the limited accumulation of physical capital, low levels of education and limited market for non-farm goods and services. The extent of rural–urban migration is also low, as indicated by the small proportion of households receiving remittance (Table 3.7).

Vulnerability and crisis management

To assess the factors causing movement into poverty, we asked respondents to report the major crises that had hit these households over the previous 10 years, and how they

Table 3.12. Productivity and profitability from rice and shrimp systems, 2001 (from authors' estimate from household survey data).

Indicators	EIZ[a]		RIZ		MIZ	
	Ninh Quoi	Minh Dieu	Phong Thanh	Phong Thanh Tay	Ninh Thanh Loi	
					Rice	Shrimp
Yield, rice equivalent (t/ha)	11.80	8.31	5.76	4.28	2.90	16.13
Summer–autumn[b]	3.79	4.71	3.56	2.46	–	9.71
Autumn–winter[b]	3.50	3.60	2.20	1.82	2.90	6.42
Winter–summer[b]	4.51	–	–	–	–	–
Gross value of production (US$/ha/year)	1381	973	674	500	339	1887
Paid-out cost (US$/ha/year)	620	367	423	346	96	550
Family income (US$/ha/year)	761	606	251	154	243	1337

[a] EIZ, early intervention zone, RIZ, recent intervention zone, MIZ, marginal intervention zone.
[b] Cropping seasons.

coped with these crises. Some 25% of the households reported facing one or more crises during the period. The major crises that caused vulnerability to livelihood were destruction of property by Typhoon Lynda (1997) and floods, death and disability of earning members, an occasional drastic fall in paddy price, death of livestock and loss of soil fertility caused by salinity, in that order of importance. The amount of loss incurred from crises varied from 1 to 10 million VND (US$66 to US$660), with a majority mentioning a loss of from 1 to 5 million VND. A majority reported that they coped with a crisis by using their own savings or by selling or mortgaging property. One in four respondents borrowed money and 17% got help from relatives. Only 38% of the people were able to recover from the losses totally, and 51% partially recovered. Twelve per cent of the households were unable to recover from their losses.

Perception of respondents of upward or downward mobility

Table 3.13 reports the opinions of the respondents on changes in economic conditions over the decade prior to the survey. Downward mobility was reported mostly in the two villages in the RIZ (in transition), where two-thirds of the households reported a deterioration in economic conditions. Upward mobility was most pronounced in the MIZ, where 60% of the households reported upward mobility. The net change was also positive in both villages in the EIZ.

The major factors behind the improvement in economic conditions in the MIZ (the shrimp-based system) were an increase in shrimp or fish production, followed by an increase in rice production (Table 3.14). In the EIZ (the rice-based system), the major factors behind upward mobility were reported as an increase in rice production, followed by an increase in employment opportunities and engagement in livestock production.

The major reasons behind downward mobility in the RIZ were a reduction in both shrimp and rice production. The closure of the sluice gates reduced the availability of brackish water needed for shrimp production. But, the acidity in the soil and the inexperience in the cultivation of modern rice varieties did not contribute to a compensation for the loss through an increase in rice production.

A significant proportion of the households in Ninh Quoi village in the EIZ also reported downward economic mobility. With the availability of fresh water, this village started growing three rice crops, which adversely affected soil fertility and caused a gradual reduction in rice yield. The reduction in rice production was reported as a major factor behind the downward mobility (Table 3.15). Other factors that contributed to the downward mobility in the village were an increase in health hazards for the farmers (increased incidence of malaria because of stagnant water in the canals) and a reduction in income from trade (the construction of sluices adversely affected inland water transport).

Table 3.13. Perceptions of changes in economic conditions in 1990–2000 (% of cases) (from sample household survey, IRRI).

Village and zone	Improved	Unchanged	Deteriorated	Net change
Early intervention zone	41.8	39.2	19.0	22.8
Ninh Quoi	34.6	40.5	24.9	9.7
Minh Dieu	50.0	37.7	12.3	37.7
Recent intervention zone	5.6	25.6	68.8	−63.2
Phong Thanh	9.6	24.9	65.5	−55.9
Phong Thanh Tay	1.9	26.3	71.8	−69.5
Marginal intervention zone				
Ninh Thanh Loi	60.4	23.4	16.2	44.2
All villages	28.6	30.5	40.9	−12.3

Table 3.14. Reasons for improvement in economic conditions (% of cases)[a] (from sample household survey, IRRI).

	Marginal intervention zone	Early intervention zone	
Reasons for improvement	Ninh Thanh Loi	Minh Dieu	Ninh Quoi
Increase in shrimp/fish production	91	3	–
Increase in rice production	23	94	75
Increase in income from trade	2	6	9
Increase in employment opportunities	7	16	27
Increase in income from livestock	1	16	23

[a] The total exceeds 100 because of multiple responses.

Table 3.15. Reasons for deterioration in economic conditions (% of cases)[a] (from sample household survey, IRRI).

	Recent intervention zone		Early intervention zone
Factors behind deterioration	Phong Thanh	Phong Thanh Tay	Ninh Quoi
Reduction in rice production	61	63	37
Reduction in shrimp/fish production	59	54	–
Reduction in natural fishing	9	21	–
Reduction in employment opportunities	12	16	15
Health hazards for earners	13	7	26
Increase in family size	2	1	17
Reduction in income from trade	5	2	15

[a] The total exceeds 100 because of multiple responses.

Discussion and Conclusions

The government intervention in water management indeed succeeded in controlling saline water intrusions into Bac Lieu Province. This had encouraged farmers to grow a second rice crop and, in some areas, a third rice crop in the eastern part of the protected area, where there is good alluvial soil and canal water became fresh before 1998.

However, management of coastal areas is not easy with regard to optimal resource use for increasing income, distributing it better and sustaining the quality of natural resources. The increase in rice production in the early intervention zone comes at the expense of a fall in the production of high-value aquatic products, from both the raising of shrimp in brackish water in fields and the capture of natural fish in canals in the recent intervention zone, where the environment was in a transition from brackish-water to freshwater ecology. Given the prevailing prices of rice and shrimp, even at the low yield (150 kg/ha) under extensive cultivation, shrimp production was many times more profitable than rice production.

The salinity control intervention had a negative impact on the livelihood of the poor. Lacking access to land and with minimal education, the livelihood of the poor depends heavily on catching natural fish and shrimp in the saline canal water. Those with marginal land were also negatively affected because they could not keep up with new technologies, and lacked capital to invest in adopting the new technologies of modern rice cultivation.

The net effect of the government investment in the construction of embankments and sluices was a substantial reduction in farm income at the transitional stage. This situation could improve over time as farmers gain experience with the new land use and cropping patterns. In the long term, the production system in the RIZ may become similar to that of the EIZ. But, because of the presence of acid sulphate soil, it is not expected that the land productivity will be as high as in the EIZ. Even if the productivity is as good as in the EIZ, farm income would still be lower than in the MIZ, where the production system was similar to that of the RIZ before the intervention.

The high value and higher profitability of shrimp production indicate that the brackish water in the coastal area is no less important a natural resource than rice lands. Findings from this study have helped the Bac Lieu provincial government to readjust the land-use policy and water management strategies that allowed intensification of rice in the east of the province and, at the same time, a shrimp-based production system in the rest of the province. This has helped reverse the downward economic mobility of the transition zone (Hoanh *et al.*, 2003).

The low yield of shrimp under present cultivation practices suggests that the productivity of the brackish-water resource could be further increased by developing and diffusing improved technologies and cultural practices. But this has to be done with caution. It has been shown that large-scale intensive shrimp farming is risky, harmful to the environment and not sustainable (see Szuster, Chapter 7, this volume). Even extensive shrimp farming in the long term may also negatively affect the environment and people's livelihood via a reduction in fruit trees, vegetation cover and homestead and livestock production (see Karim, Chapter 5, this volume). Long-term sustainability of the study area will depend upon limiting the intensification of shrimp culture and concerted action among all stakeholders to adopt integrated management of the environment. Also, shrimp raising would cause socio-economic inequality and worsen the distribution of rural income.

The economy of the study site remains at a low level of development, with land and manual labour as the dominant resources. The limited accumulation of physical capital, low quality of human capital and limited market for non-farm goods and services constrained development of the non-farm economy, which has proved to be a major pathway of poverty reduction in many countries. The government should therefore invest more for education, electrification and transport infrastructure, and develop the infrastructure for research and extension for fisheries in coastal areas, as was done in the past for rice in the deltas.

References

Carney, D. (1998) Implementing the sustainable rural livelihoods approach. In: Carney, D. (ed.) *Sustainable Rural Livelihoods: What Contribution Can We Make?* Department for International Development, London, pp. 3–23.

Chambers, R. and Conway, G. (1992) *Sustainable Rural Livelihoods: Practical Concepts for the 21st Century.* IDS Discussion Paper 296, Institute of Development Studies, Brighton, UK.

Dearden, P., Roland, R., Allison, G. and Allen, C. (2002) *Sustainable Livelihood Approaches: from the Framework to the Field.* Sustainable Livelihoods Guidance Sheets. University of Bradford, Department for International Development, UK.

Fields, G.S., Cichello, P.L., Freije, S., Menedez, M. and Newhouse, D. (2003) Escaping from poverty: household income dynamics in Indonesia, South Africa, Spain and Venezuela. In: Fields, G.S. and Pfeffermann, G. (eds) *Pathways out of Poverty*. Kluwer Academic Publishers, Boston, Massachusetts, pp. 13–34.

Foster, J., Greer, J. and Thorbecke, E. (1984) A class of decomposable poverty measures. *Econometrics* 52(3), 761–766.

Hayami, Y. and Ruttan, V.W. (1985) *Agricultural Development: an International Perspective.* Johns Hopkins University Press, Baltimore, Maryland.

Hoanh, C.T., Tuong, T.P., Gallop, K.M., Gowing, J.W., Kam, S.P., Khiem, N.T. and Phong, N.D. (2003) Livelihood impacts of water policy changes: evidence from a coastal area of the Mekong River Delta. *Water Policy: Official Journal of the World Water Council* 5(5/6), 475–488.

Lanjouw, P. and Stern, N. (1993) Agricultural change and inequality in Palanpur. In: Hoff, K., Braverman, A. and Stiglitz, J.E. (eds) *The Economics of Rural Organization: Theory, Practice and Policy.* Oxford University Press, New York, pp. 543–568.
Mellor, J. (1986) Agriculture on the road to industrialization. In: Lewis, P. and Valeriana, K. (eds) *Development Strategies Reconsidered.* Transaction Books, New Brunswick, New Jersey.
Narayan, D. (2000) *Voices of the Poor: Crying Out for Change.* Oxford University Press, New York.
World Bank (2001) *Poverty: World Development Report.* Oxford University Press, New York.

4 Social and Environmental Impact of Rapid Change in the Coastal Zone of Vietnam: an Assessment of Sustainability Issues

J.W. Gowing,[1] T.P. Tuong,[2] C.T. Hoanh[3] and N.T. Khiem[4]
[1]*School of Agriculture, Food & Rural Development, University of Newcastle, Newcastle upon Tyne, United Kingdom, e-mail: j.w.gowing@ncl.ac.uk*
[2]*International Rice Research Institute, Metro Manila, Philippines*
[3]*International Water Management Institute, Regional Office for South-east Asia, Penang, Malaysia*
[4]*Faculty of Agricultural Economics, An Giang University, An Giang, Vietnam*

Abstract

Ca Mau Peninsula, which lies at the extreme southern tip of the Mekong Delta in Vietnam, has experienced rapid environmental and socio-economic change, particularly since the 1990s when the *doi moi* (renovation) policy introduced an agenda of agriculture-led growth. The peninsula lies entirely within the zone of saline intrusion, which previously extended up to 50 km inland during the dry season, thus limiting traditional rice production to only one rainy-season crop. To promote the intensification of rice production, a plan was devised to build a series of coastal embankments and tidal sluices to control salinity intrusion. The protected area lying within Bac Lieu Province, which covers approximately 160,000 ha, is the focus for this discussion.

Data collected on environmental and socio-economic conditions within the protected area during the study period reflect a complex spatial and temporal pattern of impacts. However, the pattern can be seen as a transition between a freshwater environment supporting rice production and a brackish environment supporting shrimp production. The impacts of these changes in environmental management strategies are discussed. A strategy favouring rice production depends on the operation of sluices in such a way that a freshwater environment is maintained. Intensification of the rice production system involves relatively low risk for farmers, but results in relatively low income. On the other hand, shrimp production requires that a brackish environment be maintained and is seen to offer the potential for increased wealth, but at high levels of risk and indebtedness. Currently available evidence of impact cannot easily be extrapolated to assess long-term sustainability.

Introduction

Ca Mau Peninsula, which lies at the extreme southern tip of the Mekong Delta in Vietnam (Fig. 4.1), has experienced rapid environmental and socio-economic change, particularly since the 1990s when the *doi moi*[1] policy introduced an agenda of agriculture-led growth. The peninsula lies entirely within the zone of saline intrusion, which previously extended up to 50 km inland during the dry season, thus limiting traditional rice production to only one rainy-season crop (June–November). To promote the intensification of rice production, a plan was devised to build a series of coastal embankments and tidal sluices to control salinity intrusion (Tuong *et al.*, 2003). The first sluice in this area became operational in 1994 and the zone protected from saltwater intrusion gradually expanded westward as successive sluices were completed (Fig. 4.1). Within the protected area, the duration of freshwater conditions was extended in line with the policy to promote double or triple cropping of rice.

However, as the freshwater zone spread gradually westward, the local economy was undergoing rapid change. Profitability of the rice crop was falling and at the same time aquaculture was experiencing a dramatic boom, fuelled by technical innovations and the high local and export prices of tiger shrimp (*Penaeus monodon*). Traditional extensive systems of shrimp production based on natural recruitment of shrimp larvae were being replaced by semi-intensive monoculture production systems (Hoanh *et al.*, 2003). By 1998, tiger shrimp culture was widespread in the western part of the project area (Fig. 4.2).

The expansion of tiger shrimp culture was consistent with the official policy[2] of the central government adopted in 1998, which explicitly encouraged production for export. But shrimp farming in the protected area contradicted the land-use policy of rice intensification in the area. Despite the apparent success and popularity of shrimp farming, tidal sluices continued to be built and the freshwater zone continued to spread westward up to 2000 (Fig. 4.1). When the supply of brackish water required for shrimp production was cut off, many farmers were forced to abandon aquaculture and to convert to less profitable rice farming. Some shrimp farmers resisted and attempted to maintain favourable conditions by blocking secondary canals and pumping brackish water into their fields, but this created conflict with rice farmers, who depended on fresh water to irrigate their fields. Eventually, in 2001, the original policy emphasizing rice production was revised to accommodate extensive shrimp cultivation in the west while maintaining areas of intensive rice production in the east. Land use adjusted rapidly to this new policy and by 2002 areas of intensive rice production had shrunk back to the freshwater zone in the non-acid soils to the east (Fig. 4.2).

The background to this policy shift within the highly dynamic environmental and socio-economic circumstances of the project area is discussed by Hoanh *et al.* (2003). This shift is significant because it represents a relaxation of centralized planning in order to allow more freedom for private-sector decisions on land use. The pressure that brought about this change can be seen as the rational response by farmers to economic incentives. However, several emerging issues point to the need for continuing scrutiny and regulation of land use (and in particular shrimp farming) to ensure that short-term gain does not result in failure to achieve long-term sustainable development.

The essential requirements for sustainable development can be defined as poverty eradication, changing unsustainable patterns of production and consumption, and protecting and managing the natural resource base

[1] *Doi moi* can be translated as 'renovation' and is commonly used to describe the move from a centrally planned command economy to a market economy with accompanying democratization of social relations.

[2] Decision on the ratification of the programme of developing the export of aquatic products up to 2006; Government Decision no. 21, 1998.

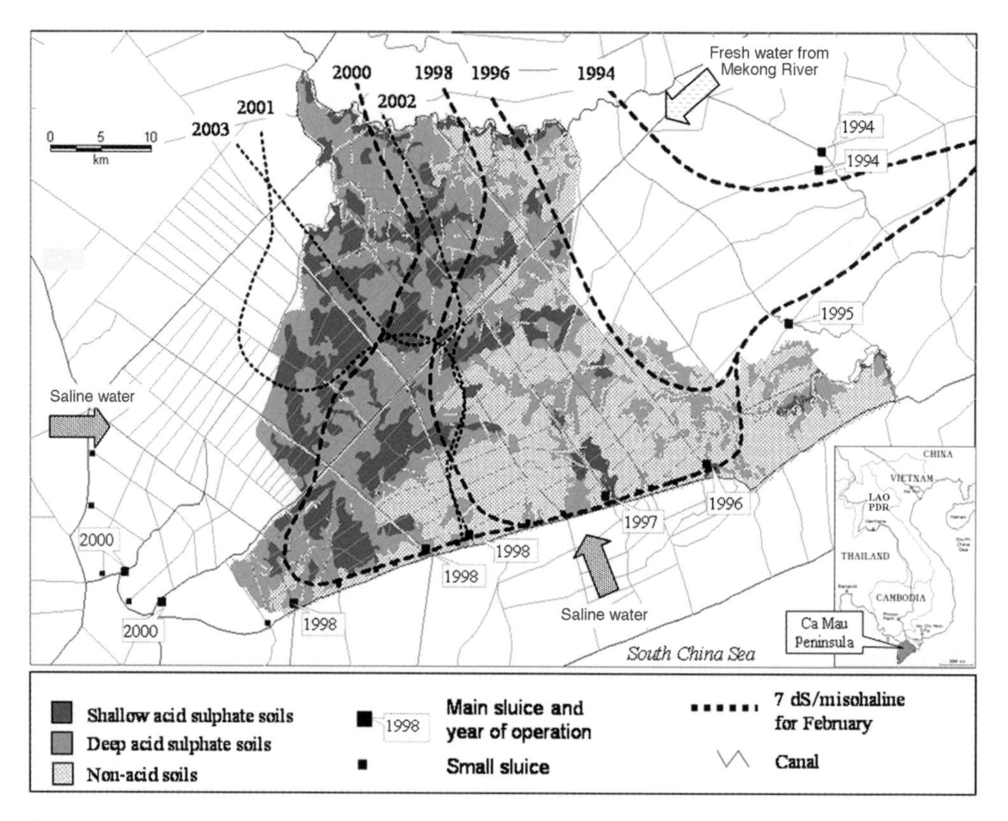

Fig. 4.1. Study area showing extent of saline intrusion at different stages of project development.

(UN, 2002). These objectives need to be understood as mutually supportive and the importance of poverty–environment linkages should be seen as the core of sustainable development strategies. Poor people are predominantly rural and predominantly dependent on natural resources for their livelihoods. They are therefore particularly vulnerable to environmental problems. They need economic growth to escape poverty, but this must be based on the sustainable use of environmental resources. Concern for environmental protection is therefore not seen as being in conflict with concern for human development. The enhancement and maintenance of environmental capital allow for the prudent use of natural resources in the short term while providing for effective protection of the environment in the long term.

The notion of *vulnerability* is critical to understanding the impacts of change. Vulnerability occurs because livelihoods are exposed to stress and are unable to cope with that stress (Adger and Kelly, 2001). Shocks (such as floods and storms) may produce stress when people lose their assets or are forced to dispose of them as a short-term survival strategy. Longer-term adverse trends (such as soil erosion, siltation or acidification) may gradually degrade assets and also create stress. The inherent fragility of poor people's livelihoods limits their adaptive options and makes them generally more vulnerable. Vulnerability therefore relates to both: (i) exposure to risk likely to affect livelihood; and (ii) weakness of the existing state that limits capacity to cope with the resulting stress.

The rapid expansion of brackish-water shrimp farming in the study area indicated farmers' preference for this production over rice. However, brackish-water shrimp farming elsewhere has been characterized by recurrent boom-and-bust cycles that cast

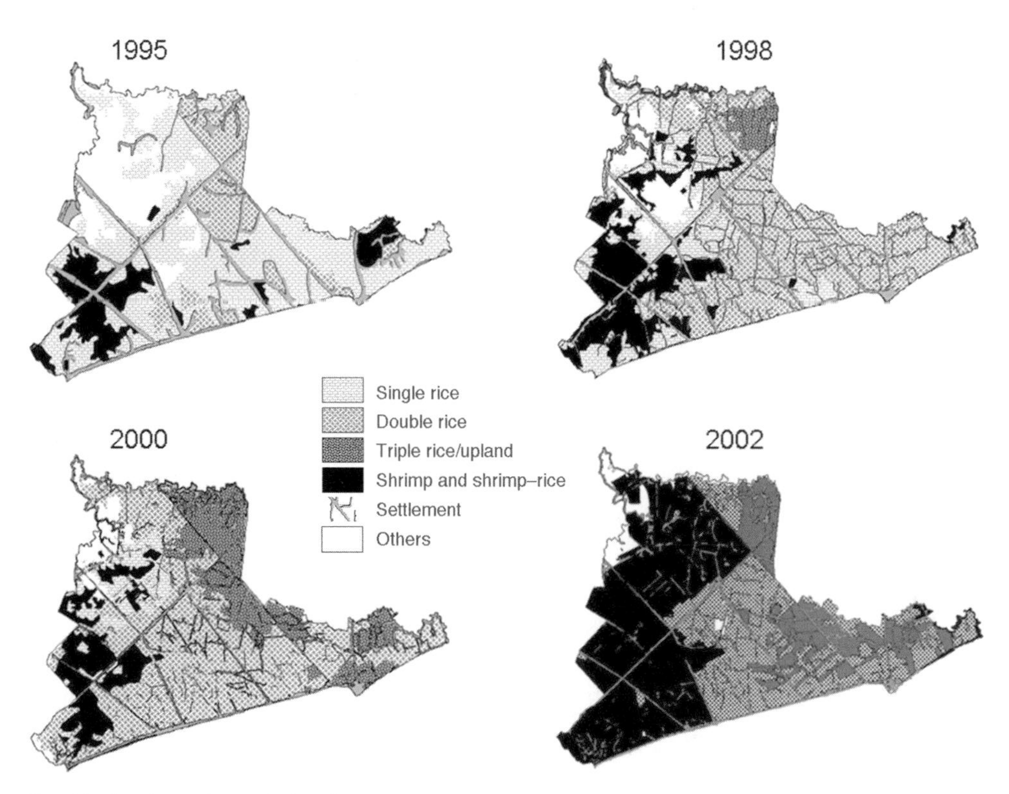

Fig. 4.2. Land-use change in the study area.

doubts on its sustainability (Primavera, 1998). Adverse environmental and socio-economic impacts have been widely reported over the past decade (Phillips *et al.*, 1993; Beveridge *et al.*, 1997; Primavera, 1998; Lebel *et al.*, 2002; Shanahan *et al.*, 2003). It is important to examine the sustainability of shrimp farming in the study area.

One of the characteristics of the study area is the presence of extensive deposits of acid sulphate soils (ASS), particularly in the western part. These soils develop as a result of the drainage of parent materials that are rich in pyrite (FeS_2), as occurs with the reclamation of brackish-water intertidal swamps. ASS are characterized by low pH (< 4), which results from the oxidation of reduced S- and Fe-bearing compounds producing acid via a number of possible chemical and biochemical pathways (Dent, 1986). The acid released in this way reacts with clay minerals to produce soil–water solutions containing high concentrations of aluminium and iron. Reclamation of ASS for agricultural use depends on leaching these toxic substances out of the root zone and has been reported to create severe acidification of the aquatic environment in Indonesia (Klepper *et al.*, 1990) and Australia (Wilson *et al.*, 1999) as well as in Vietnam (Minh *et al.*, 1997b). The acid pollution and associated changes in water chemistry can have severe impacts on fish populations (Callinan *et al.*, 1993), which are not confined to the reclaimed area. The impact on estuarine ecosystems has been reported by Sammut *et al.* (1996) and Wilson *et al.* (1999) for Australia, where mass mortalities of fish have been recorded as a result of episodic acidification, and there is also evidence of chronic long-term effects on estuarine and coastal ecosystems.

The conflict between agricultural production and environmental protection poses a question of the sustainability of development of ASS within the study area. Leachate from reclaimed ASS in the Mekong Delta shows marked peak acidity and Al concentration at the start of the rainy season (Minh *et al.*, 1997a; Tuong *et al.*, 1998), which may create shocks for the aquatic resources and *vulnerability* for the landless poor who depend a great deal on fisheries for their livelihood.

Objectives of This Study

Given the nature of the deltaic environment, strategies for both land use and water management have been equally influential in determining the nature and extent of recent environmental changes in Ca Mau Peninsula. The pre-2000 strategies, favouring rice monoculture, can be criticized (with the benefit of hindsight) for suffering from the same faults identified by Hori (2000) in a review of development planning throughout the lower Mekong basin. They failed to recognize the diversity of livelihoods of the population and did not give adequate consideration to environmental impact. The study reported here was therefore conducted between 1999 and 2003 to investigate the likely sustainability of development as it affects that part of the newly protected area lying within Bac Lieu Province (approximately 160,000 ha).

Sustainability is a multidimensional concept that is measured by reference to multiple indicators selected to represent environmental, social and economic aspects. However, it is clear that two critical issues against which recent change in Bac Lieu Province should be judged are:

- environmental sustainability, which is achieved when the productivity of the natural resource base (i.e. environmental capital) is conserved or enhanced; and
- livelihood sustainability, which is achieved when households respond and adapt to change without becoming more vulnerable.

The objective of this chapter is therefore to present an assessment of the sustainability of the development, with a particular focus on:

- the state of the environment and natural resources within the project area, and
- socio-economic conditions and the relationship between livelihoods and natural resources within the project area.

Short-term Assessment of Project Impact

Social impact: method of assessment

In recognition of the importance of livelihood sustainability, one of the work packages within the study focused on social impact assessment, that is, *assessing changes in farmers' livelihoods and farmers' resource-use strategies in coping with changes brought about by salinity intrusion protection.*

The initial activity under this work package was to establish a framework for social differentiation so that the poor could be identified as a distinct stakeholder group in order to permit an appraisal of environment–poverty linkages and any differential impact of environmental change on them. The framework adopted was based on participatory wealth-ranking exercises.

At an early stage in the project (in 2000), a quantitative survey was conducted in 14 hamlets representing different hydrological, soil and land-use regimes (Tuong *et al.*, 2003). This survey provided data required to characterize the baseline condition in terms of asset and production profiles for 350 representative households across four wealth categories. This was followed by an in-depth qualitative survey that used standard participatory rural appraisal (PRA) tools in interviews with key informants to improve understanding of livelihood strategies and to help identify the key issues affecting poor people (Chambers, 1997).

Follow-up surveys were conducted toward the end of the project (in 2003) to provide data for an assessment of short-term livelihood impacts. A combination of quantitative and qualitative surveys was again adopted within seven of the original 14 hamlets, where all households previously included in the baseline study were resurveyed. The key indicators used to measure changes in livelihoods are summarized in Table 4.1. These indicators provide some insight into the short-term impact of environmental change within the project area based on quantitative data for 1999 to 2003, which can be extended back to 1994 on the basis of insights derived from the PRA. However, the prediction of future livelihood trajectories of different stakeholders is more problematic.

Environmental impact: method of assessment

In recognition of the importance of environmental sustainability, another of the work packages within the study focused on environmental impact analysis, that is, *characterizing the changes in soil and water quality and resource use brought about by salinity protection interventions.*

The initial activity under this work package was the collection and compilation of data from existing soil and land-use maps and the water quality monitoring network in the area. Subsequently, the activities involved the actual soil and water quality monitoring carried out by the study team. By bringing these data into a GIS for spatial data management, integration and visualization, it was possible to establish an idea of the baseline condition (Kam *et al.*, Chapter 15, this volume). It also provided a means of linking the monitoring activity to modelling work based on the VRSAP[3] hydraulic model (Hoanh *et al.*, 2001, 2003).

The soil map published by IRMC[4] in 1999 at a scale of 1:50,000 provided a baseline for monitoring soil changes. Initial fieldwork was undertaken (in 2000) to verify the predictive value of the soil map at selected locations based on the previously defined sampling zones. It was found to be satisfactory, but some areas mapped in 1998 as *potential* acid sulphate soils had already developed into *actual* acid sulphate soils. A soil monitoring programme was established as part of this work package, which included intensive surveys of acidity release.

Data on water quality are subject to much more temporal variability than are soil data, and therefore presented a greater sampling problem. Data on salinity, pH and Al^{3+} from 26 locations throughout the study area were collected from 1993 to 2000, but the frequency of observations was inconsistent. A follow-up water monitoring programme was established in 2001, which included more intensive roving surveys. The aim of these activities was to provide data on short-term changes and a basis for predicting longer-term impact on soil and water quality. Particular concerns were:

- the presence of extensive deposits of acid sulphate soils (ASS) that provide a source of acidity likely to affect water quality, and
- the impact of water quality change (acidity and/or salinity) on the productivity of fisheries.

Table 4.1. Key indicators of livelihood impact.

Livelihood asset	Investigate changes in
Natural	Land ownership; total size of landholding; number of land parcels; cropping patterns; rice production; non-rice crop production; aquaculture production (shrimp and other); capture fisheries catch and seasonality, including destination of catch (home consumption or sale)
Financial	Total net HH income; contribution to total HH income from rice, shrimp, other aquaculture, livestock, employment, capture fisheries; remittances from relatives and other funds
Physical	Access to TV, radio, rowing boat, motorboat, tiller/pump/thresher
Human	Workers available per HH; age of HH head; % of female-headed HH; main occupation of HH heads

HH, household.

[3] VRSAP, Vietnam River System and Plains model.
[4] IRMC, Integrated Resource Mapping Center.

Use of the VRSAP model allowed for spatial analysis across the whole of the protected area for the period 1993–2003, but downstream impacts in the surrounding coastal zone may also be significant and these were not monitored. To analyse environmental sustainability, it is necessary to consider both wider and longer-term impacts.

Evidence of short-term impact

Data collected on environmental and socio-economic conditions within the protected area during the study period should be seen in the context of rapid change occurring at that time, which resulted in a complex spatial and temporal pattern of impacts. However, the key characteristic is the transition between a freshwater environment supporting rice production and a brackish environment supporting shrimp production, as summarized in Fig. 4.3.

A strategy favouring rice production depends on the operation of sluices in such a way that a freshwater environment is maintained. Intensification of the rice production system involves relatively low risk for farmers, but results in relatively low income. On the other hand, shrimp production requires that a brackish environment be maintained and is seen to offer the potential for increased wealth, but at high levels of risk and indebtedness.

Within the extensive areas of ASS, rice production is seriously constrained by soil conditions, but low profitability precludes the adoption of measures to ameliorate the problem. However, there is evidence that more land is brought into use after conversion to shrimp production. Inputs of lime are widely used to ameliorate acidity within the shrimp ponds, although release of acidity into the wider environment still occurs.

Other impacts on natural resources and consequently on livelihoods can also be identified. The freshwater environment favours higher production of fruits and vegetables from home gardens, and of livestock for sale and consumption, than is possible in the brackish environment. On the other hand, by sustaining freshwater conditions, the catch of wild fish from open-access canals is seen to decline.

These impacts on productivity of the natural resource base result in knock-on effects on livelihoods, but the nature of these secondary impacts varies depending on the livelihood strategy adopted by a particular

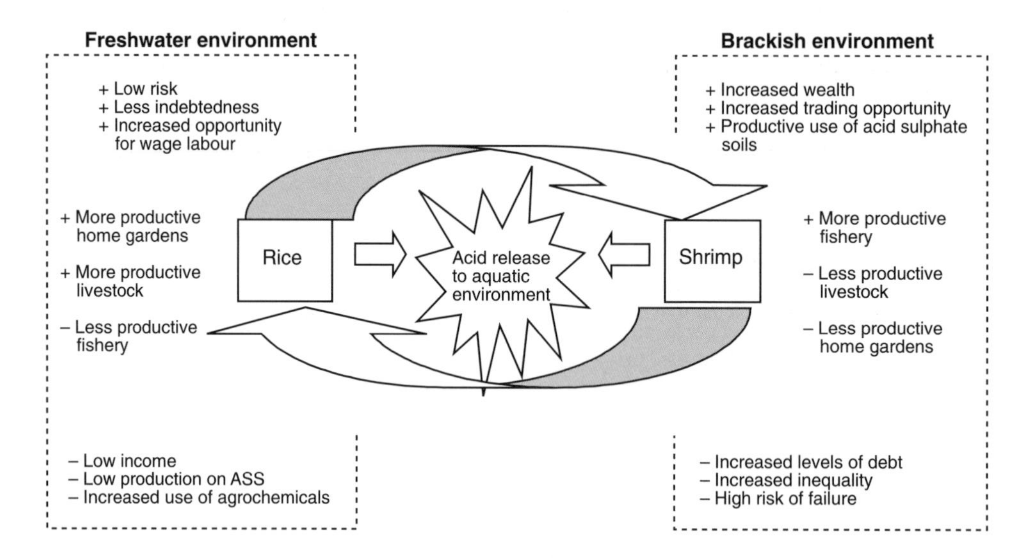

Fig. 4.3. Impacts of alternative environmental management strategies within the protected area.

household (Table 4.2). Households classified as 'very poor' were identified in five of the 14 hamlets and represent 7% of all households. They exhibit a very high proportion of functionally landless and depend on selling labour. They find more employment opportunities in rice areas than in shrimp areas. Households classified as 'poor' exist in all hamlets and represent 34% of the total population. They have some land and may benefit from conversion to shrimp production, but their asset base is low and the associated risk makes them very vulnerable to indebtedness. 'Average' and 'rich' households represent, respectively, 37% and 21% of the total.

Long-term Assessment of Sustainability

Shrimp production

Probably the most frequently voiced concern over shrimp farming is the degradation of coastal wetland ecosystems and in particular the loss of mangrove forests. Over the last 50 years, Vietnam has lost a large proportion of its original mangrove cover through various pressures, but during the last decade the principal threat was reported as shrimp aquaculture (Shanahan *et al.*, 2003). Critics have pointed to the adverse consequences of conversion of mangroves to shrimp ponds on aquatic resources and habitat and forest products, and the increased vulnerability to storms and typhoons in Thailand, the Philippines, Indonesia and Bangladesh, as well as in Vietnam. However, this criticism does not apply to the particular circumstances of the study area, where the expansion of shrimp aquaculture has occurred on land previously cleared for agriculture.

Salinization of soil and water is another major concern associated with shrimp aquaculture, particularly where shrimp ponds have expanded into rice land (Flaherty *et al.*, 1999). Shrimp production requires brackish water and its import into non-saline areas affects neighbouring rice fields and degrades water used for both agricultural and domestic needs. Again, when applied to the particular circumstances of the study area, this criticism is not valid. Salinity intrusion is a natural seasonal phenomenon within the area, which can now be managed using the recently constructed tidal sluices, and soil salinity cannot be attributed to shrimp production. There is evidence (Tran *et al.*, 1999) that salt leaching from shrimp ponds into adjacent rice fields can be significant, but no long-term buildup of salts is apparent in sequential shrimp–rice production systems (Brennan *et al.*, 2002; Phong *et al.*, 2003). This problem can be managed by appropriate zoning and the adoption of appropriate production systems.

Other environmental impacts are closely linked to the level of intensity of the shrimp production system. In this respect, there is a marked contrast between Vietnam and other major producer countries, such as Thailand, where the 'industrialization' of shrimp aquaculture has resulted in much higher intensities (Lebel *et al.*, 2002), as reflected in Table 4.3. As production becomes more intensive, economic viability becomes dependent on high levels of feed, pesticides and antibiotics

Table 4.2. Livelihood strategies.

Wealth category	Landholding (ha)[a]	Landless (%)	Livelihood strategy
Very poor (*n* = 24)	0.6	21	Mainly off-farm employment; capture fishery important; no home garden
Poor (*n* = 120)	1.3	18	50% income from own farm; also off-farm employment and fishing
Average (*n* = 131)	2.0	9	80% income from own farm
Rich (*n* = 75)	2.9	3	90% income from own farm

[a]Landholding is mean area per household available for productive use.

in order to achieve high postlarval survival rates and high growth rates. Water exchange then releases contaminated pond effluents into the wider environment. The same receiving waters generally serve as intake water for neighbouring shrimp farms and water-borne disease agents spread rapidly from farm to farm, thus encouraging the greater use of antibiotics.

Within the study area, production systems are at relatively low intensity (as also reported by Lebel *et al.*, 2002), with lime and seed (postlarvae) as the main purchased inputs. The low quality of hatchery-reared seed is a major constraint contributing to the spread of disease, high risk of failure and pressure to increase the use of antibiotics. Under present circumstances, pollution is not a major concern and the release of acidity is a more significant impact; however, there is clear evidence from elsewhere that highly intensive systems are not sustainable.

Shrimp aquaculture has been encouraged by the government of Vietnam to help raise the standard of living of rural communities. Perceived benefits include job creation during pond construction, income generation for shrimp farmers, increased trading opportunities and employment in aquaculture services (seed supply, other inputs, processing, marketing) and increased foreign exchange earnings. The potential to generate wealth undoubtedly exists, yet shrimp aquaculture has been responsible for the deterioration of livelihoods and promotion of poverty in many countries. Drawing on a series of case studies undertaken in Vietnam, Adger and Kelly (2001) concluded that poorer households are most vulnerable because: (i) they are likely to depend on a narrower range of resources and income sources; and (ii) the loss of common property management rights represents a serious erosion of their ability to cope. They found evidence of benefits from an increase in overall wealth, but increased inequality.

The high risk associated with the enterprise exposes poor people to financial ruin and landlessness. The vast majority of shrimp farmers must borrow money to pay for pond construction and purchase inputs. When the enterprise is successful, profits can be high, allowing them to pay off their debts quickly. However, the risk of failure is also high; Lebel *et al.* (2002) report a 70% rate, Shanahan *et al.* (2003) report a 70–80% rate and a survey in the study area indicates a similar risk of failure. Failure leads to spiralling indebtedness and eventually to landlessness (as reported also by Luttrell, Chapter 2, this volume), although there is as yet no direct evidence that this has occurred within the study area. Relatively poor farmers are more likely to fail, but, if their neighbours have converted from rice to shrimp farming, they have no alternative.

Landless households depend on selling labour, but evidence from elsewhere (Lebel *et al.*, 2002; Shanahan *et al.*, 2003) supports findings within the study area that shrimp farming creates fewer employment opportunities than rice farming. Environmental

Table 4.3. Comparative characteristics of typical shrimp farms (adapted from Lebel *et al.*, 2002).

Item	Study area	Ca Mau/Bac Lieu	Western Thailand	Eastern Thailand
Mean pond size (ha)	2.0	2.6	0.6	0.5
Stocking rate (per m^2)	2–3	4.8	55	66
Survival rate (%)	na[a]	11	59	51
Use of feed (% farms)	0	14	100	99
Mean production (kg/crop/ha)	150	185	4760	4460

[a]na, not available.

degradation combined with enclosure of open-access areas affects the livelihoods of poor people (particularly landless households) and adversely affects food security. Where shrimp ponds are developed on land reclaimed from open-access mangrove forest there is a particular concern, but within the study area this is not the case. However, an indirect impact on common property resources still exists because of the change in water quality and its effect on the productivity of the fishery in the canals.

Reclamation of acid sulphate soils

The conflict between agricultural production and environmental protection is the core problem and assessment of the sustainability of development of ASS within the study area leads to the question: Can the land be managed to maintain agricultural productivity while minimizing downstream impacts, or should it be returned to wetlands (Tuong *et al.*, 1998)? This requires considering processes that generate acidity (i.e. pollution source) and processes that spread acidity (i.e. pollution transport).

Figure 4.4 presents a schematic comparison of acid pollution processes in rice fields and shrimp ponds. During the dry season, the availability of brackish water allows for the filling of shrimp ponds that are limed to ameliorate the effects of acidity, but rice fields on ASS are likely to remain fallow. Shrimp ponds may lose some water to the surrounding canals via seepage (Fig. 4.4a). During the wet season, rice cropping depends on collecting direct rainfall, but can continue after the end of the monsoon by irrigating. In addition to leaching and flushing of rice fields, leaching of acidity from embankments also occurred (Fig. 4.4b).

Data collected from the study showed that the pH of the canal water in the ASS area of the study site was as low as 3 at the end of May and beginning of June, depending on the rainfall (Tuong *et al.*, 2003), but was approximately 7 in the dry season. High pH in the dry season suggests that shrimp ponds do not generate high acid pollution loads. This could also be attributed to the salinity of the brackish water in the dry season, which has high buffering capacity.

Very low pH at the beginning of the rainy season indicates that leaching and flushing of rice fields and leaching of embankments represent an environmental hazard. Evidence collected during the study indicates that the embankments of the shrimp ponds were a major source of acidity, which flushes directly into the wider environment at the start of the monsoon. This was supported by Minh *et al.* (1997a,b), who reported that acidic loading from raised beds in ASS (for growing upland crops such as sugarcane and yam) was eight times higher than that from rice fields. The formation of embankments is similar to that of raised beds and involves the excavation of acidic materials and putting them on the embankments/ raised beds. The materials oxidize during the dry season. The acidity is mobilized by rainfall. The pollution is most problematic at the beginning of the rainy season because the high loading coincides with low river discharge for dilution. Though the study did not quantify the impact of acidic pollution on fishery resources, farmers in our surveys confirmed that the diversity of fishery resources declined during the period of high acidity. Those poor farmers who relied on capture fishery also reported low catches during this period.

Acid generation from rice fields is more controllable in that it depends largely on managed flushing prior to planting and therefore depends on the timing of the cropping season. An integrated rice–shrimp system – rice cropping in the rainy season and shrimp raising in the dry season – would allow good prospects for control over acid pollution from rice fields. However, acidity release from the embankments and consequential impact on water chemistry are likely to remain a problem for many years and it is important that the operation of tidal sluices does not exacerbate the problem. These sluices may reduce neutralization by tidal inflows of estuarine water, which would otherwise occur naturally. Estimated rates of acid production and export from ASS in Australia have been reported as 100–300 kg H_2SO_4/ha/year (Sammut *et al.*,

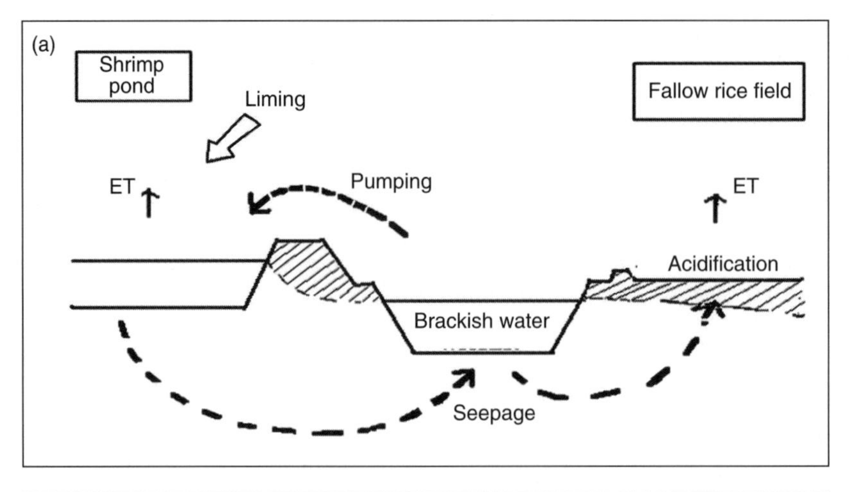

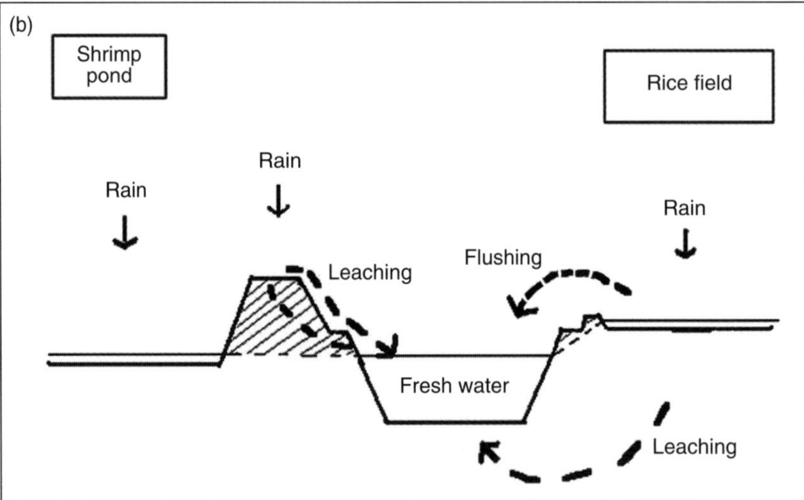

Fig. 4.4. Acid pollution processes in reclaimed acid sulphate soils in the dry season (a) and wet season (b). ET, evapotranspiration.

1996; Wilson *et al.*, 1999). Accumulation behind sluices during sluice closure, followed by sudden release as slug flow into the downstream receiving water, can be catastrophic.

Conclusions

It is clear that conditions are highly dynamic and livelihood strategies have adapted to economic, technological and policy change. But are they sustainable? The concept of sustainability requires a long-term perspective, but currently available evidence is short-term and inconclusive. The concept is also multidimensional and demands consideration of trade-offs among environmental, social and economic impacts, but information is incomplete. The analysis presented here is therefore a tentative attempt to extrapolate from what has been learned during the study by drawing upon experience elsewhere. The analysis has focused on environmental sustainability and livelihood sustainability. In making this assessment, particular consideration has been given to two critical issues that are likely to have a dominant impact on sustainability: shrimp aquaculture and the reclamation of acid sulphate soils.

Within the study area, systems of shrimp production are currently at relatively low intensity. Consequently, the environmental impact from shrimp production is not a major concern and the release of acidity from the reclamation of ASS is more significant. However, social impact is still an issue because of concern for the high risk of mass mortality from diseases and increased inequality. Effective dissemination of good practice, together with improved quality control of hatchery-reared postlarvae, can be expected to reduce the high risk associated with shrimp production. However, long-term sustainability will depend on limiting intensification and concerted action among all stakeholders to adopt integrated management of the environment.

The original development strategy, which promoted intensive rice production throughout the protected area, was rejected by many local stakeholders. Intensification of the rice production system involves relatively low risk for farmers, but results in relatively low income. Within the extensive areas of ASS, rice production is seriously constrained by soil conditions, and low profitability precludes the adoption of measures to ameliorate the problem. Consequently, there is a clear preference among many landholders to produce shrimp in this environment. Inputs of lime are widely used to ameliorate acidity within shrimp ponds and the acid pollution problem consequently decreases, but some release of acidity into the wider environment still occurs. Within the designated freshwater zone on non-acid soils, the current strategy of double-cropped rice appears to be favoured by local stakeholders and no major sustainability concerns have been identified. Elsewhere, an integrated rice–shrimp system offers the best prospects for balanced and sustainable development.

Fish have always been abundant and are considered as a commodity, like water and air, that will always be there, but there is evidence that changes within the protected area have had an impact on fishery resources. The tidal sluices have a direct impact by acting as a barrier to fish migration and an important indirect impact by controlling upstream tidal influence and salinity levels. Their impact on water quality is exacerbated by the accumulation of acid water leached from reclaimed ASS. The acid pollution and associated changes in water chemistry (e.g. high concentration of dissolved aluminium) can be expected to affect the aquatic ecosystem both within the protected area and in the downstream estuarine receiving waters. Evidence from Australia of chronic long-term effects of acid pollution on estuarine and coastal ecosystems provoked White *et al.* (1996) to raise concerns about likely adverse impacts of constructing tidal sluices. This study confirms their analysis and highlights the important connection between environmental and social impacts.

References

Adger, W.N. and Kelly, P.M. (2001) Social vulnerability and resilience. In: Adger, W.N., Kelly, P.M. and Nguyen Huu Ninh (eds) *Living with Environmental Change*. Routledge, London and New York.

Beveridge, M., Phillips, M.J. and Macintosh, D.J. (1997) Aquaculture and the environment: the supply of and demand for environmental goods and services by Asian aquaculture and implications for sustainability. *Aquaculture Research* 28, 797–807.

Brennan, D., Preston, N., Clayton, H. and Be, T.T. (2002) *An Evaluation of Rice–Shrimp Farming Systems in the Mekong Delta*. Report prepared under the World Bank, NACA, WWF and FAO Consortium Program on Shrimp Farming and the Environment, 10 pp.

Callinan, R.B., Fraser, G.C. and Melville, M.D. (1993) Seasonally recurrent fish mortalities and ulcerative disease outbreaks associated with acid sulphate soils in Australian estuaries. In: Dent, D.L. and van Mesvoort, M.E.F. (eds) *Selected Papers from the Ho Chi Minh City Symposium on Acid Sulphate Soils*. Pub. 53, ILRI, Wageningen, Netherlands.

Chambers, R. (1997) *Whose Reality Counts?* Intermediate Technology Publications, London, 297 pp.

Dent, D. (1986) *Acid Sulphate Soils: a Baseline for Research and Development*. Pub. 39, ILRI, Wageningen, Netherlands, 204 pp.
Flaherty, M., Vandergeest, P. and Miller, P. (1999) Rice paddy or shrimp pond: tough decisions in rural Thailand. *World Development* 27(12), 2045–2060.
Hoanh, C.T., Tuong, T.P., Kam, S.P., Phong, N.D., Ngoc, N.V. and Lehmann, E. (2001) Using GIS-linked hydraulic model for managing water quality conflict for shrimp and rice production in the Mekong River Delta, Vietnam. In: Ghassemi, F., Post, D., Sivapalan, M. and Vertessy, R. (eds) *Proceedings of MODSIM 2001*, International Congress on Modelling and Simulation, Canberra, Australia, 10–13 December 2001. Volume 1: *Natural Systems (Part One)*, pp. 221–226.
Hoanh, C.T., Tuong, T.P., Gallop, K.M., Gowing, J.W., Kam, S.P., Khiem, N.T. and Phong, N.D. (2003) Livelihood impacts of water policy change: evidence from a coastal area of the Mekong River delta. *Water Policy* 5, 475–488.
Hori, H. (2000) *The Mekong: its Development and Environment*. United Nations University Press, Tokyo and New York.
Klepper, O., Hatta, G. and Chairuddin, G. (1990) Environmental impacts of the reclamation of acid sulphate soils in Indonesia. *IARD Journal* 12(2), 29–34.
Lebel, L., Tri, N.H. and Saengnoree, A. (2002) Industrial transformation and shrimp aquaculture in Thailand and Vietnam: pathways to ecological, social and economic sustainability? *Ambio* 31(4), 311–323.
Minh, L.Q., Tuong, T.P., van Mensvoort, M.E.F. and Bouma, J. (1997a) Tillage and water management for riceland productivity in acid sulfate soils of the Mekong delta, Vietnam. *Soil and Tillage Research* 42(1), 1–14.
Minh, L.Q., Tuong, T.P., van Mensvoort, M.E.F and Bouma, J. (1997b) Contamination of surface water as affected by land use in acid sulfate soils in the Mekong delta, Vietnam. *Agriculture, Ecosystems and Environment* 61(1), 16–27.
Phillips, M.J., Lin, C.K. and Beveridge, M. (1993) Shrimp culture and the environment: lessons from the world's most rapidly expanding warmwater aquaculture sector. In: Pullin, R.S.V., Rosenthal, H. and Maclean, J.L. (eds) *Environment and Aquaculture in Developing Countries*. ICLARM, Manila, Philippines, pp. 171–197.
Phong, N.D., My, V., Nang, N.D., Tuong,. T.P., Phuoc, T.N. and Trung, N.H. (2003) Salinity dynamics and its implication on cropping patterns and rice performance in shrimp–rice system in My Xuyen and Gia Rai. In: Preston, N. and Clayton, H. (eds) *Rice–Shrimp Farming in the Mekong Delta: Biological and Socioeconomic Issues*. ACIAR Technical Reports No. 52e, pp. 70–88.
Primavera, J.H. (1998) Tropical shrimp farming and its sustainability. In: De Silva, S.S. (ed.) *Tropical Mariculture*. Academic Press, London, pp. 257–289.
Sammut, J., White, I. and Melville, M.D. (1996) Acidification of an estuarine tributary in Eastern Australia due to drainage of acid sulfate soils. *Marine & Freshwater Research* 47, 669–684.
Shanahan, M., Thornton, C., Trent, S. and Williams, J. (2003) *Risky Business: Vietnamese Shrimp Aquaculture Impacts and Investments*. Environmental Justice Foundation, London, 42 pp.
Tran, T.B., Dung, L.C. and Brennan, D.C. (1999) Environmental costs of shrimp culture in rice growing regions of the Mekong Delta. *Aquaculture Economics and Management* 3, 31–43.
Tuong, T.P., Minh, L.Q., Ni, D.V. and van Mensvoort, M.E.F. (1998) Reducing acid pollution from reclaimed acid sulphate soils: experiences from the Mekong delta, Vietnam. In: Pereira, L.S. and Gowing, J.W. (eds) *Water and the Environment: Innovation Issues in Irrigation and Drainage*. E. and F.N. Spon, London, pp. 75–83.
Tuong, T.P., Kam, S.P., Hoanh, C.T., Dung, L.C., Khiem, N.T., Barr, J.J.F. and Ben, D.C. (2003) Impact of seawater intrusion control on environment, land use strategies and household incomes in a coastal area. *Paddy and Water Environment* 1, 65–73.
UN (United Nations) (2002) *Report of the World Summit on Sustainable Development*. United Nations, New York, 173 pp.
White, I., Melville, M.D. and Sammut, J. (1996) Possible impacts of saline water intrusion floodgates in Vietnam's lower Mekong delta. In: Weizel, V. (ed.) *Proceedings of Seminar on Environment and Development in Vietnam*, Australian National University, Canberra, 6–7 December, 1996. Available at <http://coombs.anu.edu.au/~vern/env_dev/papers/PAP07.DOC>
Wilson, B.P., White, I. and Melville, M.D. (1999) Floodplain hydrology, acid discharge and change in water quality associated with a drained acid sulfate soil. *Marine and Freshwater Research* 50, 149–157.

5 Brackish-water Shrimp Cultivation Threatens Permanent Damage to Coastal Agriculture in Bangladesh

Md. Rezaul Karim
Urban and Rural Planning Discipline, Khulna University, Khulna, Bangladesh, e-mail: rk@bttb.net.bd

Abstract

Over the past 20 years, brackish-water shrimp cultivation in the coastal zone of Bangladesh has contributed increasingly to the national economy, but there is a lack of quantitative data on the effects of shrimp farming at the household and community levels. This chapter investigates the impact of shrimp cultivation on the environment and farmers' livelihood in a typical subdistrict of the coastal zone in Bangladesh. The variables involved with shrimp cultivation and their impact on the land are also taken into consideration and an attempt is made to analyse the patterns of land-use change that occurred between 1975 and 1999. The extensive pattern of shrimp cultivation is achieved by expansion of area rather than by intensification. The expansion of shrimp farming has resulted in decreases in crop production and many environmental problems in the form of a shortage of livestock fodder, fuel scarcity and decreases in traditional labour forces. Under the present circumstances, shrimp cultivation is no doubt beneficial for a selected group of people, but it has negatively affected the livelihoods of landless and marginal farmers, making it difficult for them to survive in the area.

Introduction

The Bangladesh coastal zone (Fig. 5.1) is a significant maritime habitat of ecologically rich and economically important natural resources. The area is situated along the largest river system running below the Himalayan Mountains. In this zone, the rivers discharge an enormous quantity of fresh water and maintain a level of salinity both on land and in sea that favours the rapid growth of a wide variety of vegetation and aquatic life.

The coastal region, especially the southwestern portion (Satkhira, Khulna and Bagerhat), is one of the most promising areas for shrimp cultivation for two major reasons (MOFL, 1997; Karim and Shah, 2001): first, its fresh- and saltwater resources are abundant in almost all seasons; second, the world's largest continuous mangrove forest, the Sundarbans, provides a food source and nursery for the offshore fishery. The mangrove forests provide a critical habitat for shrimp and other fish. Most of the shrimp culture being practised is by the extensive and improved extensive methods, known as *gher* culture. *Gher* means an enclosed area characterized by an encirclement of land

 Environment and Livelihoods in Tropical Coastal Zones (eds C.T. Hoanh, T.P. Tuong, J.W. Gowing and B. Hardy)

along the banks of tidal rivers. Dwarf earthen dykes and small wooden sluice boxes control the free entrance of saline water into the enclosed areas. In the *gher*, the sluice gates are opened from February to April to allow the entry of saline water, containing a wide variety of fish fry and shrimp postlarvae that have grown naturally to the juvenile stage in the adjacent sea and estuarine waters. This practice of natural stocking is being progressively replaced by artificial stocking of the *ghers* with only the young of specific, desired species of shrimp.

It is estimated that about 250,000 ha of land has good potential for coastal aquaculture (Ahmed, 1995). Of that, about 180,000 ha is suitable for shrimp culture (Khan and Hossain, 1996). Coastal aquaculture increased from 20,000 ha in 1994/1995 to 135,000 ha in 1996/1997, and production from 4000 to 35,000 metric tons in the same period (MOFL, 1997). The rapid expansion of shrimp farm development during the last decade, along with the adoption of extensive and improved extensive culture techniques, has caused growing concern as to its adverse effect on the coastal environment and damage to the traditional agricultural systems. The socio-economic scenarios have changed rapidly.

This chapter aims to assess the effects of the past and current situation as regards shrimp cultivation methods in relation to the coastal environment. In particular, the research aims to identify the inherent potentials and problems, as well as the emerging trends in the causes and extent of land-use changes over a period of time, with a view to understanding the implications for land-use planning and the development of more environmentally acceptable shrimp cultivation methods.

Methods

The study was conducted in 43 selected villages of Rampal Upazila (subdistrict), Bagerhat District, Khulna Division (Fig. 5.1), in 1999. Rampal Upazila covers 33,546 ha of land and is located in the centre of the western region of the coastal zone. It is situated on the Mongla River and is well connected to the Passur River. The area is close to the Sundarbans mangrove forest. The study area was selected on the basis of the following criteria: (i) the area should have extensive shrimp culture; (ii) shrimp farming should have been going on for at least 10 years in the area; (iii) the change in the physical environment should be homogeneous and representative of the situation in the coastal areas; and (iv) shrimp should be the main crop and provide the major share of income of the farmers of the area.

Information was collected from 373 respondents belonging to five landowner categories using land-use survey and questionnaire survey techniques. Field observation based on the perception of local people was carried out and recorded documents of relevant studies were examined. Changes in the land-use pattern, socio-economic conditions and the environmental situation of the area were recorded for three time periods: 1975, 1985 and 1999. Information relating to the landscape ecology in the past was collected with Retrospective Inquiry System (RIS) techniques and ground truthing was performed with the help of aerial photography and satellite images.

Results and Discussion

The percentages of major land-use categories for 1975, 1985 and 1999 are presented in Fig. 5.2. This shows that more than 80% of the land in Rampal Upazila was under rice cultivation in 1975, whereas in 1999 it decreased to less than 20%. Over this period, most of the rice land was replaced by rice–shrimp farming (rice in the rainy season and shrimp in the dry season), indicating a sharp increase in shrimp cultivation. Land occupied by other uses has not changed significantly during the study period.

Shrimp farms

Table 5.1 describes the size of shrimp *ghers* according to the household respondents. Before 1970, there was no shrimp cultivation in the study area. Most shrimp *ghers* in 1975

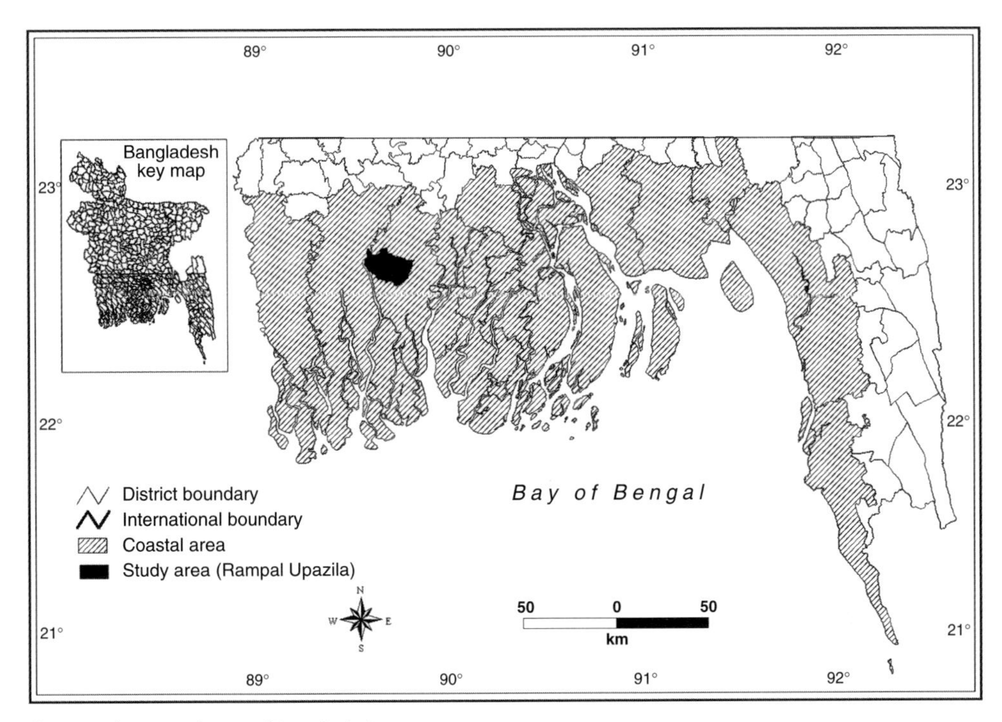

Fig. 5.1. The coastal area of Bangladesh.

were from 0.2 to 1 ha. A large number of pocket *ghers* (less than 0.2 ha) were introduced during 1975–1985 and 1985–1999 (accounting for 66.2% and 70% of the total shrimp *ghers* area, respectively). Farm areas of more than 6.47 ha represented 2.9% in 1985 and increased to 4% of the total shrimp *ghers* in 1999. Table 5.2 shows the size of shrimp *ghers* followed by paddy cultivation. The concentration of shrimp–rice farms was within the group of 0.2–1.0- and 1.0–3.0-ha categories. The percentage of farms greater than 3 ha increased during 1975–1999.

Field crops

Transplanted aman rice (rainy season) is the dominant crop in the study area. Mostly, traditional tall varieties with low yields (ranging from 1 to 3 t/ha) are cultivated. The cropping intensity of the study area was 113% in 1975 and decreased to 105% in 1985, which is much lower than the country's average (151%) (BBS, 1986a). In 1999, however, the cropping intensity in the study area was 100%. Shrimp cultivation was the only reason for the decrease in cropping intensity reported by respondents in the study area.

Yields of most of the field crops in the study area have also declined (Fig. 5.3) following the start of shrimp cultivation. Production of wheat, jute and sugarcane has been affected seriously, and now it is not possible to grow these crops because of soil salinization. The yield of the aman rice crop also declined quickly and, during the field survey in 1999, it appeared that farmers were not interested in harvesting transplanted rice crops from fields because of the very poor yield. Farmers in the study area attributed the decrease in rice yield to salinization. Though we did not have any quantitative data to support the farmers' claims, other studies mention that brackish-water shrimp farming (Rahman *et al.*, 1992) and salinity present in the soil are the major factors restricting crop production (BARC, 1990; Flowers, 1999). The soil characteristics surrounding the study area have deteriorated

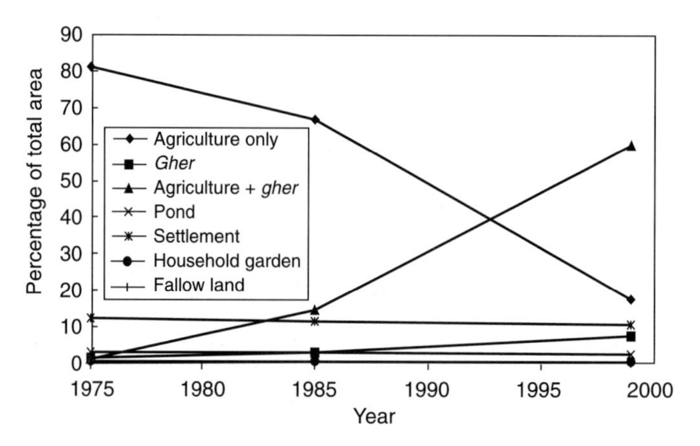

Fig. 5.2. Land uses over different years in Rampal Upazila.

Table 5.1. Size of shrimp *ghers* according to the household respondents in Rampal Upazila from 1975 to 1999 (from field survey, 1999).

Land area (ha)	Percentage of *gher* households		
	1975	1985	1999
<0.2	0	66.2	70.0
0.21–1.0	66.7	16.2	11.6
1.1–3.0	16.7	10.3	8.6
3.1–6.47	16.7	4.4	5.9
>6.47	0	2.9	4.0
Total	100	100	100

Table 5.2. Size of shrimp *ghers* followed by paddy cultivation (shrimp–rice system, SR) in Rampal Upazila from 1975 to 1999 (from field survey, 1999).

Land area (ha)	Percentage of SR households		
	1975	1985	1999
<0.2	0	36.1	41.2
0.21–1.0	75.0	28.9	25.3
1.1–3.0	25.0	27.8	22.0
3.1–6.47	0	5.2	7.9
>6.47	0	2.1	3.7
Total	100	100	100

substantially because of the gradual accumulation of salt over the years (Rahman *et al.*, 1992). Findings of these studies indicate that, once the soil becomes saline, subsequent floods and monsoon rainfalls may not leach out the salt completely and residual soil salinity is likely to increase over time.

In the study area, farmers grow some kharif vegetables such as pumpkin, brinjal (aubergine), spinach, etc, on their homestead lands. In the kharif season, which starts with the summer rains, salinity does not pose serious problems for the growth of vegetables as there is enough rain to leach out the salt, but

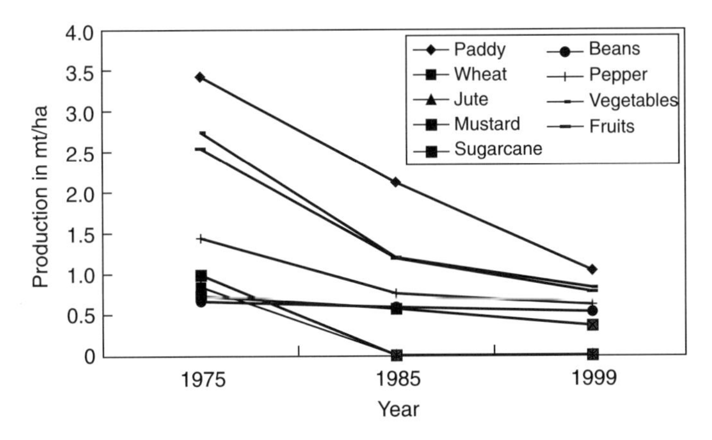

Fig. 5.3. Change in major crop yields (metric tons) in Rampal Upazila.

still production has not been satisfactory. In the rabi season from November to April, however, the intensity of soil salinity increases, which restricts most of the rabi vegetables. Small portions of land in the study area, consisting of 7% of the sampled villages (three villages), have shown that rabi crop production is affected in this way.

Table 5.3 shows that most respondents grew a large variety of vegetables in 1975. From 1975 to 1985, vegetable cultivation was seriously affected, and in 1999 the cultivation of major vegetables declined drastically. More than 80% of the sample villagers reported that vegetable cultivation was not possible because of the high soil salinity in the study area.

Vegetation

Shrimp cultivation can be considered as a land development process in which swampy forest or agricultural lands of low elevation are normally converted into shrimp ponds by embankment construction. Trees and bushes in the swampy forest or bushy land were totally cleared in preparation for shrimp farming. About 3.5% of forest and 2.4% of pasture land in the study area have been converted into shrimp ponds. Once shrimp cultivation starts, the remaining trees and vegetation also disappear fast because of high salinity and inundation. Sometimes, trees remained longer on the dykes and embankments, but they also disappeared over time because of seepage of saline water. Reeds and grasses that are used for fuel or for making mats are gradually lost because of the rise in saline levels in the waterlogged areas. The interviewees named several aquatic plants and weeds that had completely disappeared because of shrimp farming in the coastal area: durba (*Cynodon dactylon*), baju (*Tamarix troupii*), chehur (*Bauhnia vahlii*), thankuni (*Centella asiatica*), ambalisak (*Oxalis corniculata*) and kachuripana (*Eichhorina crassipes*). However, these species are still found in the non-shrimp area (Rahman *et al.*, 1992).

Table 5.4 shows the gradual encroachment of shrimp farms into the homestead areas. Before 1975, most shrimp farms (81.4%) were located farther than 500 m away from and only 1% were closer than 100 m to homestead areas. In 1999, 46% of the shrimp *ghers* were less than 10 m from homestead land, and only 2.1% were farther away than 500 m. This encroachment caused serious problems for the survival of trees in and around homesteads. A wide range of fruit trees and plant species grew abundantly before shrimp cultivation in the study area (Table 5.5). By 1985, about 60% of these trees had died. In 1999, some species (e.g. *Litchi chinensis*) almost disappeared. The interviewees reported that only some tree species, such as raintree and *sobeda*, survived to a limited extent. In some homesteads, betelnut, coconut, palmyra palm and date palms were present but did not bear fruits.

Table 5.3. Percentage of vegetable cultivation in Rampal Upazila from 1975 to 1999 (from field survey, 1999).

Scientific name	Local name	English name	Percentage				
			1975	1985	1999	1985–1999	1975–1985
Lagenaria vulgaris	Lau	Bottle gourd	98.1	83.7	50.5	–33.2	–47.6
Cucurbita maxima	Kumra	Sweet gourd	85.3	54.3	24.3	–29.9	–61.0
Amaranthes gangeticus	Lal Shak	Celery	94.4	45.2	12.3	–32.9	–82.1
Aloe indica	Sabuz Shak	Greens	95.7	46.5	19.0	–27.5	–76.7
Brassica oleracea var. *botrytis* Linn.	Ful Kopi	Cauliflower	77.5	9.4	2.9	–6.4	–74.6
B. oleracea var. *capitata* Linn.	Badha Kopi	Cabbage	79.1	7.0	1.1	–5.9	–78.1
B. oleracea	Oal Kopi	Turnip	83.7	10.2	3.5	–6.7	–80.2
Solanum melongena	Begun	Aubergine	90.1	43.6	20.1	–23.5	–70.1
S. tuberosum	Alu	Potato	86.6	27.0	10.4	–16.6	–76.2
Lycopersicon esculentum	Tomato	Tomato	91.2	24.1	6.4	–17.6	–84.8
Trichosanthes dioica	Patol	Pointed gourd	35.3	8.0	2.4	–5.6	–32.9
Luffa acutangula	Zhinga	Luffa	78.9	11.8	2.1	–9.6	–76.7
Dolichos lablab	Shim	Country bean	81.3	17.6	4.3	–13.4	–77.0
Momordica cochinchinensis	Kushi	Snake gourd	80.5	8.8	5.6	–3.2	–74.9
Abelmoschus esculentus	Dherosh	Lady's finger	79.4	14.2	9.4	–4.8	–70.1
Cucumis sativus	Sosha	Cucumber	73.3	6.1	0.8	–5.3	–72.5
Capsicum frutescens	Kathcha Morich	Green pepper	91.7	31.0	9.9	–21.1	–81.8
Vigna catiog	Borboti	Yard-long bean	75.7	3.7	2.1	–1.6	–73.5
Batatas edulischoisy	Misti Alu	Sweet potato	77.3	3.5	1.1	–2.4	–76.2
Amorphophallus campanulatus	Oal Kochu	Elephant food	79.4	5.3	0.8	–4.5	–78.6

Table 5.4. Encroachment of homestead land by shrimp farms in Rampal Upazila from 1975 to 1999 (from field survey, 1999).

Distance from homestead (m)	Percentage		
	1975	1985	1999
<10	0	0.9	46.0
10–24	0	3.4	24.1
25–50	0.3	7.8	15.8
51–100	0	4.0	5.3
101–300	0.7	5.6	3.5
301–500	17.6	19.3	3.2
>500	81.4	59.0	2.1
Total	100	100	100

In addition to its effects on vegetation clearance, decrease in pasture land and disappearance of trees, shrimp culture has had some indirect effects on vegetation. Farmers use the residues from agricultural land as fuel for cooking food, and cow dung for fuel and manure. After the introduction of shrimp cultivation, households became totally dependent on trees and forest vegetation as a fuel resource. The decrease in vegetation led to increasing soil erosion and sedimentation in the rivers and agricultural fields, which created problems within local ecosystems. The non-availability of cow dung may decrease the organic matter and fertility of the soil on the remaining agricultural land and gardens.

Livestock and poultry

Livestock and poultry play a crucial role in the traditional agricultural economy of Bangladesh. Livestock and poultry accounted for 5% of the gross domestic product (BBS, 1986b) of the country. The 1983/1984 statistics for cattle per household in the study area were 7.56, compared with 14.9 for the Greater Khulna District and 3.89 nationally. The survey result shows a sharp decline in the production of livestock and poultry in the study area (Fig. 5.4 and Table 5.6). The rate of decrease from 1985 to 1999 was nearly double that of 1975 to 1985.

It is, of course, very difficult to isolate the effects of shrimp culture from the general trends of livestock and poultry all over the country. But it was also reported that ducks

Table 5.5. Changes in numbers of trees in homestead gardens in Rampal Upazila (from field survey, 1999).

Scientific name	Local name	English name	Number of trees		
			1975	1985	1999
Mangifera indica L.	Aam	Mango	3,302	1,833	928
Artocarpus heterophyllus Lamk.	Kanthal	Jackfruit	1,369	629	197
Psidium guajava (L.) Bat	Pyara	Guava	1,752	771	603
Syzygium spp. (Wt.) Wall.	Jam	Berry	627	272	121
Bombax ceiba L.	Shimul	Silk cotton	440	160	321
Melia azadirachta L.	Nim	Neem	649	283	253
Albizzia lebbeck	Shirish	Rain tree	2,087	1,159	1,224
Borassus flabellifer L.	Tal	Palm	2,178	1,090	792
Phoenix sylvestris (L.) Roxb.	Khejur	Date	11,079	6,607	3,859
Cocos nucifera L.	Narikel	Coconut	10,525	6,115	3,953
Bambusa vulgaris Schrad	Bash	Bamboo	7,682	2,725	1,598
Areca catechu L.	Supari	Areca nut	41,960	15,625	8,603
Zizyphus jujuba Lamk.	Boroi	Plum	1,025	478	364
Eugenia malaccens	Jamrul	Star-apple	422	164	94
Citrus aurantifolia (Christ)	Lebu	Lemon	827	249	168
Litchi chinensis	Lichu	Litchi	310	75	20
Tamarindus indicus L.	Tentul	Tamarind	1,272	444	510
Musa sapientum	Kola	Banana	19,733	7,617	3,687
Achras sapota	Safeda	Sapota	342	250	193

were not allowed to move on to the shrimp farms and that poultry birds had lost their scavenging fields. This indicates that there was definitely some pressure against raising poultry and ducks in the shrimp-producing areas. The decrease in cattle population could be related to the decrease in grazing areas and the availability of food and fodder (e.g. rice straw) brought about by shrimp farming.

Table 5.6. Changes in numbers of livestock and poultry in Rampal Upazila from 1975 to 1999 (from field survey, 1999).

Type	Number of animals		
	1975	1985	1999
Cattle	2,574	1,453	585
Buffalo	691	300	92
Goats	1,289	698	257
Ducks	13,305	5,470	3,545
Chickens	8,597	5,014	3,933

Occupation and income

Table 5.7 suggests that total unemployment increased from 0% to 19% among males and from 46% to 55% among females because of the introduction of shrimp farming. These changes were highly related to the decline in agricultural employment (1975–1999) from 75% to 38% for men and from 37% to 0% for women. As shrimp farming is less labour-intensive than agriculture, with the decrease in crop cultivation, the local job market declined substantially. Before shrimp cultivation, most of the women were engaged in activities related to crop production such as harvesting crops, husking paddy, rearing livestock and poultry, cultivating vegetables, etc. Rearing of livestock became difficult because of the lack of grazing land after extensive year-round shrimp cultivation. However, a small proportion of women and children were found to be working in shrimp-processing factories, as well as in the collection of shrimp fry, because of new developments.

Table 5.8 summarizes the cost–benefit analysis of shrimp in the study area. The net annual income of the households in their own *ghers* was Tk 49,967/ha, whereas annual income from rented *ghers* was Tk 43,944/ha. Shrimp farmers who could invest around Tk 34,000/ha in their land had an opportunity to earn a net income of Tk 50,000/ha, whereas those who could not afford to lease out their land could get only Tk 6000/ha of land as lease money. As such, it can be concluded that shrimp culture is responsible for creating inequity in the society, although it had provided an opportunity for more economic use of land. On the other hand, paddy farmers had earned a net income of Tk 2700/ha with an

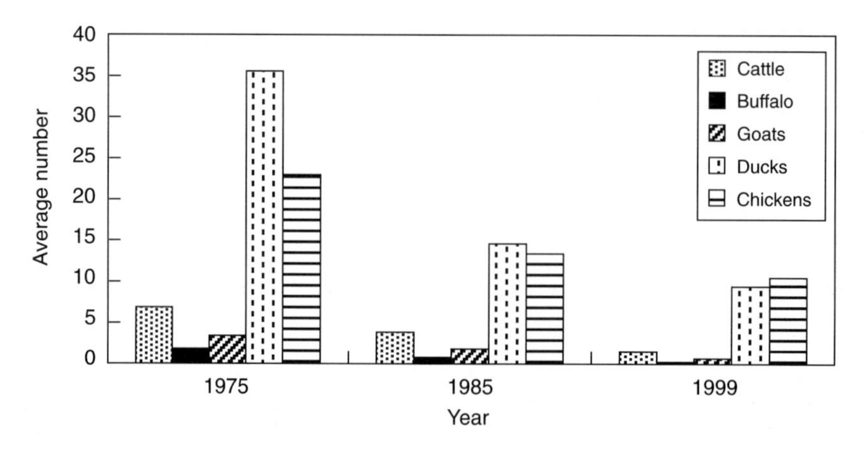

Fig. 5.4. Livestock and poultry per household in Rampal Upazila.

Table 5.7. Occupational changes in Ramzila Upazila (from field survey, 1999).

Occupation	Before shrimp culture (1975)		After shrimp culture (1999)	
	Males (%)	Females (%)	Males (%)	Females (%)
Agriculture (crop culture)	75	37	38	0
Shrimp farming/fry collection	6	0	12	23
Service	1	0	7	8
Business	6	0	10	3
Day labour	4	0	12	8
Others	8	17	2	3
Unemployed	0	46	19	55
Total	100	100	100	100

investment of around Tk 7000/ha. As a result, people are no longer interested in cultivating paddy because of its low return.

Environmental impact of shrimp farming

Brackish-water shrimp farming has altered the physical, ecological (aquatic and terrestrial) and socio-economic environment. A schematic flow diagram (Fig. 5.5) describes the various functions and interrelationships among the processes, effects and changes of each subcomponent. The practice of shrimp culture needs saline water as an input to the shrimp pond. Sluice gates are normally allowed to open two or three times when the salinity in the shrimp pond decreases and saltwater exchange from the river is necessary. As a result, heavy sedimentation from upstream water settles in the riverbed and canal bed, causing waterlogging in the shrimp ponds and on agricultural land. The shrimp-processing depot and industry drain their pollutants into the river, causing water pollution. Water in the shrimp ponds is also polluted because of the application of feed and fertilizer for the development of the shrimp. Thus, the by-products of the shrimp ponds and shrimp industry pollute water and soil and degrade the quality of the overall environment. Vegetation, crops, fish and livestock are seriously damaged by the process of shrimp cultivation, as has been discussed.

Figure 5.6 shows the interrelationship of the direct and indirect effects of shrimp farming on physical, ecological, socio-economic and environmental conditions in the study area. The problems created by shrimp cultivation are interdependent, as discussed in previous sections. The conversion of agriculture to shrimp farming created a physical and ecological imbalance, which has largely destroyed the natural ecosystem of the study area.

Conclusions

Brackish-water shrimp production in Rampal Upazila gave higher income than rice cultivation, but its expansion has had negative effects on the physical, social and natural environment. There is a need for effective planning, site selection and management of

Table 5.8. Cost–benefit analysis (in Tk) of shrimp culture in Rampal Upazila in 1999 (from field survey, 1999).

Crop	Operational cost/ha	Gross income/ha	Net income/ha
Shrimp (own *gher*)	34,033	84,000	49,967
Shrimp (rented *gher*)	40,056	84,000	43,944
Paddy	7,176	9,880	2,704

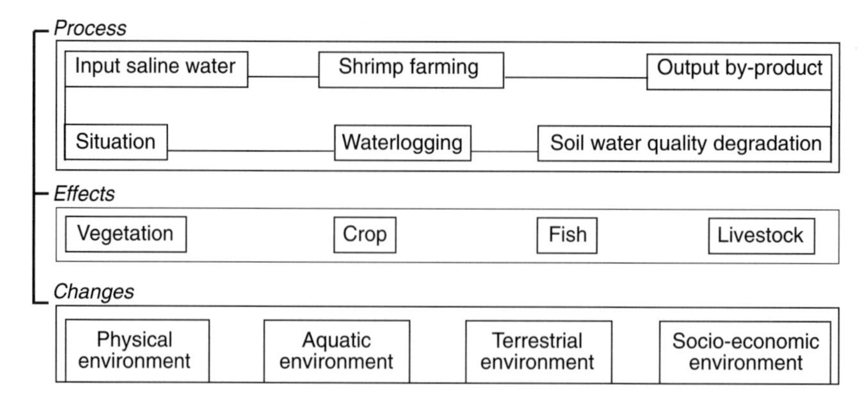

Fig. 5.5. A diagrammatic model of overall function and implications of shrimp farming.

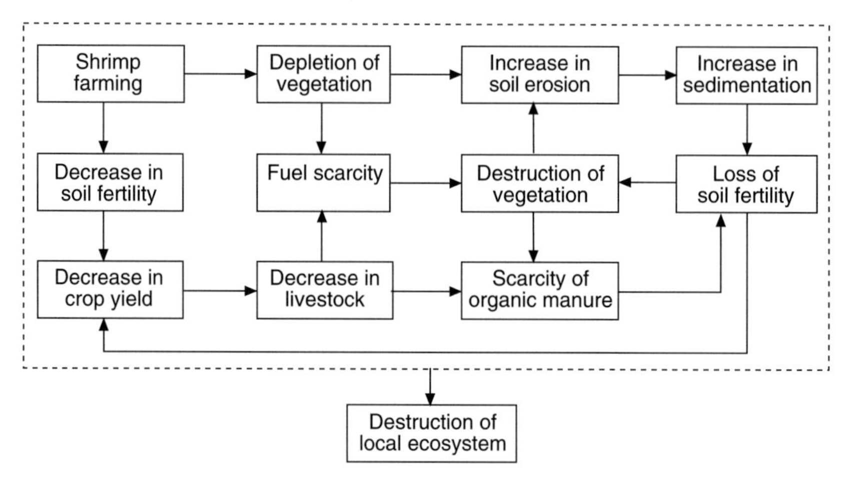

Fig. 5.6. A diagrammatic model of direct and indirect effects of shrimp farming.

shrimp farms, with due consideration given to the capacity of the environment and a comprehensive approach to sustainable development. In the economic analysis, it is important to take into account the cost of environmental degradation. Reducing the area under shrimp cultivation and raising the yield may be one way to help minimize these costs, but there may be other alternatives that require further investigation. One of the reasons for the decline in vegetation cover and biodiversity affecting livelihoods of farmers is the encroachment of shrimp ponds towards the homestead. A case can therefore be made for demarcating a buffer zone around homestead areas within which shrimp ponds should not be developed.

References

Ahmed, A.T.A. (1995) Impacts of shrimp culture on the coastal environment of Bangladesh. In: *Proceedings of the National Workshop on Coastal Aquaculture and Environmental Management*. Institute of Marine Sciences, 25–28 April 1995. Nuruddin Mahmood, University of Chittagong/UNESCO, Bangladesh, pp. 77–84.

BARC (Bangladesh Agricultural Research Council), Soil and Irrigation Division (1990) *Salinity Problems and Crop Intensification in the Coastal Regions of Bangladesh. Soil Publication No. 33.* BARC, Farmgate, Dhaka, Bangladesh, 63 pp.
BBS (Bangladesh Bureau of Statistics) (1986a) *Khulna District Statistics.* Statistical Division, Ministry of Finance and Planning, Government of the People's Republic of Bangladesh, 226 pp.
BBS (Bangladesh Bureau of Statistics) (1986b) *Report on the Bangladesh Livestock Survey 1983–84.* Bangladesh Bureau of Statistics, Reproduction, Documentation and Publishing Wing, Dhaka, Bangladesh, 175 pp.
Flowers, T.J. (1999) *Salinity and Horticulture.* School of Biological Science, University of Sussex, Brighton, UK, pp. 7–15.
Karim, M.R. and Shah, M.S. (2001) *Brackish Water Shrimp Cultivation Restricts Coastal Agriculture.* Khulna University Studies, Khulna, Bangladesh 2(1), 123–134.
Khan, Y.S.A. and Hossain, M.S. (1996) Impact of shrimp culture on the coastal environment of Bangladesh. *International Journal of Ecology and Environmental Sciences* 22(2), 145–158.
MOFL (Ministry of Fisheries and Livestock) (1997) Fisheries resources development and management. Paper presented at the National Workshop, 29 October to 1 November 1997, MOFL, Dhaka, Bangladesh.
Rahman, K., Shah, A.H. and Haque, M.A. (1992) Salinity intrusion and its effects in the south-west coastal region of Bangladesh. Paper presented at the workshop on coastal zone management in Bangladesh, 27–31 December 1992, Dhaka, Bangladesh.

6 Coastal Water Resource Use for Higher Productivity: Participatory Research for Increasing Cropping Intensity in Bangladesh

M.K. Mondal,[1] T.P. Tuong,[2] S.P. Ritu,[1] M.H.K. Choudhury,[3] A.M. Chasi,[4] P.K. Majumder,[5] M.M. Islam[6] and S.K. Adhikary[6]

[1]*Bangladesh Rice Research Institute, Gazipur, Bangladesh, e-mail: m.mondal@bdonline.com*

[2]*Crop, Soil, and Water Sciences Division, International Rice Research Institute, Metro Manila, Philippines*

[3]*Proshika Manobik Unnyan Kendra, Dhaka, Bangladesh*

[4]*HEED-Bangladesh, Dhaka, Bangladesh*

[5]*Department of Agricultural Extension, Khulna, Bangladesh*

[6]*Khulna University, Khulna, Bangladesh*

Abstract

In Bangladesh, about 1.0 million ha of coastal saline soils have been monocropped with low-yielding, traditional rice varieties during the monsoon season from June to December. Most of these lands remain fallow in the dry season because of high soil salinity and the lack of good-quality irrigation water. This research was conducted with farmers' participation to test the hypothesis that a combination of on-farm storage of surface water, to prolong freshwater availability beyond the end of the rainy season, together with the proper selection of rice varieties, can increase cropping intensity and productivity of the area. Selected farmers and local leaders were involved in the whole process, from designing the new cropping systems to managing, testing and evaluation. In the wet season, the traditional rice varieties were replaced by short-duration, high-yielding varieties (HYV), which can be harvested earlier, about 1.5 months before traditional varieties. This opened up opportunities for early establishment (in mid-November) of short-duration HYV of rice during the dry season. River water was directly used for irrigation of the dry-season crop up to mid-February. Beyond this time, river water became too saline for irrigation purposes. Before it became too saline, river water was taken in through sluices in the first week of February and conserved in on-farm canal networks. The stored water was used to irrigate rice from mid-February to the end of March. The new cropping system increased annual rice yield by two- to threefold and farmers' profits by 1.5- to twofold compared with the farmers' traditional system, and with no apparent negative effect on the environment. The technology was taken up at a fast pace, indicating that farmers preferred it to shrimp farming. Principles of the technology can be applied to other monsoon, deltaic coastal areas.

Introduction

More than 30% of the cultivable land in Bangladesh is in the coastal area. About 1.0 million ha of arable lands are affected by varying degrees of salinity. Farmers grow mostly low-yielding, traditional rice varieties during the wet season. Most of the lands remain fallow in the dry season (January–May) because of soil salinity and the lack of good-quality irrigation water (Karim *et al.*, 1990; Mondal, 1997). Crop yields, cropping intensity, production levels and people's quality of livelihood are much lower in this region than in other parts of the country, which have enjoyed the fruits of modern agricultural technologies based on high-yielding varieties, improved fertilizer and water management and improved pest and disease control measures (BBS, 2001). At the same time, food demand in the area is increasing with the steady increase in human population.

During recent years, commercial shrimp farming became very attractive in the coastal region of the country. Small and medium farm-holders often lease out their agricultural lands to wealthy people, who have converted large tracts of traditional rice lands to shrimp farms. This conversion sometimes induced environmental degradation and social unrest, and it may not deliver sustainable benefits to small farm-holders (Majid and Gupta, 1997). Supplying farmers with alternative production systems with high land and water productivity is crucial for food security, enhancing farmers' livelihood and sustaining the environment of the coastal zone.

Experiences elsewhere proved that the use of short-duration, high-yielding rice varieties (HYV), the effective use of rainwater and proper crop scheduling that matches crop water requirements with the water supply and quality dynamics can increase the cropping intensity and productivity of rice lands in coastal areas (Tuong *et al.*, 1991; My *et al.*, 1995). Mondal (1997) successfully grew HYV in the wet season in the coastal zone areas of Bangladesh. Mondal (2001) also showed that river water in Khulna District remained suitable for irrigation far into the dry season until mid-February. We hypothesized that, if the river water could be taken into and stored in on-farm canal networks before it became too saline, freshwater availability could be prolonged adequately to irrigate an additional crop of HYV of rice grown after wet-season rice, thus increasing the cropping intensity of the coastal land and improving the socio-economic status of the resource-poor farmers in the coastal areas of Bangladesh. This chapter describes the processes and outcomes of farmer participatory research (2001–2004) to develop a new, rice-based cropping system at a typical site in the south-western coastal area of Bangladesh, which was carried out to test the above hypothesis. It also discusses possible refinements of the system and the implications of the findings for resource management for improving farmers' livelihood in coastal areas.

Methodology

Characteristics of the study site

The study was carried out at Kismat Fultola village under Batiaghata Upazila (subdistrict, the smallest administrative unit in Bangladesh), Khulna District (Fig. 6.1). The site has many typical characteristics of the agricultural land of the south-western coastal zone of the country. The area has two distinct seasons: a rainy season from June to October and a dry season from November to May (Fig. 6.2). Formerly, the soil was very saline because of the intrusion of seawater during the dry season. To increase agricultural production, in the 1960s the area was included in the government's Coastal Embankment Project (CEP), in which designated areas (polders) were surrounded by dykes or embankments, separating them hydrologically from the main river system and offering protection against tidal floods, salinity intrusion and sedimentation (Islam, 2005). Thanks to the CEP project, crops have been saved from salinity and flooding, and wet-season rice yields at some places increased by 200–300% (Nishat, 1988, as cited in Islam, 2005). Nevertheless, most rice soils in the area are still moderately saline (EC_e = 4–8 dS/m) to saline (EC_e = 8–16 dS/m).

The area of the present study lies on the banks of the Kazibachha River (Fig. 6.1). The river has a diurnal tidal regime. During flood tide (daily high tide), the water level could be 1–3 m higher than the land surface, thus offering opportunities for gravity irrigation in both the wet and dry seasons, provided the salinity level is not too high. Water of the river is fresh most of the time during the rainy season, with an average electrical conductivity mostly below 1.0 dS/m from July to December and below 4.0 dS/m from mid-January to mid-February (Fig. 6.3). However, the average monthly salinity of the river water can reach a maximum of about 20 dS/m at the end of the dry season, and this high salinity level makes the river water unsuitable for irrigation during March to June.

The farmers' common practice is to grow rainfed, low-yielding and long-duration local rice varieties in the wet season (transplanted in July and harvested in December, Fig. 6.4), with an annual average yield of only 2.0–2.5 t/ha. Inadequate rainfall (Fig. 6.2) and the lack of good-quality irrigation water restrict crop cultivation during the dry season, but about 40% of the farmers grow sesame during March to May on residual soil moisture. However, the crop is often damaged by unpredictable high rainfall that may occur in May (Fig. 6.2) at the reproductive stage. The farmers reported that, out of 30 years of sesame cultivation, they were

Fig. 6.1. Location map of the study site (Kismat Fultola village, Khulna District).

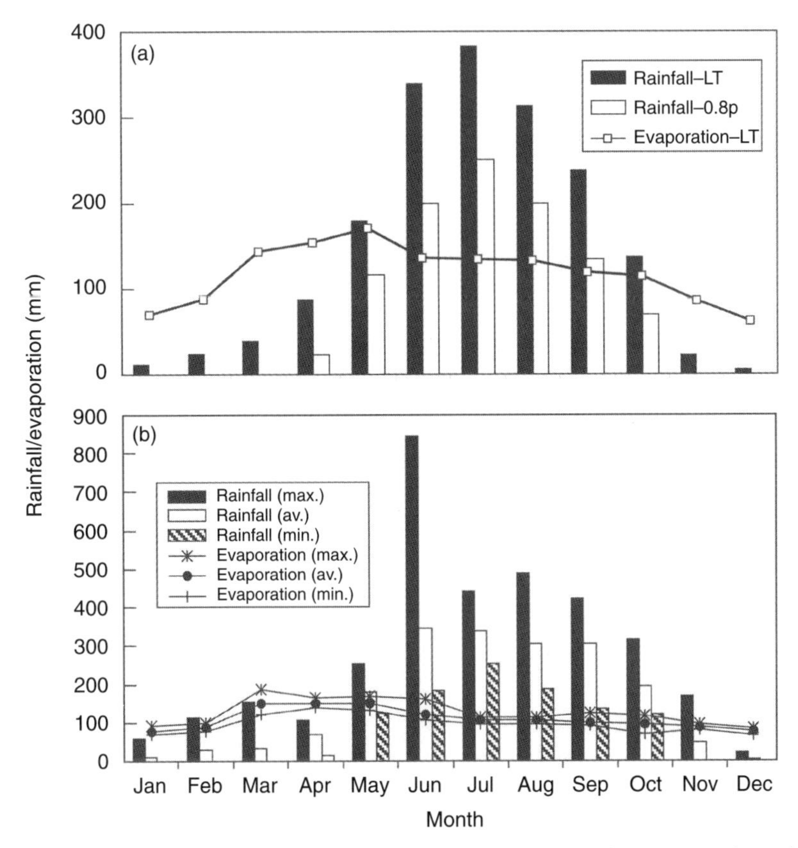

Fig. 6.2. Rainfall and evaporation pattern in Khulna District, 1902–1984 (a) and 1998–2003 (b). LT, long-term average values; 0.8p, values corresponding to 80% probability of exceedence. Source of the long-term data, FAO (1988).

successful in only 10 years and complete damage occurred in another 10. Sesame harvested in the remaining 10 years was partly damaged/rotten and was poor in quality, with a consequent low market price.

Conceptualizing a new cropping system and water management strategy

A new cropping system (Fig. 6.4), with a rainfed rice crop during the wet season and an additional irrigated rice crop during the dry season (from mid-November to mid-April), was conceptualized to increase cropping intensity of the study area. The conceptualization followed the principles described by Tuong *et al.* (1991). It critically assessed agro-hydrological factors in matching the availability of good-quality water and the water requirement of the new rice cropping pattern. The new cropping pattern made use of two principal components, described below.

1. An innovative water management strategy. Rice cultivation in the area after October needs irrigation water. River water is suitable for irrigation until mid-February. To extend irrigation water availability beyond mid-February, river water can be taken into and stored in the on-farm canal network before the water becomes too saline. The stored water can be used to irrigate the dry-season rice until the harvest. For a given area, the larger the amount of the on-farm storage, the longer the irrigation period can be extended.

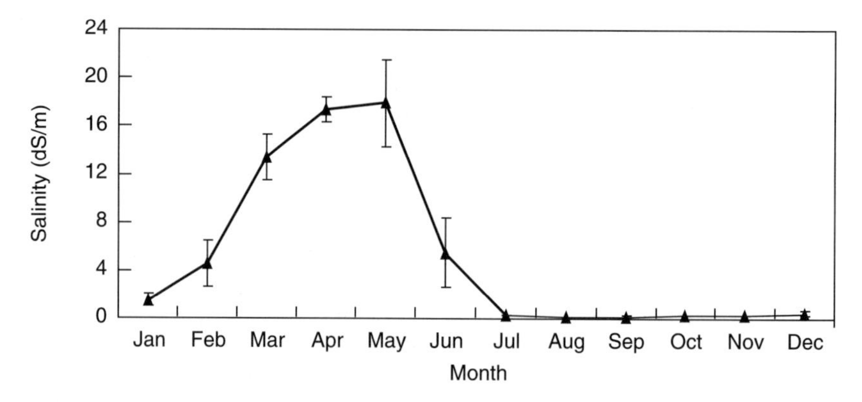

Fig. 6.3. Average monthly salinity of Kazibachha River measured at Kismat Fultola village, Khulna District, 1997–2004. Vertical and capped bars indicate standard error of the means of eight monthly values; each monthly value was the average of 10–20 daily values.

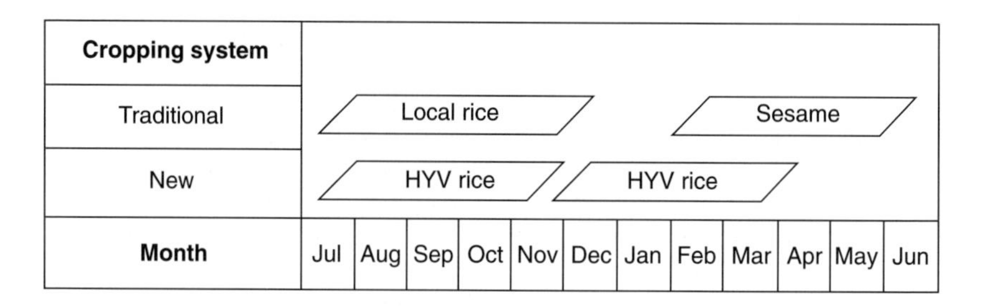

Fig. 6.4. Traditional and new cropping systems at Kismat Fultola, Khulna District. HYV, high-yielding varieties.

2. Use of short-duration HYV of rice. The use of short-duration varieties is critical for increasing the cropping intensity of rainfed lowland rice (Tuong *et al.*, 1991). It was perceived that HYV of rice in the wet season not only gave higher yields (Mondal, 1997) but also reduced the growth span of wet-season rice compared with traditional varieties, and facilitated the advancement of dry-season rice establishment. Early establishment and the use of short-duration HYV in the dry season allowed the crop to be harvested before water salinity reached damaging levels in April and May. We aimed at harvesting the dry-season crop as early as possible to reduce irrigation requirements during the dry season and to avoid salinity damage later in the season.

The participatory process

In 2001, driven by the need for more rice, some farmers in the study area transplanted dry-season rice in February and used river water for irrigation at the end of the cropping season. This resulted in total crop failure because the river water was too saline. In early 2002, we organized a series of meetings with farmers, local leaders and government and non-government organizations (GO-NGO) to investigate the reasons for the 2001 crop failure and to discuss the newly conceptualized cropping system and water management. Twenty-nine farmers agreed to test the new cropping system in the same location where they experienced failure of rice in 2001.

The first step in the farmer–researcher collaboration was a 3-day, hands-on training on cultivating HYV of rice. Details on the participatory experiments (see below) were discussed with individual farmers before each cropping season from 2002 to 2004. Neighbouring farmers, local leaders, social workers, GO-NGO officials, journalists, educators, donors and women's communities were invited to join the field days to participate in evaluating the performance of HYV of rice and the new cropping system during each season.

In the dry season of 2003, we helped organize a farmer-run workshop so that participating farmers could share their research experiences with the extension personnel, researchers, social workers, leaders and elite educators, GO-NGO officials, donors and women's communities. A workshop was organized in June 2004, focusing mainly on extension personnel, journalists and policymakers, to acquaint them with the technologies for wider dissemination.

The experiment

A 3.5 ha area, provided with a sluice gate (a gate that can be opened at high tide for water intake as well as at low tide for drainage) at the south and a flash gate (for water intake only) at the north, was selected for this participatory experiment. The area was divided among 29 participating farmers. In the 2002 wet season, researchers recommended farmers to try 15 HYV of rice developed by the Bangladesh Rice Research Institute (BRRI) (Table 6.1 – hereafter, all varieties/lines will be referred to by their abbreviations shown in Table 6.1). In the 2003 wet season, the advanced line BR6110 was added. Each farmer selected two to five varieties/lines from the recommended ones and grew them on their own land within the experimental field. In addition, all the tested, wet-season rice varieties/lines were grown by three participating farmers in a contiguous area of about 0.2 ha inside the experimental field, named the 'crop museum'.

Farmers transplanted 30-day-old seedlings in the last week of July, when soil salinity was adequately reduced by rainfall. After the harvest of rainy-season crops, farmers transplanted the dry-season rice crop around mid-December. All participating farmers planted variety BD28 in the dry season. In addition, six farmers also grew BD36 in the 2002/3 dry season and BD29 in the 2003/4 dry season. All rice varieties used in the dry season were non-aromatic and photoperiod-insensitive (Table 6.1). The growth duration of BD28 and BD36 was similar, but BD29 had a longer duration.

Rainfall supplied most of the water required for wet-season rice crops. During the occasional droughts in the rainy season, supplemental irrigation was provided during flood tide through the flash gate and irrigation canal networks where water was gravity-fed into the fields. This method of irrigation was also employed to supply water for the dry-season crops from December to mid-February, when the salinity level of the river water remained below 4.0 dS/m. During the last intake, as much water as possible was stored in the canals and in the fields. Water stored in the canals was used to irrigate rice from mid-February to the end of March with the help of a low-lift pump.

Monitored parameters

Climate and water resources

Rainfall and Class-A pan evaporation were measured daily at the study site during the experimental period. Groundwater level was measured weekly from June 2002 to May 2004 at Kismat Fultola, at seven observation wells installed at regular intervals for up to 1 km distance perpendicular to the river, and at a well 500 m south of the experimental site, and about 50 m away from the riverbank. Groundwater salinity in the observation wells was measured weekly by using a portable electrical conductivity meter.

Water salinity of the Kazibachha River was measured daily as far as 500 m north of the intake sluice of the study site during ebb (daily low tide) and flood tide using a portable electrical conductivity meter.

Table 6.1. Characteristics of high-yielding varieties/lines of rice grown in the experimental field (2002–2004) at Kismat Fultola village, Khulna District, Bangladesh (from BRRI, 2004).

Variety/line	Abbreviation	Growth duration (days)	Bred for salinity tolerance?	Other properties
BR10	BR10	150	No	Non-aromatic, slightly photoperiod-sensitive
BR11	BR11	145	No	Non-aromatic, slightly photoperiod-sensitive
BR22	BR22	150	No	Non-aromatic, photoperiod-sensitive
BR23	BR23	150	Yes, low tolerance	Non-aromatic, photoperiod-sensitive
BR25	BR25	135	No	Non-aromatic, photoperiod-insensitive
BRRI dhan30	BD30	145	No	Non-aromatic, slightly photoperiod-sensitive
BRRI dhan31	BD31	140	No	Non-aromatic, photoperiod-insensitive
BRRI dhan32	BD32	130	No	Non-aromatic, photoperiod-insensitive
BRRI dhan33	BD33	118	No	Non-aromatic, photoperiod-insensitive
BRRI dhan34	BD34	135	No	Aromatic, photoperiod-insensitive
BRRI dhan37	BD37	140	No	Aromatic, photoperiod-sensitive
BRRI dhan38	BD38	140	No	Aromatic, photoperiod-sensitive
BRRI dhan39	BD39	122	No	Non-aromatic, photoperiod-insensitive
BRRI dhan40	BD40	145	Yes, medium tolerance	Non-aromatic, photoperiod-sensitive
BRRI dhan41	BD41	148	Yes, medium tolerance	Non-aromatic, photoperiod-sensitive
BR6110-10-1-2[a]	BR6110	145	–	Non-aromatic, photoperiod-sensitive
BRRI dhan28	BD28	140	No	Non-aromatic, photoperiod-insensitive
BRRI dhan29	BD29	160	No	Non-aromatic, photoperiod-insensitive
BRRI dhan36	BD36	140	No	Non-aromatic, cold-tolerant, photoperiod-insensitive

[a] BR6110-10-1-2 was added in 2003.

Monthly values at ebb and flood tides were taken as the average of daily measurements.

Soil salinity

Topsoil salinity of Kismat Fultola village was measured at 15-day intervals in the dry season and at 30-day intervals in the wet season from May 2002 to April 2004. At each sampling time, 17 soil samples were taken randomly from the 0–15 cm soil layer of the experimental field to make a composite sample. Soil salinity of the composite sample was determined by the saturation extract method (ASA and SSSA, 1982).

Yield assessment

Areas of 5 m^2 each, from four different corners and one from the middle of the plot, were harvested from seven to 16 farmers' fields to estimate grain yield, and values were later adjusted to 14% moisture content (BRRI, 1993). This procedure was followed to determine grain yield of the experimental fields as well as that of fields of farmers who did not originally participate in the experiment, but later adopted the new cropping system and used the recommended varieties on their own farms (uptake fields).

Results and Discussion

Agro-hydrological conditions

Figure 6.2b shows monthly rainfall and evaporation at the study site from 1998 to 2003. They did not deviate very much from the long-term (1902–1984) values (Fig. 6.2a) and confirmed that rainfall could supply adequate water to meet crop evapotranspiration from June to October. The dynamics of groundwater depth at two monitoring points at the study site are presented in Fig. 6.5. The water table at 1 km distance from the river was nearer to the ground surface than that closer to the river, suggesting that the further the distance from the riverbank, the more difficult is the drainage.

The groundwater salinity levels are shown in Fig. 6.6. The salinity of the groundwater at the study site was about 4 dS/m throughout the year, but the salinity at the point 500 m south of the study site reached a maximum of 12 dS/m. This implies that groundwater salinity could vary greatly within short distances. The use of groundwater remains a risky venture, owing to high

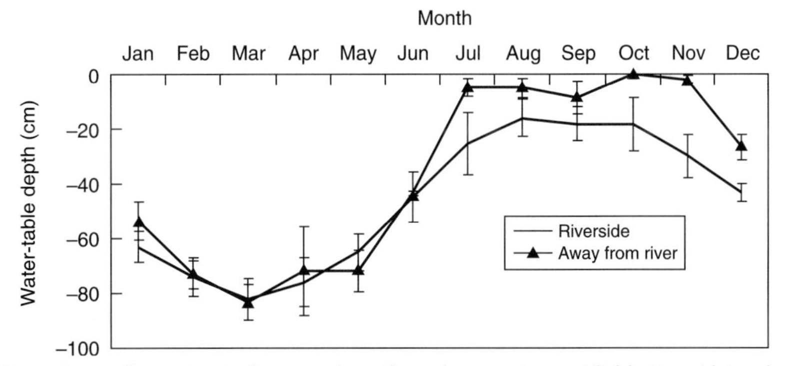

Fig. 6.5. Groundwater fluctuation in the coastal aquifer at the experimental fields (riverside) and at 1 km from the river, Kismat Fultola, Khulna District, 2002–2004. Vertical and capped bars indicate standard deviations of the means of three monthly values; each monthly value was the average of four weekly values.

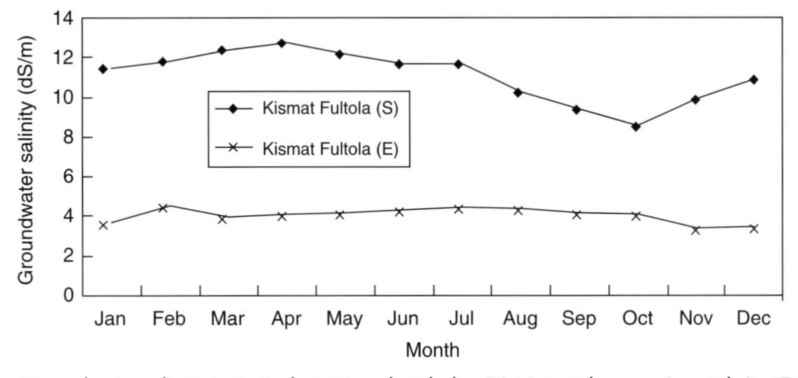

Fig. 6.6. Groundwater salinity in Batiaghata Upazila, Khulna District, at the experimental site (E) and at 500 m south of the site (S), 1996–2003. Both sites are 50 m from the river bank.

spatial variability and the possible intrusion of saline water from the river into the coastal aquifers if the water level of the aquifers is lowered because of excessive withdrawal of water for irrigation.

The dynamics of the salinity of the river water during the dry season (Fig. 6.7) followed the same pattern as that observed from long-term records (Fig. 6.3). Salinity of the river water started to increase in the second half of February and reached maximum values in April/May. There were significant differences in salinity levels of the river water during ebb and flood tides. The average salinity at flood tides reached 4 dS/m in mid-February, but remained below 4 dS/m during ebb tides until the beginning of March.

Variation in soil salinity between cropped and fallow lands

A typical soil salinity pattern of the rice soils at Kismat Fultola is shown in Fig. 6.8. Topsoil salinity in fallow lands varied from 5.0 to 12.0 dS/m in the dry season and remained below 4.0 dS/m in the wet season (when traditional rice is grown under rainfed conditions). But when rice was grown in the dry season by using river water, the dry-season soil salinity level decreased considerably from January to March. The decrease in soil salinity during the dry season under the new cropping system compared with the traditional rice–fallow system was due to the leaching of salts via irrigation water.

Rice grain yields

High-yielding rice varieties in the wet season

Yield of the non-aromatic HYV grown in the wet season varied from 4.0 to 5.0 t/ha (Fig. 6.9) and the aromatic rice varieties yielded about 3 t/ha. The difference in yield was due to the lower yield potential of the aromatic rice varieties (BRRI, 2004). Among the non-aromatic HYV, BD33 and BD39 had the shortest growth duration (about 120 days, Table 6.1), but yielded as well as the varieties with 140–150 day duration (Fig. 6.9). Many tested genotypes yielded more than BR23 and BD41, which were especially bred for salinity tolerance, indicating that salt tolerance is not an absolute requirement for rice grown during the rainy season. This is supported by the low soil salinity level during the rainy season at the study site (Fig. 6.8). All tested HYV yielded much higher than the farmers' traditional varieties. This was attributed to the higher yield potential and harvest index of the HYV (BRRI, 2004), and probably the higher nutrient input and supplementary irrigation in the new cropping system.

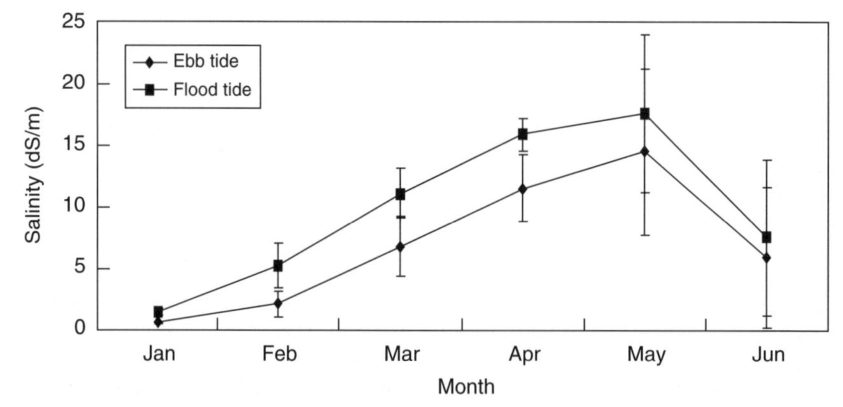

Fig. 6.7. Variation in water salinity of the Kazibachha River during ebb and flood tides at Kismat Fultola, Khulna District (average of 2000, 2003 and 2004). Vertical and capped bars indicate standard error of the means of three monthly values; each monthly value was the average of 10–15 daily values.

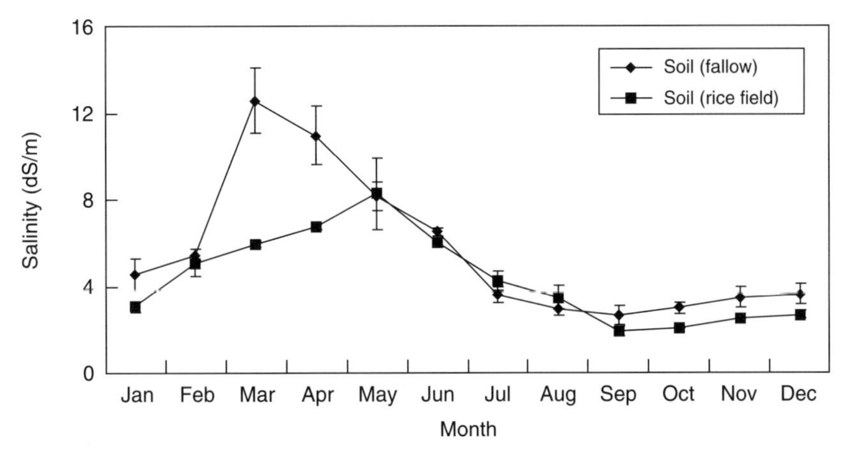

Fig. 6.8. Soil salinity in fallow and rice lands at Kismat Fultola, Khulna District, 1998–2004. Vertical and capped bars indicate standard error of the means of seven monthly values; each monthly value was the average of two measurements.

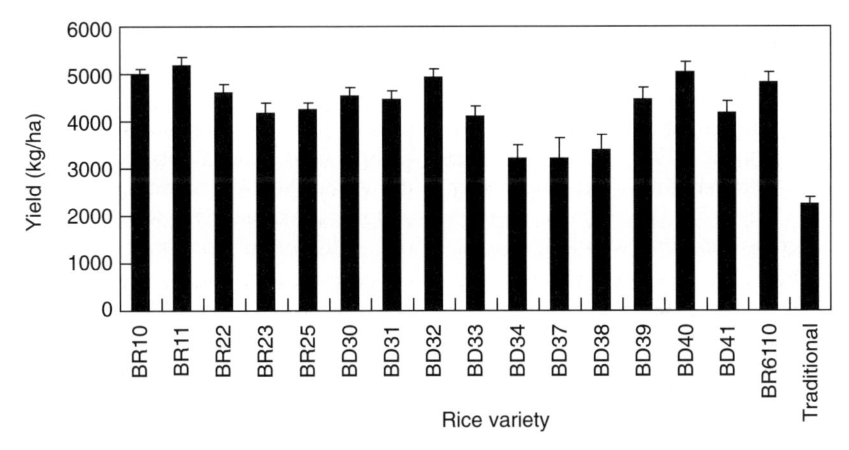

Fig. 6.9. Rice yield in the wet season (average of 2002 and 2003) at Kismat Fultola, Khulna District. Vertical and capped bars indicate standard error of the 2-year mean yields of 6–13 farmers' fields.

High-yielding rice varieties in the dry season

The average yield of BD28 in the 2001/2 and 2002/3 dry seasons was about 3000 kg/ha (Fig. 6.10). This was lower than its potential yield (BRRI, 1993). The main reason was the lack of farmers' experience in managing HYV at the outset, and also an unusual cold spell just after transplanting rice in the 2002/3 dry season (data not shown). The lack of farmers' experience was reflected in the large variation in rice yield of the sampled plots, from as low as 1000 kg/ha to as high as 4625 kg/ha, with about 40% of the experimental area having an average yield of 4300 kg/ha.

BD28 yielded more in the 2003/4 dry season than in the previous year. Average yields of BD28 and BD36 were about 3167 and 3630 kg/ha, respectively (Fig. 6.10). It is interesting to note that neighbouring farmers who used BD28 in the 2003/4 dry season obtained much higher grain yield (4425 kg/ha) than the farmers participating in the research (Fig. 6.10). Among them, five farmers obtained 6314 kg/ha with BD29 and 5326 kg/ha with BD36. The higher yield of BD29 could be attributed to its longer duration (Table 6.1). We do not have a full explanation for the higher yields in the 'uptake fields', but we postulate that the rice fields of the 'uptake' farmers were away from the river and may have had less salinity than the experimental field.

Changes in annual rice production and farmers' income

Farmers in the coastal region usually harvest about 2.0–2.5 t/ha of rice grain per year and earn a benefit of about Tk 15,000–20,000/ha (US$1 = Tk 60, Table 6.2). For farmers who grow sesame during the dry season, total annual grain production in rice equivalence in good years is about 3.0–4.5 t/ha (Table 6.3). By adopting the new cropping system and water management approach, participating farmers can harvest 6.0–8.5 t/ha of rice grain per year.

The additional profit from growing HYV instead of traditional rice varieties in the wet season was Tk 10,000/ha. In the dry season, even with the low average yield obtained in 2002/3, the participating farmers still obtained a profit of about Tk 7500/ha. In total, net yearly benefits from the new cropping system were Tk 32,500–37,500/ha, with a 50–100% increase over the traditional cropping system.

About 40% of the participating farmers have landholdings of less than 1.0 ha and the rest have about 1.0–2.0 ha per family. Almost all farmers have to depend on these land resources for their entire livelihood. Most expenditures are usually met from income from rice. Only about 5–10% of the total land owned by the participating farmers was used in this project, yet about 40% of their total rice production came from within the 3.5 ha project area, where cropping intensity has increased from 100% to 200%.

Environmental impact

Throughout this new system, coastal water resources were productively used for rice cultivation without hampering or disturbing the environment. Integrated pest management (IPM) techniques were adopted for the control of insects and pests, resulting in little use of pesticides despite the enormous pest attack normally experienced during the wet season in the Batiaghata area. In the experimental field, soil salinity in the dry season decreased (Fig. 6.8) because of the use of relatively non-saline river water for dry-season rice cultivation. Conservation of relatively

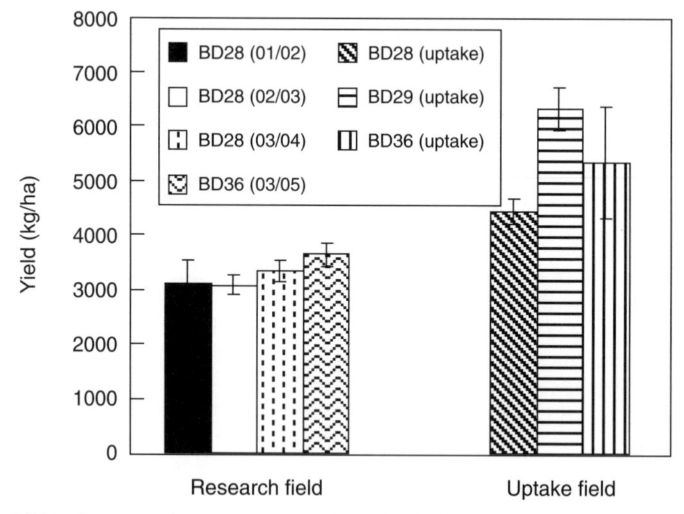

Fig. 6.10. Rice yield in the research (2001–2004) and uptake fields (2003–2004) in the dry season at Kismat Fultola, Khulna District. BD, BRRI dhan; vertical and capped bars indicate standard error of the means of 3 years, 6–19 farmers per year from research fields and 2–17 farmers from uptake farmers' fields in the 2003/4 season.

Table 6.2. Cost–benefit comparison between the traditional and newly developed cropping system using local and non-aromatic HYV of rice.

	Item	Traditional cropping	New cropping
Wet season	Production cost (Tk[a]/ha)	5,000	10,000
	Rice yield (t/ha)	2.0–2.5	5.0–6.0
	Total income (Tk/ha)	20,000–25,000	35,000–40,000
	Profit (Tk/ha)	15,000–20,000	25,000–30,000
Dry season	Production cost (Tk/ha)		15,000
	Rice yield (t/ha)		3.0
	Total income (Tk/ha)		22,500
	Profit (Tk/ha)		7,500
Annual	Rice yield (t/ha)	2.0–2.5	8.0–9.0
	Production cost (Tk/ha)	5,000	25,000
	Profit (Tk/ha)	15,000–20,000	32,500–37,500

[a] US$1 = Taka 60 (approx.).

Table 6.3. Cost–benefit comparison between traditional rice–sesame and the modified rice–rice cropping system using HYV of rice.

	Item	Traditional cropping	New cropping
Wet season	Production cost (Tk[a]/ha)	5,000	10,000
	Rice yield (t/ha)	2.0–2.5	5.0–6.0
	Total income (Tk/ha)	20,000–25,000	35,000–40,000
	Profit (Tk/ha)	15,000–20,000	25,000–30,000
Dry season	Production cost (Tk/ha)	900	15,000
	Sesame[b]/rice yield (t/ha)	0.728	3.0
	Total income (Tk/ha)	9,318	22,500
	Profit (Tk/ha)	8,418	7,500
Annual	Rice and sesame yield (t/ha)	2.7–3.3	8.0–9.0
	Production cost (Tk/ha)	5,900	25,000
	Profit (Tk/ha)	23,418–28,418	32,500–37,500

[a] US$1 = Taka 60 (approx.).[b] Adopted from Mondal (1997), assuming the best yields of sesame.

non-saline river water in natural canals benefited the health of farm animals in the locality, as animals usually depend on canal water for drinking during the dry season. Increased sedimentation owing to the use of river water for supplemental irrigation during the wet season may help improve soil fertility. The microclimate of the area is considerably improved in the dry season because of crop cultivation using stored irrigation water.

Conclusions and Recommendations

This study showed that soil salinity during the wet season was low enough to grow modern rice varieties, even if they are not specifically bred for salt tolerance. In addition, HYV rice can be grown in the dry season using river water stored in on-farm canal networks before it becomes too saline. The stored water is used for irrigating rice in the latter part of the dry season. The modified water management and cropping practices increased the productivity of the coastal saline rice lands and farmers' income by 50–100% over the traditional farmers' practice.

The participatory approach – involving multilevel stakeholders, from farmers to local government officials and policymakers at all stages of the project – used in this study accelerated the uptake of the technology by neighbouring farmers. Rice production during the dry season in coastal ecosystems improves Bangladesh's food security and enhances farmers' livelihoods via increased income. It also helps improve the environment by reducing salt accumulation during the dry season and ameliorating the microclimate via vegetation cover. These are important factors in farmers' preference of the technology over shrimp farming, as evidenced by the fast-paced adoption of the technology, while shrimp area has not increased in the study area.

Many rivers and canal networks lie in the coastal region, especially in the polder areas of coastal Bangladesh, where non-saline river water and rainfall can be conserved for crop production. These offer a good opportunity for improving land productivity and the livelihood of resource-poor farmers in these areas. In our study, the turnaround time between the wet-season crop harvest and dry-season crop establishment is more than a month. There are still more opportunities to advance crop establishment and therefore harvesting of the dry-season crop, which would further reduce the amount of water needed for supplementary irrigation at the end of the dry season. There is also a possibility of dry direct seeding of the rainy-season crop to advance its establishment and harvest even further (My *et al.*, 1995), followed by an earlier dry-season crop. Alternatively, the last river water intake can be extended by about 2 weeks if river water can be pumped into the canal systems at the ebb tide with lower salinity, instead of relying solely on gravity for the intake of river water at the flood tide. This of course will entail more investment. Cost–benefit analysis needs to be performed, and social acceptance of these changes needs to be further studied.

As the rice farmers have no control over open-water bodies (i.e. natural canals, where they can conserve rainwater and non-saline river water) and sluice gates, they cannot properly plan for rice cultivation in the dry season. New management practices should be established to allow farmers to control the sluice gates and open water bodies to irrigate their farms without harming the interests of those who rely on fisheries in the open water for their livelihood. Canal networks need to be excavated to maximize water storage, and water balance studies have to be carried out to determine the maximum rice area that can be irrigated using stored water.

Our study focused on the coastal area of Bangladesh, but the success of the technology at the study site and similar success in Vietnam (Tuong *et al.*, 1991) indicate that the same principles can be applied to increase cropping intensity and improve farmers' livelihoods in many coastal areas in tropical deltas with similar characteristics.

References

ASA (American Society of Agronomy) and SSSA (Soil Science Society of America) (1982) *Methods of Soil Analysis, Part 2*. Madison, Wisconsin, pp. 225–260.

BBS (Bangladesh Bureau of Statistics) (2001) *Statistical Yearbook of Bangladesh*. Statistical Division, Ministry of Planning, Dhaka, Bangladesh.

BRRI (Bangladesh Rice Research Institute) (1993) *Modern Rice Cultivation* (*Adunik Dhaner Chas*, in Bangla). BRRI, Joydebpur, Gazipur, Bangladesh.

BRRI (Bangladesh Rice Research Institute) (2004) *Modern Rice Cultivation* (*Adunik Dhaner Chas*, in Bangla). BRRI, Joydebpur, Gazipur, Bangladesh.

FAO (Food and Agriculture Organization of the United Nations) (1988) *Land Resources Appraisal of Bangladesh for Agricultural Development*. Report 3(1), FAO, Rome.

Islam, R. (2005) Managing diverse land uses in coastal Bangladesh: institutional approaches. Paper prepared for the international conference on 'Environment and Livelihoods in Coastal Zones: Managing Agriculture-Fishery-Aquaculture Conflicts', 1–3 March, 2005, Bac Lieu, Vietnam.

Karim, Z., Hussain, S.G. and Ahmed, M. (1990) *Salinity Problems and Crop Intensification in the Coastal Regions of Bangladesh*. Soils Publication No. 33, Soils and Irrigation Division, BARC, Farmgate, Dhaka, Bangladesh, pp. 1–20.

Majid, M.A. and Gupta, M.V. (1997) *Research and Information Needs for Fisheries Development and Management*. Proceedings of National Workshop on Fisheries Resources Development and Management in Bangladesh, 29 October–1 November 1995. MOFL/BOBP/FAO/ODA, pp. 160–177.

Mondal, M.K. (1997) Management of soil and water resources for higher productivity of the coastal saline ricelands of Bangladesh. PhD thesis. University of the Philippines, Los Baños, Philippines.

Mondal, M.K. (2001) *Development of Suitable Salinity Management Techniques and Their Environmental Impact Assessment on the Coastal Ecosystem of Bangladesh*. Final report submitted to the Bangladesh Agricultural Research Council (BARC), Dhaka, Bangladesh.

My, T.V., Tuong, T.P., Xuan, V.T. and Nghiep, N.T. (1995) Dry seeding rice for increased cropping intensity in Long An Province, Vietnam. In: Denning, G.L. and Xuan, V.T. (eds) *Vietnam and IRRI: A Partnership in Rice Research*. International Rice Research Institute, Manila, Philippines, and Ministry of Agriculture and Food Industry, Hanoi, Vietnam, pp. 111–122.

Tuong, T.P., Hoanh, C.T. and Khiem, N.T. (1991) Agro-hydrological factors as land qualities in land evaluation for rice cropping patterns in the Mekong Delta. In: Deturck, P. and Ponnamperuma, F.N. (eds) *Rice Production on Acid Soils of the Tropics*. Institute of Fundamental Studies, Kandy, Sri Lanka, pp. 23–30.

7 Coastal Shrimp Farming in Thailand: Searching for Sustainability

B. Szuster
Department of Geography, University of Hawai'i at Manoa, Honolulu, Hawaii, USA, e-mail: szuster@hawaii.edu

Abstract

Shrimp farming in Thailand provides a fascinating example of how the global trade in agricultural commodities can produce rapid transformations in land use and resource allocation within coastal regions of tropical developing nations. These transformations can have profound implications for the long-term integrity of coastal ecosystems, and represent a significant challenge to government agencies attempting to manage land and water resources. Thailand's shrimp-farming industry has suffered numerous regional 'boom and bust' production cycles that created considerable environmental damage in rural communities. At a national scale, these events were largely masked, however, by a shifting cultivation strategy and local adaptations in husbandry techniques. This chapter outlines the need to upgrade planning systems, improve water supply infrastructure and enhance extension training services within coastal communities to address ongoing systemic environmental management problems within the Thai shrimp-farming industry.

Introduction

In Lewis Carroll's *Through the Looking Glass*, the Red Queen tells Alice that 'in this place it takes all the running you can do to keep in the same place'. This phrase has been used to illustrate a variety of natural and social phenomena (Van Valen, 1973) and it also aptly describes the history of shrimp farming in coastal Thailand. This nation has been a leading global producer of farmed shrimp since 1992 (FAO, 2002), but the industry has been plagued by persistent environmental problems stemming from a combination of natural resource degradation and associated viral disease outbreaks (Szuster and Flaherty, 2002).

Environmental problems have created widespread crop failures throughout Thailand, but a predicted national-level collapse in farmed shrimp production has not occurred (Dierberg and Kiattisimkul, 1996; Vandergeest *et al.*, 1999). This chapter traces the development of shrimp farming in Thailand. It argues that the Thai aquaculture industry has for many years managed to avoid a national-scale collapse in shrimp production by shifting farm sites and modifying husbandry techniques. Although these strategies have succeeded in maintaining overall production levels, underlying problems related to planning, infrastructure and social organization continue to exist and threaten the long-term sustainability of coastal shrimp farming in Thailand.

History of Coastal Shrimp Farming in Thailand

Thailand possesses over 2700 km of coastline and a tropical climate ideal for farming marine species such as shrimp. Basic biophysical factors are important, but the presence of appropriate aquaculture technologies, low agricultural wages and the availability of government tax incentives were also critical in supporting the development of shrimp farming in Thailand (Kongkeo, 1995). The shrimp-farming industry has gone through several distinct developmental phases that are reviewed below to illustrate the ability of Thai aquaculturalists to innovate and respond to changing environmental and socio-economic conditions.

Pre-expansion phase (1930–1971)

Shrimp farming was probably introduced to coastal Thailand by Chinese immigrants during the 1930s (Tookwinas, 1993). These early farms (called *thammachaat* or 'natural farms') applied traditional techniques that involved flooding low-lying coastal paddy fields during the dry season. Wild shrimp within the seawater were captured during this process and retained until the paddy fields were drained prior to planting a wet-season rice crop. Enclosures were large (30 ha or more) and a limited amount of daily water exchange was provided by natural tidal flows. Traditional farming techniques were inherently a polyculture because seawater provided the entire supply of shrimp seed and farmers exerted no control over species composition. Tidal flows also provided naturally occurring food organisms to sustain the captured shrimp during the culture period. Traditional shrimp farming requires no special technical skills or infrastructure, and input costs are minimal due to the use of naturally occurring seed stock and food. Yields are low (approximately 200 kg/ha/year), due to an absence of stocking control and poor survival rates. Species cultured in this manner include the banana shrimp (*Penaeus merguiensis*), Indian white shrimp (*P. indicus*), school shrimp (*Metapeneaus monoceros*) and black tiger shrimp (*P. monodon*).

Simple, traditional shrimp-farming techniques were modified during the late pre-expansion phase. New shallow shrimp ponds were constructed within coastal mud flats and numerous salt farms were also converted to shrimp production after World War II. This conversion was largely motivated by depressed salt prices, and more than 50% of the salt farms in the upper Gulf of Thailand region had been converted to shrimp production by the late 1960s (Csavas, 1994). Pond enclosures are still large in modified traditional systems (5 ha or more), but yields can reach 400 kg/ha/year by exercising limited control over fry stocking, improving water management and applying manure or chemical fertilizers to induce algal blooms. Only a small number of modified traditional shrimp farms continue to operate in Thailand today because of the modest harvests associated with this culture system (Pillay, 1997). Environmental damage associated with traditional shrimp farming is generally limited, but a large amount of intratidal land is needed for pond enclosures and this requirement can affect coastal habitats such as mangrove.

Early expansion (1972–1987)

Shrimp farming continued in a largely traditional form in Thailand until the early 1970s, when the Thai Department of Fisheries began experimenting with semi-intensive monoculture techniques (Katesombun, 1992). This modified culture system provided higher yields and was widely adopted in coastal areas already supporting traditional shrimp farms. Black tiger shrimp were the focus of semi-intensive husbandry experiments because this species possesses a high export value and is able to grow quickly under artificial conditions (Pillay, 1990). Another factor was the successful production of hatchery-raised black tiger shrimp postlarvae by Thai personnel trained in Japan. This breakthrough, in conjunction with the development of improved feeds and husbandry techniques, set the stage for a dramatic expansion of shrimp farming in coastal Thailand (Liao, 1992).

Semi-intensive shrimp culture uses pond enclosures that are smaller than traditional farms (1–8 ha), but this system provides significantly higher yields (up to 1000 kg/ha/year). This increase in productivity is gained through the use of hatchery-raised fry, supplementary feeding and a limited degree of mechanical water management provided by low-lift axial flow pumps to supplement the water exchange provided by tidal action (Tookwinas and Ruangpan, 1992). Construction costs are higher due to the need to construct levees or dykes, but the pond enclosures are more uniform, which provides additional control over the grow-out environment (Pillay, 1993). The quality of effluent released into the surrounding environment by semi-intensive operations is usually poorer than that of traditional farms, and overall environmental impacts are more pronounced due to the increased culture intensity (Miller, 1996, unpublished Masters thesis). However, coastal habitats such as mangrove can be less affected by semi-intensive operations if sited in supra-tidal areas that possess soil and drainage characteristics that are better suited to aquaculture (Menasveta, 1997).

Shrimp boom (1988–1995)

The introduction of the semi-intensive culture systems to Thailand was quickly followed by the development of intensive farming techniques (called *phattana* or 'developed farms'). This technology was introduced from Taiwan and features smaller ponds (0.16–1.0 ha), the use of hatchery-raised shrimp fry at high stocking densities, mechanical aeration, prepared feeds, fertilizers, chemicals and antibiotics (Flaherty and Karnjanakesorn, 1995). Average farm yields increased to as much as 2000 kg/ha/year from 1987 to 1999 and total annual farmed shrimp production also skyrocketed (Fig. 7.1). In addition to favourable biophysical conditions and the availability of Taiwanese intensive farming technology, the Thai 'shrimp boom' was supported by Thai government policies and international development agencies such as the Asian Development Bank (Flaherty and Vandergeest, 1998). These institutions provided significant assistance to the emerging shrimp aquaculture sector in areas such as financial support to potential farmers, aquaculture research and extension services and infrastructure construction in coastal areas (e.g. roads and canals).

Intensive culture techniques were first

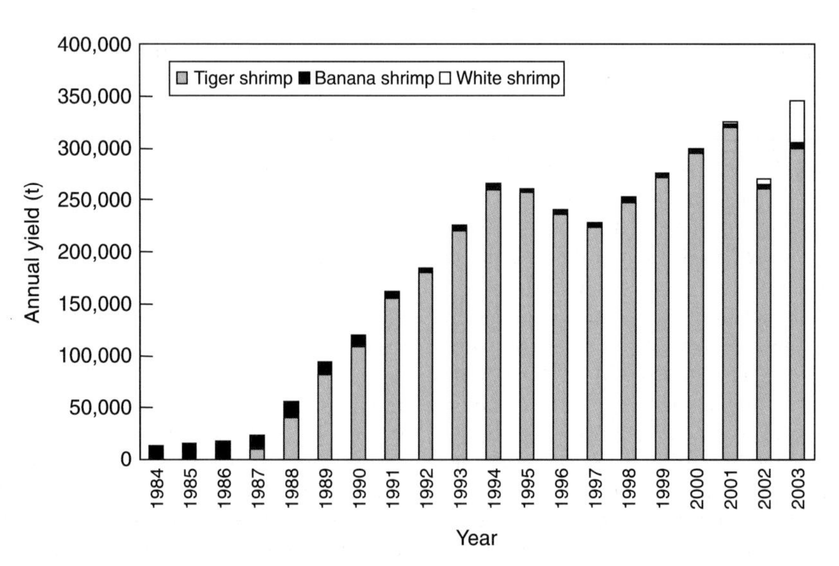

Fig. 7.1. Farmed shrimp production in Thailand, 1984–2003.

adopted in coastal areas surrounding the Upper Gulf of Thailand from 1987 to 1989. Most of these farms were small (less than 5 ha) and constructed on existing traditional shrimp farms, salt pans or wetlands. More than 80% were abandoned after only a few years of operation as a result of environmental degradation, viral disease problems and a lack of experience with intensive shrimp husbandry techniques (Jenkins, 1995). Derelict shrimp ponds became a common sight in Samut Prakarn and Samut Sakhon provinces at this time, with pond bottom soils contaminated by salt and chemical residues (Beveridge and Phillips, 1993). Although a small number of abandoned shrimp ponds were used to grow lower-value species such as fish or crab, many remained idle or were converted to non-agricultural land uses such as housing or manufacturing.

After the collapse of shrimp farming in the Upper Gulf of Thailand region, the geographic focus of shrimp farming moved to the eastern and southern coastal regions. These areas generally possess better soil and water supply characteristics than sites near Bangkok, and a large number of new farms were constructed during the early 1990s (Jory, 1996). The Thai government also recognized the emergence of serious environmental problems in the shrimp aquaculture sector during the early 1990s (e.g. mangrove destruction, soil degradation, water pollution) and responded by supporting research into revised aquaculture techniques and natural resource management practices (Kongkeo, 1997). Initiatives supported by the Thai government and agencies such as the World Bank and the Food and Agriculture Organization of the United Nations included training and extension services for small-scale aquaculturalists, research into water recycling and zero-discharge farming techniques, mangrove reclamation in abandoned shrimp-farming areas, seawater irrigation systems and research into shrimp disease and the breeding of domesticated species (MIDAS Agronomics, 1995).

There is no doubt that the management strategies, extension programs and infrastructure projects described above provided immediate benefits to Thailand's shrimp farmers. However, their success in bringing long-term sustainability to the aquaculture industry is debatable. Total annual shrimp production continued to rise through 1995, but this increase was largely as a result of new farm construction in eastern and southern Thailand (Fig. 7.2). Improved management was not a major factor (Flaherty *et al.*, 2000). This shifting cultivation strategy was initially quite successful in maintaining national production levels, but, by 1995, most of the coastal land suitable for shrimp farming was already in production or abandoned. Total annual harvests declined in 1996 as a result of viral disease problems in eastern and southern Thailand (Department of Fisheries, 2002), and this led several observers to suggest that a national crash in shrimp production could be imminent (Flaherty and Karnjanakesorn, 1995; Dierberg and Kiattisimkul, 1996). Coastal shrimp production did, in fact, continue to fall during the latter half of the 1990s, but an important innovation allowed the shrimp boom to continue. Low-salinity shrimp-farming techniques were developed during the mid-1990s (Ponza, 1999, unpublished Masters thesis), and these would allow intensive shrimp farming to make a dramatic return to the Upper Gulf of Thailand region.

Low-salinity shrimp farming (1996–2002)

As crop failures became commonplace throughout coastal Thailand, the shrimp-farming industry began to search for alternatives to maintain production (Kaosa-ard and Pednekar, 1996). Crop failures in coastal areas during the early 1990s had a serious economic impact on novice shrimp farmers, who generally possessed very little aquaculture experience. They responded to this crisis in several ways. Some farmers attempted to raise additional shrimp crops, but this strategy simply compounded husbandry or environmental management mistakes that had led to the initial crop failures. Other farmers switched to safer crops such as freshwater fish or crab that involve fewer risks, but also smaller potential profits. Many individuals

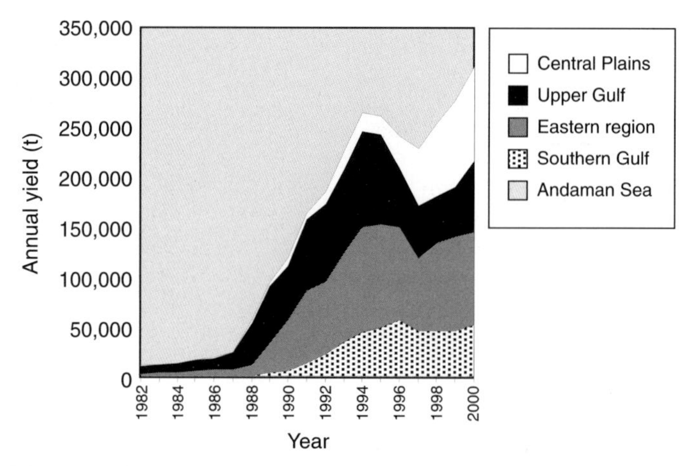

Fig. 7.2. Farmed shrimp production in Thailand by region (t), 1982–2000.

simply abandoned aquaculture and turned to off-farm employment in order to repay the large debts incurred from shrimp farming (Banpasirichote, 1993).

Relocating shrimp farms inland from disease-prone coastal areas also emerged as a response to widespread crop failures (Flaherty and Vandergeest, 1998). During the mid-1990s, farmers in Samut Prakarn and Chachoengsao provinces discovered that tiger shrimp could be grown in seasonally saline watersheds located near the coast (Szuster, 2001, unpublished PhD dissertation). Typically, these areas could support only a single dry-season shrimp crop because brackish water is unavailable during the rainy season. However, the development of low-salinity husbandry techniques allowed two, or even three, shrimp crops to be raised within a single calendar year (Flaherty *et al.*, 1999). Low-salinity farming techniques are generally similar to those used in coastal operations, but, while coastal farms use seawater to fill and replenish pond enclosures, low-salinity farms combine fresh water with hypersaline water purchased from coastal salt pans or saltwater concentrate operations (Miller *et al.*, 1999). Freshwater inputs are also used to offset evaporation and seepage losses over the grow-out period, and this can reduce salinity levels to near zero by harvest unless supplementary saline water or bagged salt is applied. Harvest on low-salinity farms occurs earlier than on coastal operations as a result of falling salinity levels and the negative effect this has on shrimp health and development (Ponza, 1999). Given the shorter culture period and suboptimal growing environment, shrimp from low-salinity farms tend to be smaller and of poorer quality than shrimp produced in coastal areas.

The success of low-salinity techniques in seasonally saline areas was soon noticed by rice farmers in freshwater regions located further inland from the coast (Pongnak, 1999). Rice farmers realized that the high potential profits derived from shrimp production could easily offset the costs associated with trucking salt water to their land (Szuster *et al.*, 2003). Development opportunities were limited only by basic site suitability criteria, such as relatively flat terrain, suitable soils and a reliable source of water (Flaherty *et al.*, 1999). These factors led a large number of rice farmers in central Thailand to convert irrigated paddy fields into shrimp ponds during the latter half of the 1990s (Committee on Inland Shrimp Farming, 1998). It is difficult to estimate the extent of shrimp farming within freshwater areas of central Thailand during the late 1990s because information collected by the Royal Department of Fisheries does not distinguish between freshwater and brackish production sources. However, surveys conducted by the Thai Department of Land Development and the Department of

Fisheries suggest that low-salinity farms operating within freshwater areas could have accounted for as much as 40% (or approximately 100,000 t) of Thailand's total farmed shrimp output in 1998 (Limsuwan and Chanratchakool, 1998). Although it is difficult to assess the accuracy of this estimate, its magnitude alone indicates that the expansion of low-salinity shrimp farming within inland freshwater areas of central Thailand masked a very serious collapse of brackish-water shrimp production in coastal regions (Fig. 7.2).

The expansion of low-salinity shrimp farming into freshwater rice-growing areas of the Chao Phraya River Delta initially proceeded with little overt government support or scrutiny. This low profile disappeared when the Thai print media became sharply critical of low-salinity shrimp farming and the potential environmental damage this activity could produce within Thailand's most important agricultural region (Arunmart and Ridmontri, 1998). Many analysts suggested that low-salinity shrimp farms could produce soil salinization, water pollution and increased competition between agriculture and aquaculture for fresh water (Miller *et al.*, 1999; Pongnak, 1999). Following a rancorous debate between pro- and anti-shrimp-farming groups in the media, and the hasty completion of environmental impact studies, the Thai government banned shrimp farming within non-coastal provinces on the basis of a recommendation from the National Environment Board (Srivalo, 1998). Provincial governors in coastal provinces were subsequently instructed to identify and map brackish-water areas (where shrimp farming would be permitted) and freshwater zones (where shrimp farming would be restricted). In spite of the 1998 ban on low-salinity shrimp farming, the practice continued relatively uninterrupted over the following 2 years. Harvests may even have increased between 1999 and 2001 as the Thai government encouraged farmers to increase harvests to take advantage of high market prices stemming from a catastrophic collapse of shrimp production in Latin America (Bangkok Post, 2000a). The Thai shrimp-farming industry also lobbied strenuously for a reversal of the ban on shrimp farming in freshwater areas during this period (Bangkok Post, 1999). Although government sources indicated that the restriction on this practice would be relaxed (Boyd, 2001), opposition from environmental groups, soil scientists within Thailand's Land Development Department and His Majesty King Bhumibol convinced the National Environment Board to reaffirm the ban (Samabuddhi, 2001). Progress on enforcing the ban has been slow, however, and a recent survey conducted by the Thai government suggests that many low-salinity shrimp farms continue to operate within freshwater areas (Samabuddhi, 2003).

Recent developments (2003–2004)

Thai government policies have certainly restricted the expansion of low-salinity shrimp farming within freshwater regions of the country (*The Nation*, 2001). This constraint has not prevented shrimp farmers from attempting to maintain overall production levels in the face of unresolved environmental problems and more rigorous quality restrictions imposed by major importers (Boonchote, 2003). Two recent strategies to maintain production are the conversion of low-salinity shrimp ponds to freshwater prawn culture in Nakhon Pathom and Suphanburi provinces (Szuster *et al.*, 2003), and the importation of the Pacific white shrimp (*Penaeus vannamei*), native to Latin America (NACA, 2003). Converting low-salinity shrimp farms to freshwater prawn production is a positive development that addresses soil salinization concerns, but the introduction of non-native shrimp species to Thailand is troubling. The Pacific white shrimp is popular with Thai shrimp farmers for its high yields and disease resistance characteristics. Existing Thai government regulations restrict the importation of this exotic species to research purposes only, but this control has been widely ignored in the interest of maintaining national shrimp production levels (Wangvipula, 2002). Farmers appear to have no difficulty in obtaining seed stock and the Thai Shrimp Farmers

Association estimates that up to 60% of all harvests in 2004 could be white shrimp (*Bangkok Post*, 2004). The importation and use of exotic shrimp could, however, disrupt host biotic communities or transfer exotic pathogens to native species such as the black tiger shrimp (Arthington and Bluhdorn, 1996). The Taura syndrome virus (TSV) is endemic in the Pacific white shrimp and has produced massive crop losses in Latin America (FAO, 1997). Although no cases of TSV transference to native shrimp species in South-east Asia have been documented to date, this pathogen may have already infected wild shrimp populations in Taiwan (NACA, 2003). Viral transmission pathways include the use of infected broodstock or seedstock, pond effluent disposal, pond flooding, shrimp escapes, transport to processing facilities and sediment or solid waste disposal (Lightner, 1996).

Strategies to Enhance Sustainability

Aquaculture planning

Thailand's coastal shrimp-farming industry finds itself at a critical juncture. In the face of trade sanctions by major overseas trade partners in Europe and America and growing domestic public scrutiny, many industry leaders and government decision-makers are aware that the sector must undergo a major transformation to achieve future success. An industry that is infamous for its 'slash-and-burn' approach to coastal land use, narrow focus on short-term profits and disregard for environmental regulations must now embrace stability, sustainability and regulation to secure access to international markets. This transformation can be achieved only through the development and implementation of formal planning structures and processes. Aquaculture zoning and other forms of integrated management have been proposed to enhance the sustainability of coastal shrimp farming in Thailand (Tookwinas, 1999), but this has been attempted in only a limited number of local pilot projects such as the recent Coastal Habitats and Resource Management initiative (CHARM, 2003). There is little doubt that aquaculture zoning on a nationwide basis could protect environmental resources, minimize land-use conflicts and maximize shrimp production by siting farms in areas best suited for aquaculture. The Thai Department of Fisheries has carried out aquaculture suitability studies within major shrimp-farming areas in southern and eastern Thailand, and this information could support aquaculture zoning and other forms of integrated management. Thailand's Land Development Department (1999) has also investigated environmental conditions in coastal areas and recommended that shrimp farm construction be restricted to areas possessing soil parent materials with a conductivity of 2 mS/cm or greater (measured at 1.5 m below the surface). This action could mitigate potential impacts related to soil salinization and restrict shrimp farms to less productive agricultural areas where saline sediment lies relatively close to the surface. Although a complex regulatory environment will probably slow the emergence of effective aquaculture planning in Thailand (MIDAS Agronomics, 1995; Flaherty *et al.*, 2000), the identification of specific shrimp farm zones could impose a degree of stability on the industry and provide a focus for infrastructure and training programmes that are needed to support sustainable shrimp-farming practices.

Water supply infrastructure

Clean, plentiful water supplies are essential for successful aquaculture in the same sense that fertile soils represent the basis for agricultural crop production. The development of aquaculture water supply systems in Thailand, however, has been largely unplanned and has evolved by adapting the existing infrastructure, originally designed to support rice cultivation (Braaten and Flaherty, 2000). Exceptions to this include a small number of large, government-supported seawater irrigation projects that provide separate water supply and waste treatment infrastructure for coastal shrimp farmers in eastern and southern Thailand

(Tookwinas and Yingcharoen, 1999). Small irrigation canals serve as the only option for both water supply and wastewater disposal in all other shrimp-farming areas. The water pollution and disease transference implications of this practice are obvious, but the uncoordinated development of shrimp farming in Thailand has left most small-scale farmers with no alternative to the existing water distribution system (Beveridge and Phillips, 1993). Farmers accept that their water supplies are contaminated and have attempted to mitigate disease concerns by reducing water inputs during the culture period, applying antibiotics and switching to species such as the Pacific white shrimp with different disease resistance characteristics. Although these strategies can provide some short-term relief, viral disease outbreaks remain a constant threat that is magnified by the presence of inadequate water supply systems and sub-standard water management practices (Szuster *et al.*, 2003).

As the lead government agency charged with the responsibility for constructing and maintaining Thailand's irrigation infrastructure, the Royal Irrigation Department has traditionally focused on the needs of agriculture in general, and wet rice paddy production in particular. Aquaculture has a very low profile within the Royal Irrigation Department. The water supply needs of shrimp farmers are viewed as the responsibility of the Department of Fisheries. This division of responsibilities ignores the fact that aquaculture is a dominant water user group in many coastal areas (Szuster, 2001). More effective management of water supply infrastructure within coastal Thailand is needed, and should be a priority in areas specifically zoned for aquaculture. A telling example of the current lack of coordination between Thai government agencies with respect to aquaculture water supplies is the recently completed Bangpakong River dam in eastern Thailand. This dam was planned and constructed by the Royal Irrigation Department to limit natural saltwater tidal flows in the Bangpakong River during the dry season (Kasetsart University, 1994). Dry-season saline intrusion in upstream areas limits irrigation opportunities and agricultural production, but is a defining ecological characteristic of low-gradient river systems in South-east Asia. The presence of seasonally saline flows in the Bangpakong River also facilitated the development of low-salinity shrimp farms that can be found more than 100 km upstream of the Gulf of Thailand. Aquaculture water use, however, was not given a high priority by the Royal Irrigation Department during the planning of the Bangpakong River dam project (Kasetsart University, 2000). This led to a significant investment in water supply infrastructure that not only ignored the needs of shrimp farming, but also seriously affected the hydrology and natural ecology of the Bangpakong River ecosystem (*Bangkok Post*, 2000b). Improved water supply systems are a fundamental requirement for the evolution of sustainable shrimp-farming practices, but additional infrastructure spending does not guarantee success. Improved consultation and cooperation among the Thai government agencies responsible for water resource management are also needed to effectively plan new infrastructure investments and manage existing water supply systems to the benefit of both aquaculture and agriculture.

Social organization

In contrast to the significant corporate presence within India and Latin America, the Thai shrimp-farming industry is dominated by small, independent owner-operators (Flaherty *et al.*, 1999). Shrimp farmers are typically former agriculturalists, fisherfolk or small-scale business investors with little or no previous background in aquaculture (Flaherty *et al.*, 2000). Most operations are managed as family farms, with little or no assistance from hired farm labour or professionals such as biologists or veterinarians. Access to credit is also limited because of the modest scale of the operations and the financial risks associated with shrimp farming (Vandergeest *et al.*, 1999). A small number of shrimp farms are owned by outside investors who lease land within farming communities to construct larger operations (CORIN, 2000), but this situation is unusual in most areas. Several large corporate farms were constructed in Thailand during the 1990s (primarily in the southern region), but most of these operations subsequently closed as a result of environmental degradation, viral disease outbreaks and unsatisfactory contract-farming arrangements (Boonchote, 2003; Vandergeest *et al.*, 1999). The corporate presence in the Thai shrimp aquaculture industry is now largely restricted to the manufacture of shrimp feed and processing for market rather than to the grow-out phase. A small number of government shrimp-farming projects, such as the Kung Kraben Bay Royal Initiative, have also been developed. These projects typically involve several hundred small family farms organized to take advantage of the communal water supply and waste treatment infrastructure. These government projects are expensive to construct, however, and less than 5% of all shrimp farms currently have access to seawater irrigation facilities (Tookwinas and Yingcharoen, 1999).

The small-scale and highly mobile nature of shrimp farming in Thailand has allowed the industry to increase national production in spite of serious economic and environmental challenges. This organizational structure possesses inherent weaknesses, however, that must be addressed if the industry is to become more stable and sustainable. Substantial effort has gone into the creation of best management practices and a voluntary Code of Conduct for shrimp farmers (Boyd, 1999; FAO and NACA, 2000). The development of these initiatives stems from an acceptance that many existing shrimp farm management practices are detrimental to environmental quality and even potentially hazardous to human health (Boyd, 2003). Major proponents of the Code of Conduct include the Thai Department of Fisheries and major aquaculture industry groups (e.g. the Thai Marine Shrimp Farmers Association, the Thai Frozen Foods Association, the Thai Food Processors Association and Global Aquaculture Alliance). Transferring this information on improved husbandry and environmental management practices to a large cadre of independent-minded rural farmers has proved to be difficult (CORIN, 2000). Training courses and manuals have been developed (Prompoj, 2002; Wailailak University and the Thai Shrimp Network, 2002), but only a relatively small number of farmers will be enrolled in Code of Conduct implementation pilot projects over the next 5 years (Tookwinas, 2002). A large number of farms will not qualify for certification under the Code of Conduct because their operations are too small to support essential infrastructure such as treatment ponds. Many other farmers may not participate because they do not recognize the financial benefit of implementing the code (Ampornpong, 2002, unpublished MSc thesis).

These difficulties highlight the need for both improved government extension services at the local level and support for the organization of local shrimp-farmer groups. Duplicating the infrastructure and techniques used by neighbours is a common management strategy adopted by many Thai shrimp farmers (Flaherty *et al.*, 1999). Seminars provided by feed or chemical sales agents have also been used in the past to spread knowledge of husbandry techniques in rural areas. Relatively few farmers have attended Thai government-sponsored aquaculture seminars or belong to local cooperative shrimp-farmer groups (Miller *et al.*, 1999). A small number of organizations such as the Surat Thani Shrimp Farmers Club have emerged to deal with common husbandry, environmental management and socio-economic concerns, but it is notable how few cooperative shrimp-farmer groups exist at the local level in Thailand. Most shrimp farmers also have limited contact with staff from the Department of Fisheries and many are proud of their ability to achieve success 'on their own' with little or no outside assistance (Miller *et al.*, 1999). This attitude has been attributed to weak group cohesion within Thai society that limits cooperation and the spread of knowledge outside of an individual's immediate close relations (CORIN, 2000). It is important that this attitude be overcome. Future improvements in sustainability depend on better cooperation between government agencies and rural communities, and greatly improved collaboration among local groups of shrimp farmers. More emphasis on capacity building within local government extension services and community shrimp-farmer organizations is also required. Initiatives such as the best management practices and the Code of Conduct are useful management tools, but their success in improving the overall environmental performance of the shrimp-farming industry will largely depend on supplementary measures that support effective farm-level social organization and cooperation.

Conclusions

The evolution of shrimp aquaculture in Thailand has been characterized by a constant search for unexploited areas to replace farm sites degraded by poor husbandry practices, pollution or disease. Over the past 15 years, the focus of farming activities has shifted from the Upper Gulf of Thailand to eastern and southern Thailand, and finally inland to freshwater areas of central Thailand. Local technical innovations such as low-salinity culture techniques supported a move into freshwater areas, and non-native Pacific white shrimp have recently been introduced to areas suffering from viral disease problems in the traditional black tiger shrimp crop. Shifting cultivation strategies and technical innovations have been successful to the extent that national farmed shrimp production has increased over the past 10 years (FAO, 2004), but environmental concerns associated with water pollution, land salinization, land-use conflicts and disease transference remain unresolved (Szuster and Flaherty, 2002). This situation presents a serious challenge to the shrimp-farming industry because few opportunities exist for expansion in coastal areas and the ban on low-salinity culture has removed expansion opportunities within freshwater regions (Szuster *et al.*, 2003). The shifting cultivation strategy that supported the Thai 'shrimp boom' appears to have run its course, and serious consideration must now be given to measures that will allow the industry to create a sustainable future within existing shrimp-farming areas.

The issues we have noted as requiring critical attention largely relate to group cooperation and social organization. Indeed, it is reasonable to suggest that a large majority of the strictly technical concerns preventing the evolution of more sustainable shrimp cultural practices have been resolved over the past 20 years (Boyd and Clay, 1998). This includes advances in husbandry practices, domesticated broodstock supplies, disease resistance, siting criteria, water supply infrastructure and waste treatment techniques. Structural concerns within the Thai natural resource management system were overlooked during the shrimp boom years, but the end of this gold rush mentality has highlighted the need for stability, regulation and sustainability in all regions supporting aquaculture. Issues such as aquaculture zoning, water resource management and capacity building within local extension services and farmers' organizations can no longer be ignored. Ultimately, the future of the Thai shrimp-farming industry depends on how effectively these issues are managed because the short-term strategies of the past will clearly not sustain the shrimp-farming industry in the future.

References

Arthington, A.H. and Bluhdorn, D.R. (1996) The results of species interactions resulting from aquaculture operations. In: Baird, D.J., Beveridge, M., Kelly, L. and Muir, J.F. (eds) *Aquaculture and Water Resources Management*. Blackwell Scientific Publications, Oxford, UK, pp. 114–139.

Arunmart, P. and Ridmontri, C. (1998) Shrimp farm ban planned: move to protect the Chao Phraya basin. *Bangkok Post*, 30 May 1998.

Bangkok Post (1999) Shrimp farmers say ban order is unfair. *Bangkok Post*, 1 September 1999.

Bangkok Post (2000a) Bullish time for shrimps. *Bangkok Post*, 20 September 2000.

Bangkok Post (2000b) Bangpakong dam proves disastrous. *Bangkok Post*, 11 April 2000.

Bangkok Post (2004) High yields of white shrimp send production soaring in Thailand. *Bangkok Post*, 7 December 2004.
Banpasirichote, C. (1993) *Community Integration into Regional Industrial Development: a Case Study of Khlong Ban Po, Chachoengsao*. Background Report for the Chulalongkorn University Social Research Institute 1993 Year-end Conference, Jomtien, Thailand, 10–11 December 1993.
Beveridge, M.C.M. and Phillips, M.J. (1993) Environmental impact of tropical inland aquaculture. In: Pullin, R.S.V., Rosenthal, H. and Maclean, J.L. (eds) *Environment and Aquaculture in Developing Countries*. International Center for Living Aquatic Resources Management (ICLARM), Manila, Philippines, pp. 213–236.
Boonchote, V. (2003) Boom for black tiger farms over. *Bangkok Post*, 8 December 2003.
Boyd, C. (1999) *Codes of Practice for Sustainable Shrimp Farming*. Global Aquaculture Alliance, St Louis, Missouri.
Boyd, C. (2001) Inland shrimp farming and the environment. *World Aquaculture* 321, 10–12.
Boyd, C. (2003) The status of codes of practice. *World Aquaculture* 342, 63–66.
Boyd, C. and Clay, J. (1998) Shrimp aquaculture and the environment. *Scientific American* 278, 58–65.
Braaten, R. and Flaherty, M. (2000) Hydrology of inland brackish water shrimp ponds in Chachoengsao, Thailand. *Aquacultural Engineering* 23(4), 295–313.
CHARM (2003) *Overall Workplan of the Coastal Habitats and Resources Management Project (CHARM)*. The European Union and Government of the Kingdom of Thailand, Bangkok, 25 November 2003.
Committee on Inland Shrimp Farming (1998) *Environmental Impact of Shrimp Farming in Freshwater Area*. Ministry of Science, Technology and Environment, Bangkok, Thailand.
CORIN (2000) *Shrimp Farming Experiences in Thailand: a Continued Pathway for Sustainable Coastal Aquaculture*. Report submitted to the Network of Aquaculture Centres in Asia-Pacific by the Coastal Resources Institute (CORIN). Prince of Songkhla University, Hat Yai, Thailand.
Csavas, I. (1994) Coastal aquaculture in Thailand. *FAO Aquaculture Newsletter* 7, 2–5.
Department of Fisheries (2002) *Fishery Production Statistics 2000*. Fishery Statistics and Information Subdivision, Division of Fishery Economics, Bangkok, Thailand.
Dierberg, F. and Kiattisimkul, W. (1996) Issues, impacts, and implications of shrimp aquaculture in Thailand. *Environmental Management* 20, 649–666.
FAO (1997) *Review of the State of World Aquaculture*. Inland Water Resources and Aquaculture Service, Fishery Resources Division. Food and Agriculture Organization of the United Nations, Rome.
FAO (2002) *Code of Conduct for Responsible Fisheries*. Food and Agriculture Organization of the United Nations, Rome.
FAO (2004) *Fishstat Plus Fisheries Statistical Database*. Fishery Information, Data and Statistics Unit. Food and Agriculture Organization of the United Nations, Rome.
FAO and NACA (2000) *Aquaculture Development Beyond 2000: the Bangkok Declaration and Strategy*. Conference on Aquaculture in the Third Millennium, 20–25 February 2000. Food and Agriculture Organization of the United Nations (FAO) and Network of Aquaculture Centres in Asia-Pacific (NACA), Bangkok, Thailand.
Flaherty, M. and Karnjanakesorn, C. (1995) Marine shrimp aquaculture and natural resource degradation in Thailand. *Environmental Management* 19, 27–37.
Flaherty, M. and Vandergeest, P. (1998) Low salt shrimp aquaculture in Thailand: goodbye coastline, hello Khon Kaen! *Environmental Management* 22(6), 817–830.
Flaherty, M., Vandergeest, P. and Miller, P. (1999) Rice paddy or shrimp pond: tough decisions in rural Thailand. *World Development* 12, 2045–2060.
Flaherty, M., Szuster, B. and Miller, P. (2000) Low salinity shrimp farming in Thailand. *Ambio* 29(3), 174–179.
Jenkins, S. (1995) *Key Researchable Issues in Sustainable Shrimp Aquaculture in Thailand*. ACIAR and CSIRO, Sydney, Australia.
Jory, D.E. (1996) Marine shrimp farming in the Kingdom of Thailand: Part 1. *Aquaculture Magazine* 22(3), 97–106.
Kaosa-ard, M. and Pednekar, S. (1996) *Environmental Strategy for Thailand*. Thailand Development Research Institute, Bangkok, Thailand.
Kasetsart University (1994) *Data Evaluation for the Bangpakong River Basin Development Project, Main Report*. Department of Aquatic Resources Engineering, Kasetsart University and the National Economic and Social Development Board, Bangkok, Thailand.

Kasetsart University (2000) *Planning and Management for Water Resources and Land Use in the Bangpakong River Basin. Volume 1, General Information.* Kasetsart Institute of Research and Development and the Office of the Eastern Seaboard Development Committee, Bangkok, Thailand.
Katesombun, B. (1992) Aquaculture promotion: endangering the mangrove forests. In: *The Future of People and Forests in Thailand After the Logging Ban.* Project for Ecological Recovery, Bangkok, Thailand, pp. 103–122.
Kongkeo, H. (1995) *How Thailand Became the World's Largest Producer of Cultured Shrimp. Aquaculture Towards the 21st Century.* Proceedings of the INFOFISH International Conference on Aquaculture, Colombo, Sri Lanka.
Kongkeo, H. (1997) Comparison of intensive shrimp farming systems in Indonesia, Philippines, Taiwan and Thailand. *Aquaculture Research* 28(10), 789–796.
Land Development Department (1999) *Zoning Maps for Shrimp Farming in Freshwater Areas.* Ministry of Science, Technology and Environment, Bangkok, Thailand.
Liao, I.C. (1992) Marine prawn culture industry in Taiwan. In: Fast, A.W. and Lester, L.J. (eds) *Marine Shrimp Culture: Principles and Practices.* Elsevier, Amsterdam, pp. 653–675.
Lightner, D.V. (1996) *A Handbook of Shrimp Pathology and Diagnostic Procedures for Disease of Cultured Penaeid Shrimp.* World Aquaculture Society, Baton Rouge, Louisiana.
Limsuwan, C.S. and Chanratchakool, P. (1998) *Closed Recycle System for Sustainable Black Tiger Shrimp Culture in Freshwater Areas.* Proceedings of the Fifth Asian Fisheries Forum, 11–14 November 1998, Chiang Mai, Thailand.
Menasveta, P. (1997) Intensive and efficient shrimp culture – the Thai way – can save mangroves. *Aquaculture Asia* 2(1), 38–44.
MIDAS Agronomics (1995) *Pre-Investment Study for a Coastal Resources Management Program in Thailand.* The World Bank, Washington, D.C.
Miller, P., Flaherty, M. and Szuster, B. (1999) Inland shrimp farming in Thailand. *Aquaculture Asia* 4(1), 27–32.
NACA (2003) *Asian Regional Status of Culture, Movement and Management of* P. vannamei. Network of Aquaculture Centres in Asia (NACA), Bangkok, Thailand.
Pillay, T.V.R. (1990) Asian aquaculture: an overview. In: Mohan, J. (ed.) *Aquaculture in Asia.* Asian Fisheries Society, Indian Branch, Department of Aquaculture, College of Fisheries, Mangalore, Karnataka, India, pp. 31–42.
Pillay, T.V.R. (1993) *Aquaculture: Principles and Practices.* Cambridge University Press, Cambridge, UK.
Pillay, T.V.R. (1997) Economic and social dimensions of aquaculture management. *Aquaculture Economics and Management* 11, 3–11.
Pongnak, W. (1999) Case study on the impact and conflict in using the nation's freshwater land resources for farming penaeid shrimp. *Greenline* 4 (June–September), 6–15.
Prompoj, W. (2002) *Quality Shrimp Production Using the Code of Conduct in Thailand.* Department of Fisheries, Bangkok, Thailand. (In Thai.)
Samabuddhi, K. (2001) Farmers realizing ban likely to stay. *Bangkok Post,* 20 September 2001.
Samabuddhi, K. (2003) Inland prawn farms thrive despite ban. *Bangkok Post,* 20 February 2003.
Srivalo, P. (1998) Cabinet bans inland shrimp farms. *The Nation,* 8 July 1998.
Szuster, B. and Flaherty, M. (2002) Cumulative environmental effects of low-salinity shrimp farming in Thailand. *Impact Assessment and Project Appraisal* 20(2), 189–200.
Szuster, B., Molle, F., Flaherty, M., and Srijantr, T. (2003) Socioeconomic and environmental implications of inland shrimp farming in the Chao Phraya Delta. In: Molle, F. and Srijantr, T. (eds) *Perspectives on Social and Agricultural Change in the Chao Phraya Delta.* White Lotus Press, Bangkok, Thailand, pp. 177–194.
The Nation (2001) Prawn ban stays intact. *The Nation,* 6 October 2001.
Tookwinas, S. (1993) *Intensive Marine Shrimp Farming Techniques in Thailand.* Proceedings of the First International Symposium on Aquaculture Technology and Investment Opportunities, Riyadh, Saudi Arabia.
Tookwinas, S. (1999) Coastal planning of shrimp farming: carrying capacities, zoning and integrated planning in Thailand. In: Smith, P.T. (ed.) *Towards Sustainable Shrimp Culture in Thailand and the Region.* Proceedings of a workshop held at Hat Yai, Songkhla, Thailand, 28 October–1 November 1996.
Tookwinas, S. (2002) *Assistance and Issues in the Implementation of the Code of Conduct for Shrimp*

Aquaculture. Report prepared for the World Bank, Network of Aquaculture Centres in Asia-Pacific, World Wildlife Fund and Food and Agriculture Organization of the United Nations consortium on shrimp farming and the environment. Published by the consortium, Bangkok, Thailand.
Tookwinas, S. and Ruangpan, L. (1992) Marine shrimp culture and mangroves. *Thai Fisheries Gazette* 45, 953–966.
Tookwinas, S. and Yingcharoen, D. (1999) Seawater irrigation system for intensive marine shrimp farming. *Aquaculture Asia* 4(3), 33–38.
Vandergeest, P., Flaherty, M. and Miller, P. (1999) A political ecology of shrimp aquaculture in Thailand. *Rural Sociology* 64(4), 573–596.
Van Valen, L. (1973) A new evolutionary law. *Evolutionary Theory* 1, 1–30.
Wailailak University and the Thai Shrimp Network (2002) *Techniques to Help Shrimp Farmers Overcome Problems Through Alternative Farming Practices*. Proceedings of a workshop held 17 August 2002. Nakhon Sri Thammarat, Thailand. (In Thai.)
Wangvipula, R. (2002) Switch to west white shrimp breed urged. *Bangkok Post*, 20 July 2002.

8 Tracing the Outputs from Drained Acid Sulphate Flood Plains to Minimize Threats to Coastal Lakes

B.C.T. Macdonald,[1] I. White,[1] L. Heath,[1] J. Smith,[2] A.F. Keene,[3] M. Tunks[4] and A. Kinsela[2]

[1] *Centre for Resource and Environmental Studies, Australian National University, Canberra, Australia, e-mail: ben.macdonald@anu.edu.au*

[2] *School of Biological, Earth and Environmental Sciences, University of New South Wales, Sydney, Australia*

[3] *School of Environmental Science and Management, Southern Cross University, New South Wales, Australia*

[4] *Tweed Shire Council, New South Wales, Australia*

Abstract

Drainage of acid sulphate flood plains for agriculture and urban development has led to the acidification of coastal waterbodies, major fish kills and other environmental effects in the subtropical areas of eastern Australia. These have produced problems for local governments and conflicts in communities. Here, we trace the effects of drainage from an acidified, subtropical flood plain on water quality and sediments in an estuarine lake, Cudgen Lake, in northern New South Wales. This shallow, brackish lake was once a renowned fish and prawn nursery. The local government has monitored water quality in drains and the lake since 1990. This has revealed episodic discharge events with pH as low as 2 and concentrations of dissolved iron and aluminium that are toxic to gilled organisms. Highly acidic waters were found to accumulate in drainage channels after rains following long dry periods. These were then discharged into the lake, causing major fish kills, low benthic organism biodiversity and infestations of acid-tolerant reeds and disease-bearing mosquitoes. High concentrations of arsenic, mercury and lead were found to have accumulated in iron monosulphides in the lake sediments, together with massive amounts of aluminium and iron. Ion ratios show that these metals were mobilized by drainage from the acidified soils within the flood plain. Although the lake sediments at present represent a sink for these metals, disturbance and oxidation of the monosulphides in the sediments could release major contaminants into the lake. The heavy metals sequestered in the lake sediments could partly explain the low benthic biodiversity. These results provide pointers for enhancing the conservation and fishery values of the lake.

Introduction

Pressure is increasing to develop and drain coastal lowlands to reduce flooding and waterlogging, particularly in the Asia–Pacific region. Many of these coastal flood plains contain iron pyrite (FeS_2), with concentrations sometimes averaging over 3.5% (w/w),

 Environment and Livelihoods in Tropical Coastal Zones (eds C.T. Hoanh, T.P. Tuong, J.W. Gowing and B. Hardy)

deposited under brackish, reduced conditions during and following the last sea-level rise. Drainage promotes oxidation of sulphides to sulphuric acid, acidifying groundwater and accelerating mineral weathering (Dent, 1986). Discharge from drained coastal flood plains following flooding can export hundreds of tonnes of acidic products into coastal streams, with devastating effects on fisheries and coastal communities (Sammut *et al.*, 1996).

The well-studied primary environmental impacts of acid drainage are partly due to the presence of dissolved metals, particularly aluminium and iron, but also other heavy metals (Willett *et al.*, 1993; Tin and Wilander, 1994; Åström and Björklund, 1995; Sammut *et al.*, 1996). However, the fate of these exported metals has been less well studied. There is potential for them to be sequestered in iron monosulphide (FeS)-rich sediments in receiving waterbodies (Billon *et al.*, 2001), with the risk that, under certain conditions, metals can be remobilized, causing secondary environmental effects.

Many sulphidic coastal flood plains in eastern Australia have been drained for agriculture. Massive fish kills have followed (Easton, 1989), generating conflicts, particularly between fishers and farmers. Often, local governments have jurisdiction over these drained flood plains and it falls to them to minimize conflicts and environmental, economic and social impacts. In this chapter, we describe collaborative work between researchers and local government aimed at providing information to governments and their communities on the processes occurring in drained acid sulphate flood plains discharging into coastal lakes, on the potential risks of secondary impacts and on possible remediation measures.

Cudgen Lake Study Area

This study was conducted in the subtropical Cudgen Lake catchment on the east coast of Australia (28°20′S, 153°29′E) in northern New South Wales. Average annual precipitation is 1620 mm, with 70% of rainfall occurring from December to May, and average annual evaporation is 1100 mm. The Cudgen Lake flood plain was formed following sea-level rise 12,000 years ago (Roy, 1973; Chappell, 1991) and has soils rich in reduced sulphides, such as pyrite. The area of acid sulphate flood plain is 20 km^2 and the total Cudgen Lake catchment area is 100 km^2. Figure 8.1 shows the extent of acid sulphate soils within the catchment. Cudgen Lake is a shallow (1–2 m), weakly tidal (tidal range 30–50 mm), brackish lake that covers 160 ha (Fig. 8.1). It has restricted drainage to the sea through Cudgen Creek, which discharges through an interbarrier depression.

Drainage of the flood plain

The Cudgen flood plain was once an extensive wetland. European settlers began to clear the flood plain for agricultural production in 1869. Drainage of the flood plain began in 1907, primarily for dairying. Mechanization greatly accelerated the rate of drainage from the 1940s to the 1980s. There are multiple land uses within the Cudgen catchment, including grazing, sugarcane production, tea tree, tropical fruits, tea, wetland conservation areas, and urban, recreation and transport infrastructure. In 1995, Cudgen Lake and its surrounding area was declared a state nature reserve because of its fisheries and conservation values.

The Cudgen Lake flood plain is the most extensively drained coastal flood plain in NSW, with a drainage density of 70 m/ha. Drainage and resort developments have caused widespread oxidation of sulphides within the flood plain and have led to conflicts within the catchment community, which have generated problems for the local government, the Tweed Shire Council.

Environmental impacts of flood plain drainage

Cudgen Lake was once a renowned nursery for fish and prawns. However, following massive drainage developments in the late 1980s, major fish kills occurred in the lake in May/June and December 1991. An addi-

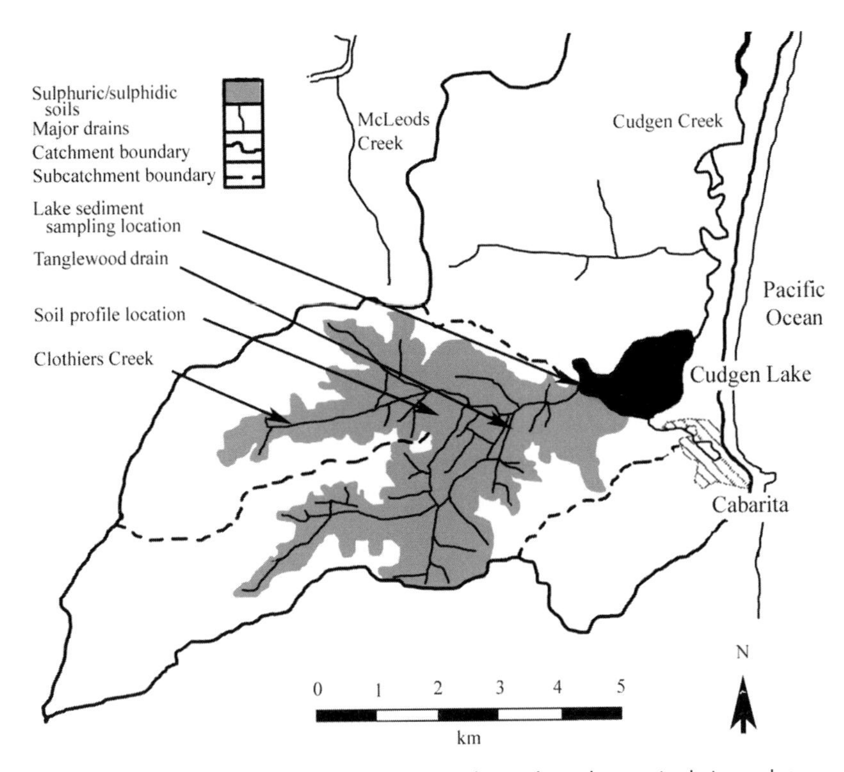

Fig. 8.1. Cudgen Lake study site showing the lake, its catchment boundary, main drains and streams, and the distribution of acid sulphate soils in the flood plain (from Macdonald *et al.*, 2004).

tional major fish kill occurred in August 1998, when intense rains followed a prolonged dry period. These fish kills generated widespread community concern, particularly among recreational and commercial fishers.

Monitoring by the Tweed Shire Council following the 1987 fish kills showed highly acidic water (pH 2.5), with large concentrations of dissolved aluminium (approx. 60 mg/l) being discharged into the lake (White *et al.*, 1997). The continued discharge of oxidation products from the flood plain has severely affected water quality, compromised the conservation value of the reserve and changed the lake's ecology. In addition, significant scalded areas, devoid of vegetation, have developed in the catchment, thus reducing grazing and cropping productivity.

Following acidification of the lake, the acid-tolerant spike-rush (*Eleocharis* spp.) and sedge (*Schoenoplectus litoralis*) have become the dominant vegetation. These have restricted fishing and recreation activities, such as sailing and boating. In the acidified areas of the lake, benthic diversity is also low and reed-infested areas are devoid of benthic life. The acidification of the lake also threatens the health of the local community. The mosquito larva *Aedes vigilax*, responsible for carrying Ross River Fever, an infection caused by a member of the *Togaviridae* family of viruses, occurs in acidified areas of the lake.

The conservation and fisheries values of Cudgen Lake, the extensive disturbance of acid sulphate soils and the presence of major scalds within its flood plain have led to the identification of the Cudgen Lake flood plain as one of 13 acid sulphate soil 'hot-spot' priority areas by the NSW Acid Sulphate Soil Management Advisory Committee.

Measurements

Soil profile and water quality samples were collected from the flood plain, from drains and from Cudgen Lake. Measurements of pH and electrical conductivity (EC) were made in the field. Soil water content, soluble and exchangeable cations (1:5 extracts), organic carbon, acidity and pyrite content, and aluminium, iron, chloride and sulphate content were determined in the laboratory (Macdonald *et al.*, 2004). Cudgen Lake sediment samples were collected from near the inlet of Clothiers Creek, which drains the 20 km^2 acid sulphate flood plain (Fig. 8.1). Two 0.4 m sediment cores were extracted and frozen with liquid nitrogen for transport back to the laboratory. A 1 m core was extracted using a sediment gouge, and sediment descriptions, field pH and redox were measured immediately. Bulk density of the upper sediments, reactive iron, total S and C, organic S and C, FeS and FeS_2 were determined at 0.02-m intervals. A separate 0.4-m sediment core was dried and crushed for total elemental analysis using concentrated HNO_3 that extracts metals from all but the most weathering-resistant minerals. Sediment porewaters were sampled using diffusion-controlled dialysis membrane samplers, 'peepers' (Hesslein, 1976). Peepers were removed after 10 days' equilibration and the porewaters analysed for metals, cations and nutrients (Macdonald *et al.*, 2004).

Results and Discussion

Flood plain soil properties

The upper 0.25 m of the flood plain soil profile had greater than 60% organic C and the top 0.5 m contained sulphide oxidation products, with aluminium and iron, followed by sodium, as the dominant soluble and exchangeable cations, and pH as low as 2.7. This acidic layer had bulk densities as high as 1.2 t/m^3 and volumetric water contents around 0.35 m^3/m^3. The exchangeable Al and Fe species represent a store of about 5×10^5 moles H^+/ha of acid products in the flood plain topsoil. Below 0.5 m was unoxidized blue–grey sulphidic clay gel with pH from 5.5 to 7, organic C less than 2%, bulk density around 0.4 t/m^3 and volumetric water contents around 0.9 m^3/m^3. The predominant soluble and exchangeable cations in this sulphidic layer are the base cations magnesium, calcium and sodium. Their concentrations are greater in the unoxidized layer than in the topsoil. The soil profile illustrates the oxidation processes that occur within these coastal flood plain soils. The oxidation of pyrite ultimately can be written as (Dent, 1986)

$$\begin{aligned} &FeS_2 + 15/4H_2O \rightarrow Fe(OH)_3 \\ &+ 2SO_4^{2-} + 4H^+ \end{aligned} \tag{8.1}$$

The acidified porewaters react with common estuarine clay minerals such as illite to release metal ions, principally Al^{3+}, K^+, Fe^{2+}, Na^+, Mg^{2+} and Ca^{2+} (White *et al.*, 1997):

$$\begin{aligned} &(K_{0.5}Na_{0.36}Ca_{0.05})(Al_{0.45}Si_{3.46})\,O_{10}(OH)_2 \\ &+ 7.41H^+ + 2.59H_2O \rightarrow 0.5K^+ + 0.36Na^+ \\ &+ 0.05Ca^+ + 0.3Mg^{2+} + 0.25Fe\,(OH)_3 \\ &+ 1.95Al^{3+} + 3.46H_4SiO_4 \end{aligned} \tag{8.2}$$

These acid-weathering products increase the EC of porewaters. Higher-valence cations exchange with lower-valence cations on clay exchange sites in the acidified topsoil, building up a store of acid products. The acid porewaters also release heavy metals associated with clays or other minerals into the soil porewaters, which may be transported into the topsoil or exported into adjacent waterways. Drainage of the Cudgen flood plain has transformed the soil from a sink for reduced sulphur, metals and carbon to a transient store, from which they are mobilized by rainfall into the flood plain drains.

Flood plain drain-water quality

After rainfall, water quality in the flood plain drains deteriorates and pH decreases, whereas EC, concentrations of Fe and Al, and the ratio of SO_4^{2-}:Cl, increase (Fig. 8.2). Provided Cl^- is a conservative ion, increasing SO_4^{2-}:Cl^- indicates transport of SO_4^{2-} from the soil profile to drains (Sammut *et al.*, 1996). The relation between Al concentration and pH in Fig. 8.2 is

$$\log_{10}[\mathrm{Al}] = 3.53 - 0.68\ \mathrm{pH} \quad (8.3)$$

The mean dissolved Al concentration in the Tanglewood drain in Fig. 8.2 is 28 mg/l and that for iron is 12 mg/l. This Al concentration is greater than the average concentration for 147 acid mine drainage sites (20 mg/l) (Singh *et al.*, 1997). The mean annual discharge into Cudgen Lake is 6×10^4 Ml (Heath *et al.*, 2001). The Tanglewood drain appears to be representative of other drains within the catchment, so that the approximate mean annual discharge to the lake is 1680 t of dissolved Al and 720 t of dissolved Fe. The peak discharge of oxidation products occurs during the recession phase of a flood event, when groundwater inputs are at a maximum (Sammut *et al.*, 1996; White *et al.*, 1997).

Lake sediment characteristics

The Cudgen Lake sediment core had three distinct layers: a black monosulphidic layer, a pyritic grey sandy clay layer and a pyritic grey clay gel layer (Fig. 8.3). The pH of the sediments was neutral but increasing at depth, whereas SO_4 and dissolved Fe were present in the top two layers but decreased when going into the gel layer (Fig. 8.3).

The sedimentation rate at the sampling site since European settlement was estimated to be 3 mm/year and is composed of an iron monosulphidic-rich facies. This surface layer of the lake sediments has high organic C, S and monosulphides (FeS) to a depth of 0.19 m. Drainage of the flood plain acid sulphate soils has increased the supply of SO_4^{2-} and Fe to the lake sediments. Their concentrations decrease with depth because of reduction and mineralization (Fig. 8.3). Reduced S in the underlying sandy clay layer (to a depth of 0.55 m) is principally pyrite (FeS_2), and, in the clay gel layer below 0.55 m, pyrite concentration is approximately 1.5%. This layer is comparable to the pyritic clay gel layer underlying the flood plain.

The monosulphide layer has elevated concentrations of Al, As, Hg (Fig. 8.4) and Pb that exceed ANZECC (Australia and New Zealand Environmental Conservation Council) guideline trigger values. These metals can accumulate in sediments by co-precipitation with FeS, adsorption to FeS or direct formation of metal sulphides. Sulphate reduction and the formation of a monosulphide layer provide a lake sediment sink for metals sourced from the flood plain acid sulphate soils.

Table 8.1 shows the approximate total mass of metals within the monosulphide layer of Cudgen Lake. This is based on assumptions of an average layer depth of 0.2 m, bulk densities of 0.124 t/m^3 and half of the 160 ha lake being covered by these sediments. There is a large store of metals within the lake system and these sediments will probably continue to accumulate because of inputs from the flood plain soils. If these lake sediments were allowed to oxidize, through dredging, by resuspension or by variations of lake pH due to acid drainage, stored metals could be released into the water column. The continued very low level of benthic organism diversity in the areas of Cudgen Lake prone to acidification may be partly due to the accumulation of heavy metals within the lake sediments.

Conclusions

We have shown here how the drainage of sulphidic flood plains can redistribute sulphur and metals through the drainage system into the sediments of receiving waterbodies. Flood plain acid sulphate soils, once a sink of sulphur and metals, have become a source, and the drainage network is the conduit through which dissolved species are exported into

Table 8.1. The mean total weight (t) of acid-extractable metals in the top 0.2 m monosulphide sediment layer in Cudgen Lake (from Macdonald *et al.*, 2004).

Metal	Total weight (t)
Al	101,700
Ce	60
As	200
Co	30
Cr	37
Cu	41
Fe	401,600
V	270
Pb	100

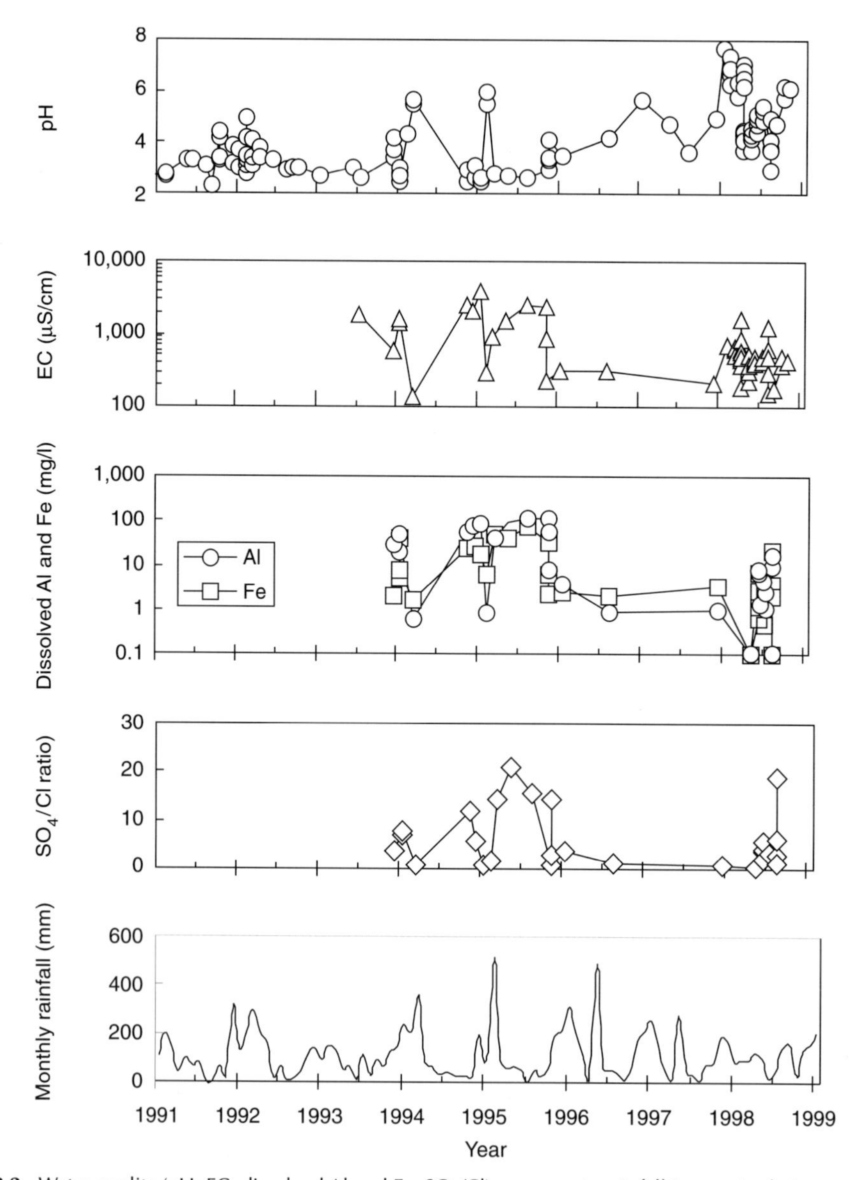

Fig. 8.2. Water quality (pH, EC, dissolved Al and Fe, SO_4/Cl) response to rainfall in a main drain, Tanglewood drain, discharging into Cudgen Lake.

downstream lakes and estuaries. Increased sulphate and organic matter inputs into these waterbodies have enhanced monosulphide formation and metal sequestering in the upper lake sediments. These accumulations of labile sulphides pose a potential threat to estuarine environments if sediments are disturbed or lakes drained. They may also explain the low benthic organism diversity in areas of the lake prone to acidification.

It is clear that the continuing discharge of acidified products into Cudgen Lake will continue to affect the conservation and fisheries value of the lake unless this problem is examined. Work is under way to cap scalded areas in the flood plain with clean fill. In

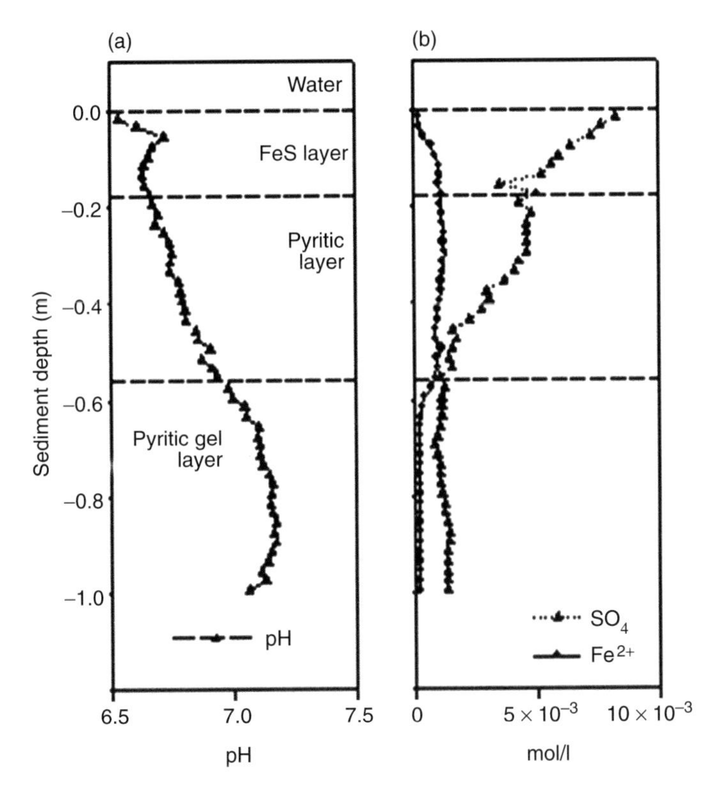

Fig. 8.3. Profiles of pH (a) and dissolved iron and sulphate (b) in Cudgen Lake sediments (from Macdonald *et al.*, 2004).

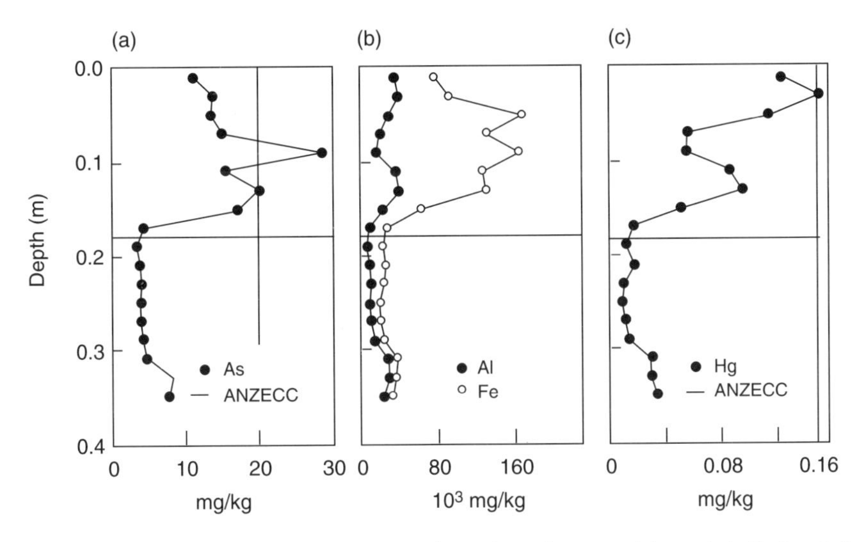

Fig. 8.4. Concentrations of arsenic (a), aluminium and iron (b) and mercury (c) metals in Cudgen Lake sediments and the ANZECC trigger value for contaminated sediments (from Macdonald *et al.*, 2004).

addition, ways to decrease drainage density are being examined and lime dosing of drainage discharge has been tested. A useful corollary of this work is the suggestion that artificially constructed wetlands may be useful for treating discharge from drained acid sulphate soils by trapping acid products and dissolved metals in monosulphide layers.

Acknowledgements

This work was supported by the Australian Research Council under grants LP0219475 and DP0345145 and by the NSW ASSPRO initiative.

References

Åström, M. and Björklund, A. (1995) Impact of acid sulfate soils on stream water geochemistry in western Finland. *Journal of Geochemical Exploration* 55, 163–170.

Billon, G., Ouddane, B. and Boughriet, A. (2001) Chemical speciation of sulfur compounds in surface sediments from three bays (Fresnaye, Siene and Authie) in northern France, and identification of some factors controlling their generation. *Talanta* 53, 971–981.

Chappell, J. (1991) Late Quaternary environmental changes in eastern and central Australia, and their climatic interpretation. *Quaternary Science Reviews* 10, 377–390.

Dent, D. (1986) *Acid Sulfate Soils: a Baseline for Research and Development*. ILRI: Wageningen, Netherlands.

Easton, C. (1989) The trouble with the Tweed. *Fishing World* 3, 58–59.

Heath, L.C., Beavis, S. and White, I. (2001) A GIS hydrology system for managing acid sulfate soils. In: Vertessy, R. (ed.) *Congress on Modelling and Simulation, MODSIM 2001*. Australian National University, Canberra, 10–13 December 2001.

Hesslein, R.H. (1976) An *in situ* sampler for close interval pore water studies. *Limnology and Oceanography* 21, 912–914.

Macdonald, B.C.T., Smith, J., Keene, A.F., Tunks, M., Kinsela, A. and White, I. (2004) Impacts of runoff from sulfuric soils on sediment chemistry in an estuarine lake. *Science of the Total Environment* 329, 115–130.

Roy, P.S. (1973) Coastal geology of the Cudgen area, north coast of New South Wales. *Records of the Geological Survey of New South Wales* 17, 41–52.

Sammut, J., White, I. and Melville, M.D. (1996) Acidification of an estuarine tributary in eastern Australia due to drainage of acid sulfate soils. *Marine and Freshwater Research* 47, 669–684.

Singh, B., Harris, P.J. and Wilson, M.J. (1997) Geochemistry of acid mine waters and the role of microorganisms in such environments: a review. In: Auerswald, K., Stanjek, H. and Bigham, J.M. (eds) *Soils and Environment: Soil Processes from Mineral to Landscape Scale*. Catena Verlag, Reiskirchen, Germany, pp. 159–192.

Tin, N.T. and Wilander, A. (1994) Chemical condition in acidic water in the Plain of Reeds, Vietnam. *Water Research* 29, 1401–1408.

White, I., Melville, M.D., Wilson, B.P. and Sammut, J. (1997) Reducing acidic discharges from coastal wetlands in eastern Australia. *Wetlands Ecology and Management* 5, 55–72.

Willett, I.R., Melville, M.D. and White, I. (1993) Acid drainwaters from potential acid sulfate soils and their impact on estuarine ecosystems. In: Dent, D.L. and van Mensvoort, M.E.F. (eds) *Selected Papers of the Ho Chi Minh City Symposium on Acid Sulfate Soils*. ILRI, Wageningen, Netherlands, pp. 419–425.

9 From Conflict to Industry – Regulated Best Practice Guidelines: a Case Study of Estuarine Flood Plain Management of the Tweed River, Eastern Australia

I. White,[1] M. Melville,[2] B.C.T. Macdonald,[1] R. Quirk,[3] R. Hawken,[3] M. Tunks,[4] D. Buckley,[4] R. Beattie,[5] L. Heath[1] and J. Williams[6]

[1] *Centre for Resource and Environmental Studies, Australian National University, Canberra, Australia, e-mail: ian.white@anu.edu.au*

[2] *School of Biological, Earth and Environmental Sciences, University of New South Wales, Australia*

[3] *NSW Canegrowers, Condong Sugar Mill, New South Wales, Australia*

[4] *Tweed Shire Council, New South Wales, Australia*

[5] *NSW Sugar Milling Co-operative Limited, New South Wales, Australia*

[6] *NSW Department of Primary Industries, New South Wales, Australia*

Abstract

Different sectors of society claim rights to use valuable coastal ecosystem resources. Conflicts over their use are therefore inevitable. In eastern Australia, government-encouraged development and drainage of coastal flood plains, principally for agriculture, resulted in accelerated oxidation of acid sulphate soils and export of toxic acidic drainage to coastal streams. Major impacts on infrastructure, ecology, fisheries and aquaculture resulted. In 1987/1988, all gilled organisms were killed in 23 km of the Tweed River estuary by acid outflows from canelands. This generated major conflicts among fishers, environmentalists and sugarcane producers farming the flood plain. Here, we describe the evolution of a collaborative learning approach to coastal flood plain management involving cane farmers, local government and researchers, and the institutional response to this fish kill. Existing knowledge in Australia was inadequate and efforts centred on providing information and options for better management and regulation of sulphidic estuarine areas and on mitigating impacts on downstream ecosystems. Farmers, researchers and local government officers working collaboratively generated information on the properties and management of sulphidic flood plains under the highly variable rainfall conditions common in Australia. This provided options for management that were rapidly translated into practice and underpinned mandatory best management guidelines for the NSW sugar industry. Increases in productivity and decreases in acid water discharge have resulted. Essential features of the collaborative partnership are analysed and the institutional response, which led to the adoption of Australia's national strategy for the management of acid sulphate soils, is described.

 Environment and Livelihoods in Tropical Coastal Zones (eds C.T. Hoanh, T.P. Tuong, J.W. Gowing and B. Hardy)

Introduction

Global challenges of coastal zone management

Coastal ecosystems are both immensely valuable (Costanza *et al.*, 1997) and valued by their dependent communities. Despite this, they continue to degrade on a global scale because of natural and human-induced changes (Kremer and Crossland, 2002). Of human pressures, cropping, grazing and megacities in coastal catchments have the largest impacts (WRI, 2000), mainly because of enhanced discharges of nitrogen, carbon and sediment from changes in coastal land and water regimes (GESAMP, 2001). The principal drivers for exports into coastal areas are river and groundwater discharge, population density, agricultural activities and catchment area (Buddemeier *et al.*, 2002; Kremer and Crossland, 2002).

Sustainable resource use and maintenance of coastal system functions are enormous challenges. Some of the key challenges are (Kremer and Crossland, 2002):

- improving the availability and accessibility of resource and environmental information,
- fostering participatory approaches to coastal zone management,
- developing wise use options and agreements (context-specific best management practices),
- ensuring that planning and management cope with change, and
- developing policies that take into account risks and vulnerability.

The supply of reliable information, participatory approaches and the adoption of context-specific best management practices are of central importance to sustainable coastal zone management, as we shall attempt to demonstrate in this chapter.

Information on coastal ecosystems

Some believe that it is not possible to wait while knowledge gaps in coastal ecosystems are investigated. Instead, action learning or adaptive management has been proposed as the way forward (Jiggins, 2002). Adaptive management of complex situations assumes that mistakes due to incorrect information can be identified, through rigorous monitoring, and corrected (Dovers and Mobbs, 1999). This presupposes linear processes in which the consequences of wrong actions can be reversed readily. Estuaries appear to behave in a non-linear, hysteretic manner (Harris, 1999), that is, after relatively small changes they exhibit dramatic collapses that are not easily reversed.

Information on coastal ecosystem processes is patchy, even in intensively managed areas. The general paucity of data is evident in Australia. Relatively little research has been undertaken on the processes, consequences of environmental change or impacts of human activities in coastal eastern Australia, despite the concentration of population there (Thom, 2002). We argue here that the hysteretic nature of coastal acid sulphate soil systems means that reliable information is of fundamental importance to improved environmental outcomes. The effective transfer of information to bring about changes in land management presents challenges to conventional models of knowledge transfer (Esman and Uphoff 1984; Roling, 1988; Curtis, 1998).

Participatory approaches to coastal zone management

Participatory approaches to natural resource management are a key to successful coastal management (Thom and Harvey, 2000). Problems arising from the past management of coastal resources in eastern Australia have partly resulted from the plethora of conflicting visions and disparate goals among protection, rehabilitation, economic development and regional employment growth, as well as from the inheritance of past legislation and administrative goals (Thom, 2002) and the failure to involve communities. The establishment of a shared vision through involvement of all sectors (Chambers, 1983) is an essential step in reducing conflict in coastal areas. Substantial evidence shows that participation through local organizations can produce broad-based management changes (Chambers, 1983; Esman and Uphoff 1984; Roling, 1988; Curtis, 1998; Ashby, 2003). Here, we describe the evolution of such an approach.

Coastal stewardship, wise practice options and conflict resolution

Many different sectors of society claim right of access to and use of resources in coastal regions (UNESCO, 2002a). As a consequence, a number of government agencies and authorities have frequently conflicting responsibilities for the coastal zone. As a result, no specific agency or institution is responsible for their management, despite the ever-increasing demand for finite resources and space (Healthy Rivers Commission, 2000). Conflicts over their use are often inevitable. Some even claim that conflict is a key driver for change.

In any conflict resolution, the first steps are to describe the nature and cause of the conflict, to identify and to bring together all the stakeholders in order to try to reach a consensus or compromise agreement. To reach agreement, there needs to be a process or mechanism where conflicting parties are able to address and resolve conflicts. Independent, impartial, outside parties, such as universities, can assist in developing agreements (UNESCO, 2002a).

Coastal stewardship has been proposed as one way of reducing conflicts by promoting ownership. It involves voluntary compliance, strong commitment and willing participation in the sustainable use of coastal resources (UNESCO, 2002b). The challenges in coastal stewardship are to inform, educate, motivate and empower communities to become managers and custodians of their environment. We provide in this work examples of this model.

Wise coastal practices are actions, tools, principles or decisions that contribute significantly to the achievement of sustainable development (embracing environmental, social, cultural and economic considerations) of coastal areas. Several characteristics of wise practices are particularly relevant to conflict prevention and resolution. These include participatory processes, consensus building, effective and efficient communication processes, capacity building and the need to respect traditional frameworks (UNESCO, 2002a). Wise practice agreements bring together all stakeholders, including governments, in a framework of voluntary compliance (UNESCO, 2002b). There are increasing concerns, particularly among regulators, however, that, without strong underpinning regulations, voluntary compliance agreements contain no effective mechanisms to address persistent breaches. This is an important factor in the case study discussed here, in which we illustrate the development of mandatory, context-specific wise practice agreements.

Recent developments in multi-agent-based simulations (MABS), coupled with role-playing games, provide powerful tools for studying interactions between societies and their environment (Bousquet *et al.*, 2002; Perez *et al.*, 2003). They have the potential to greatly reduce conflict over natural resource management and resource allocation; however, they were developed long after the work described here commenced.

In this chapter, we discuss issues involved in addressing some of these challenges. We describe a case study of a partnership developed to improve the sustainability of the acidified, coastal flood plain of the Tweed River in eastern Australia. The cooperative learning approach to coastal flood plain management described here evolved over the last 15 years. The partnership includes local government, cane farmers and their industry and academic institutions, and successfully addresses some of the important challenges outlined above. We first outline briefly issues concerned with coastal flood plain development in Australia.

Coastal Flood Plain Development in Eastern Australia

Coastal flood plains were the first regions in Australia developed for agriculture following European settlement. They have favourable temperatures and plentiful, young fertile soils, associated with large wetlands and low areas with high water tables (King, 1948).

While coastal flood plains were valuable drought refuges for cattle, waterlogging was a major constraint to increased cropping and improved pastures. Governments of all political persuasions encouraged land drainage. Landowner-initiated drainage in New South Wales (NSW), eastern Australia, dates back to the 1820s. The attendant increased investments in flood plain agriculture following drainage, however, required protection from frequent floods and storm surges, which involved the re-engineering of flood plain hydrology (Thom, 2002). As a result, the time scale for inundation of many backswamp areas was reduced from about 100 to around 5 days (White *et al.*, 1997).

Most coastal flood plains are now, on average, drier and export drainage waters at a much greater rate through more efficient drainage canals. Increased rates of drainage discharge, with accompanying sediment and chemical loads, often exceed the natural assimilative capacity of receiving waters. While naturally occurring, periodic droughts also dried out coastal flood plains (Lin and Melville, 1993) and natural drainage channels were much less efficient at delivering run-off to receiving waters than were constructed canals (Sammut *et al.*, 1996).

Institutional arrangements for flood plain drainage

Frequent floods in coastal NSW, particularly in the 1950s, led to joint Australian Commonwealth–NSW state government flood mitigation schemes. These schemes required participation from local governments and led to the construction of major levees, floodgates, retention basins and large primary canals on most of the river systems in NSW. These are now the responsibility of local governments.

The Drainage Act of 1904 established Drainage Unions, composed of local farmers who manage their secondary tributary drains feeding into main drains or directly into rivers. Drainage Unions had the authority to tax local landowners for maintenance and improvement of secondary drains and to carry out drainage work without prior state or local government approval. The major NSW water reforms of 2000 abolished many Drainage Unions, although some have been replaced by Drainage Boards, and their functions transferred to local governments.

Farm-level tertiary field drains, feeding into secondary drains, were constructed and maintained by individual landholders. Thousands of kilometres of drains have been constructed in coastal NSW. Many primary, secondary and tertiary drains are protected by one-way floodgates preventing tidal ingress of brackish water into low-lying backswamps. Floodgates impede fish passage to feeding and breeding areas (Sammut *et al.*, 1995). Figure 9.1 shows the high density of secondary and tertiary drains feeding into the Tweed River in eastern Australia.

Until the introduction in 1997 of the model provisions for Local Environment Plans (LEPs), there was little constraint on the construction of tertiary drains. Constructed drains and flood mitigation structures permitted widespread grazing, dairying, tea tree (*Leptospermum* sp.) and sugarcane industries to be established. Almost all of this government-encouraged development was carried out without recognition of the presence of acid sulphate soils or of the potential impacts of their drainage.

Acid sulphate soils in Australia

Many coastal flood plains throughout the world have Holocene-age soils (< 10,000 years old) containing iron sulphide minerals (Dent, 1986), mostly rich in the mineral pyrite, although in some regions, monosulphides are important (Bush *et al.*, 2004). Acid sulphate soils were recognized in Europe over 250 years ago (Pons, 1973). In Australia, their existence was appreciated relatively recently (Woodward, 1917; Teakle and Southern, 1937; Walker, 1972; Willett and Walker, 1982). Prior to the release of NSW acid sulphate soil risk maps (Naylor *et al.*, 1995), acid sulphate soils did not offi-

Fig. 9.1. Networks of secondary and tertiary drains on the upper flood plain of the Tweed River, northern NSW, eastern Australia. The river is at far left.

cially exist in Australia. As recently as 2002, some Australian states still denied their existence.

Walker (1972), in his comprehensive study, specifically warned of the dangers of continuing to drain sulphidic coastal flood plains. His advice was ignored. Preoccupation with the problems of inland agriculture in eastern Australia discouraged serious attention to issues in coastal areas. Detrimental impacts of acid sulphate soils in Australia include corrosion of engineering infrastructure (White *et al.*, 1996), massive fish kills and fish diseases (Brown *et al.*, 1983; Noller and Cusbert, 1985; Hart *et al.*, 1987; Easton, 1989; Callinan *et al.*, 1993; Sammut *et al.*, 1995), dramatic changes in stream ecology (Sammut *et al.*, 1995, 1996), blooms of harmful cyanobacteria (Dennison *et al.*, 1997) and emissions of sulphur dioxide (Macdonald *et al.*, 2004a). Acidic outflows can also threaten aquaculture (Simpson and Pedini, 1985).

Exported acidic materials have been found to be sequestered in the sediments of shallow receiving waters, from where they are available for subsequent remobilization (Macdonald *et al.*, 2004b). Thousands of tonnes of acidic products have been discharged from eastern Australian flood plains in single flood events (Sammut *et al.*, 1996; Wilson *et al.*, 1999).

Figure 9.2 shows an early attempt at identifying potential coastal areas of acid sulphate soils in Australia (White *et al.*, 1996). It is estimated that there are over 40,000 km^2 of acid sulphate soils, with some in every state of Australia. The current widespread recognition, policies and strategies on acid sulphate soils in Australia stemmed from a single incident.

Fish kills on the Tweed River

Heavy rains in early 1987 following a prolonged dry period strongly acidified the entire 23 km of the Tweed River estuary in north-eastern NSW, which drains to the sea at Tweed

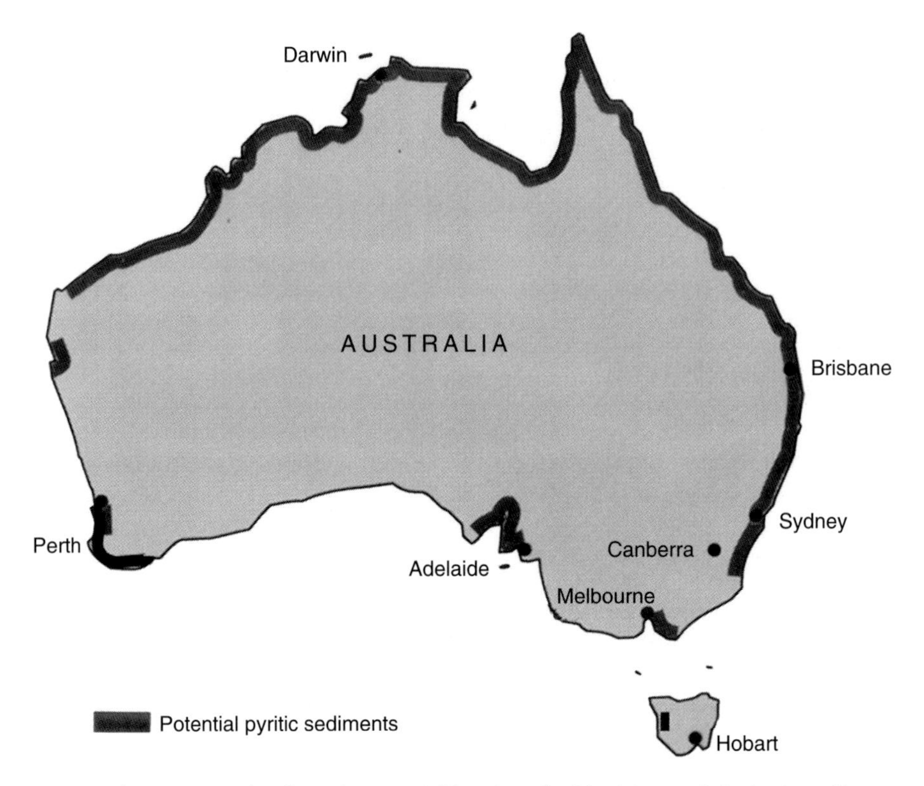

Fig. 9.2. An early attempt at identifying the potential location of acid sulphate soils in the Australian coastal zone from the distribution of mangroves (from White *et al.*, 1996).

Heads (see Fig. 9.3). The dissolved aluminium completely clarified the normally turbid estuary and all gilled organisms were killed (Easton, 1989). Possibly more than 1000 t of dissolved and colloidal aluminium was discharged in the event and the river remained almost sterile for a further 18 months.

The region surrounding the Tweed River is one of the fastest growing population centres in Australia because of its climate and highly prized natural amenities. It is a centre for tourism and recreational fishing. The fish kills on the Tweed attracted major media coverage and local attention because nearly 70% of the commercial fish species caught in Australia spend part of their lives in estuaries.

Over 80% of the 11,000 ha Tweed River flood plain is used for producing sugarcane. Many reasons were advanced for the fish kills and sugarcane farming was implicated in most of them. At first, pesticides were suspected. It was noticed that mosquito larvae had not been killed by the event. This precluded pesticides and led to the identification of acidic, aluminium-rich drainage from acid sulphate soils, which made up much of the canelands (Easton, 1989).

The 1987 fish kills on the Tweed generated serious conflicts between, on the one hand, commercial and recreational fishers, tourism operators, oyster farmers and environmentalists and, on the other, the sugarcane industry. Livelihoods on both sides were at risk. An initial public meeting of 500 locals ended in acrimony with threats to blow up or vandalize tidal floodgates on drains and inundate farmlands with brackish estuarine water, and counterthreats to shoot trespassers who interfered with drainage structures. One week later, the parties reconvened with cooler heads and elected representatives from each stakeholder group. This body became the Tweed River Advisory Committee (TRAC).

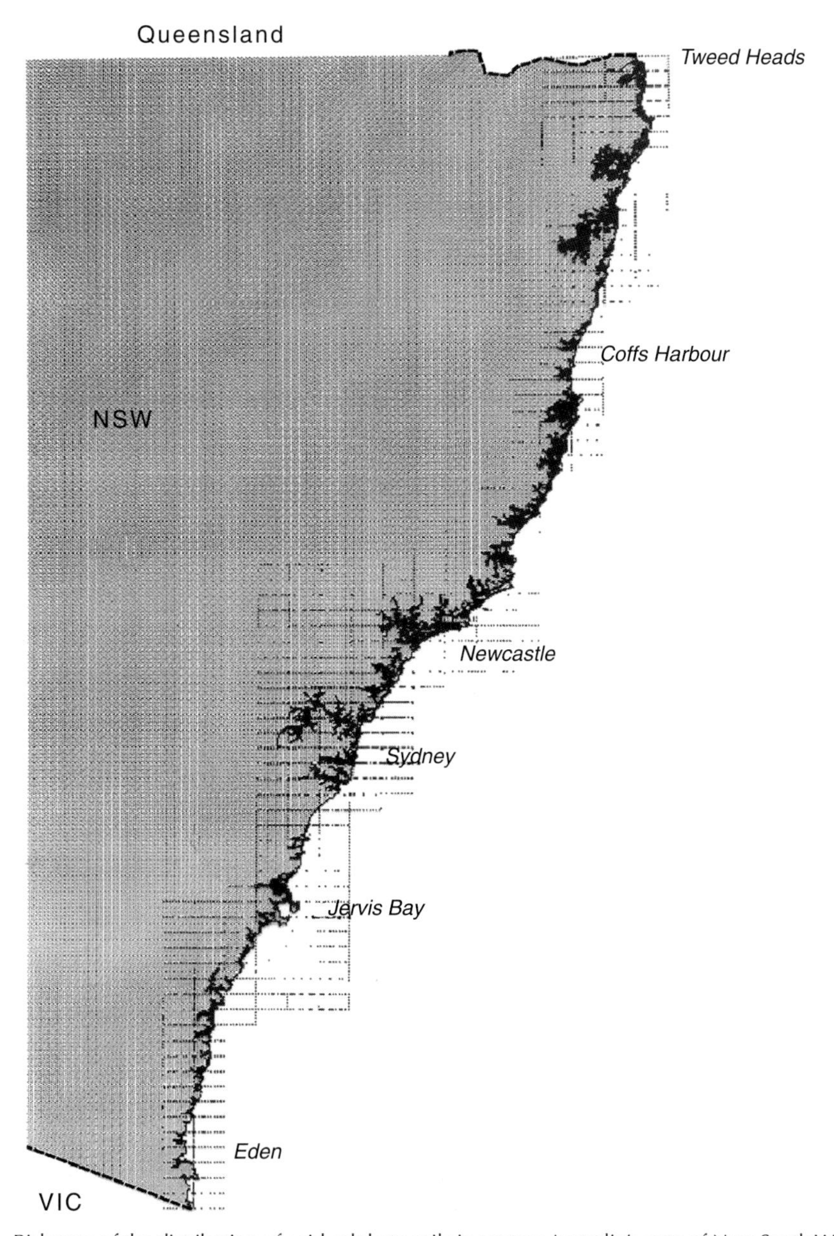

Fig. 9.3. Risk map of the distribution of acid sulphate soils in eastern Australia's state of New South Wales. The Tweed River discharges into the ocean at Tweed Heads (from Naylor *et al.*, 1995).

Evolution of a Cooperative Learning Partnership

In an attempt to find a formal way forward, the local government, Tweed Shire Council, convened a meeting of key representatives from TRAC in 1989. The meeting included farmers and fishers, oyster farmers, representatives from the sugar industry, local government and NSW state resource agencies, and researchers from university and federal research organizations. At the time, only three scientists in Australia had significant research experience in acid sulphate soils.

The aim of the Tweed meeting was to determine the cause of the fish kills and to explore possible solutions. Despite the animosity, the ultimate goals were to develop a shared vision for management and use of the flood plain and to initiate procedures to address the drainage water quality problems. The then six state government agencies with responsibilities for estuarine management had conflicting agendas and different opinions of the causes of the event. State agencies were wary of issues connected with coastal development which was, and continues to be, a major and controversial political question. Tweed Shire Council helped foster the development of a local acid sulphate soil committee, composed of the conflicting parties, local government and researchers, in 1990. It was finally agreed that information on the cause of the fish kills was incomplete and that research on the processes causing water quality problems should be carried out.

Adequacy of understanding

Further research is often proposed as a way to postpone difficult choices and decisions in natural resource management. In 1990, when research on the Tweed began, there was a considerable body of knowledge external to Australia about acid sulphate soils and some of the general principles of their management were well established (Dent, 1986). However, significant knowledge gaps existed, particularly in the Australian context with its extremely variable climate and hydrology. One of the principal issues was the comparative contribution of natural and farming-related processes to acid discharge. Adaptive management was not considered since the effects of a single day's drainage excavations of acid sulphate soils can have impacts that last for decades (White *et al.*, 1997).

A number of factors made the Tweed River flood plain particularly attractive to researchers. There is essentially a single land use on the sulphidic flood plain – sugarcane production – and the sugarcane industry is a well-structured cooperative, and therefore results of research are readily transferable. A single local government, Tweed Shire Council, managed the catchment, and the council was determined to find solutions. Finally, the connection between land use and estuarine water quality seemed straightforward.

Knowledge gaps

In 1989, most previous work on acid sulphate soils had focused on improving their agronomic performance (Dent, 1986), mainly due to experience in the Netherlands and in areas where off-site environmental impacts were discounted. Little attention had been paid to the downstream and off-site impacts of developing acid sulphate soils for agriculture, although Walker (1972) was an exception to this. Existing knowledge had focused on acid generation from the continuing oxidation of pyrite. To stop oxidation, air must be prevented from entering the sulphidic horizon, usually by reflooding the soil or by raising the groundwater level above the sulphidic horizon. Reflooding was expected to reverse the acidification processes by reducing oxidized sulphate back to sulphides through the microbially catalysed oxidation reaction involving organic matter (Dent, 1986):

$$\begin{aligned}&4FeOOH(s) + 4SO_4^{2-} + 9CH_2O + 8H^+ \\ &\rightarrow 4FeS(s) + 9CO_2 + 15H_2O\end{aligned} \quad (9.1)$$

Reflooding of sulphidic flood plains is still the most common management strategy recommended for acid sulphate soils but is anathema to coastal landowners. Previous work also suggested that lengthy, detailed and expensive soil surveys were required to determine the distribution of acid sulphate soils in coastal catchments.

Armed with this set of beliefs, research began on the Tweed in early February 1990. Monitoring by the Tweed Shire Council had identified the McLeods Creek (a re-engineered secondary drainage canal) catchment as one of the acid-exporting 'hot spots' on the Tweed River. It was clear from even preliminary measurements that the flood plain did not conform to conventional wisdom. Soil profiles showed that the surface soil had a pH above 4, making this technically not an acid sulphate soil despite obvious indicators of sulphide oxidation in the subsoil (Fig. 9.4).

One feature of the farmed flood plain was the complex, dense and seemingly over-constructed farm-level drainage system excavated without engineering drainage design (see Fig. 9.1). It was clear that an understanding of the hydrology, mostly ignored in previous studies, and its relation to the soil stratigraphy, groundwater and drain water quality were essential to understanding the processes.

Distribution of acid sulphate soils

Determination of the spatial extent of acid sulphate soils in the Tweed River flood plain was fundamentally important to improved management. A conventional soil survey would have taken too long and would have been too expensive. A geomorphic approach to mapping based on the conditions necessary for sulphide accumulation in estuarine sediments was developed. Over the last 6500 years, sea levels have remained fairly constant, the eastern Australian land surface has been tectonically relatively stable over that time and fluvial sediment accumulation there has not been large (Lin and Melville, 1993).

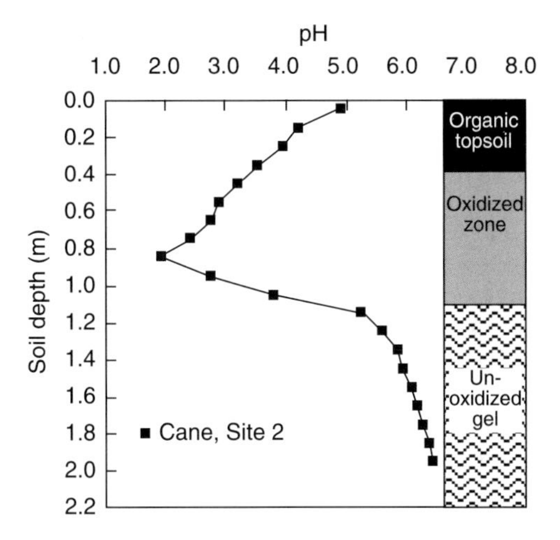

Fig. 9.4. Profile of soil pH in an acid sulphate soil in a caneland on the Tweed River flood plain taken from a cane farm on Fig. 9.1.

Based on this, the top of the sulphidic horizon in Australian Holocene coastal sediments should be close to where it was last formed, at about mean high-tide level (about 1 m Australian Height Datum, in eastern Australia). Because this horizon has been buried by a variable thickness of alluvium, it was predicted that Holocene coastal sulphidic sediments in eastern Australia would be found at sites with surface elevations of less than about 5 m AHD (Melville *et al.*, 1990, 1993; White and Melville, 1993). This prediction provided the basis for producing acid sulphate soil risk maps for the entire state of NSW (see Fig. 9.3, Naylor *et al.*, 1995). These maps have been both remarkably accurate and useful planning tools and form the basis for Local Environment Plans, Development Control Plans and Development Assessments that control developments in acid sulphate soil areas in NSW.

Cane farmers' response

Initial discussions between researchers and cane farmers in the McLeods Creek catchment and sugar mill officers during the preliminary site visit were less than encouraging. The cane industry rejected the suggestion that acidification of streams was attributable to their soils, some suggesting instead that it was road gravel that caused the problem. Rejection is a readily identified first stage in many areas of natural resource management (see Fig. 9.5). Frequently, many do not progress beyond this stage.

The industry wanted to lower water tables further by up to 1 m to improve cane production since waterlogging was perceived to be a main impediment to cane production. Cane farmers were divided on whether researchers should be invited into the catchment. The strong opposition was understandable since the prevailing wisdom was that, to control acid discharges, low-lying areas should be reflooded with brackish estuarine water, thereby eliminating cane production and farmers' livelihoods and property values. In addition, farmers felt they were being blamed unfairly for acidic

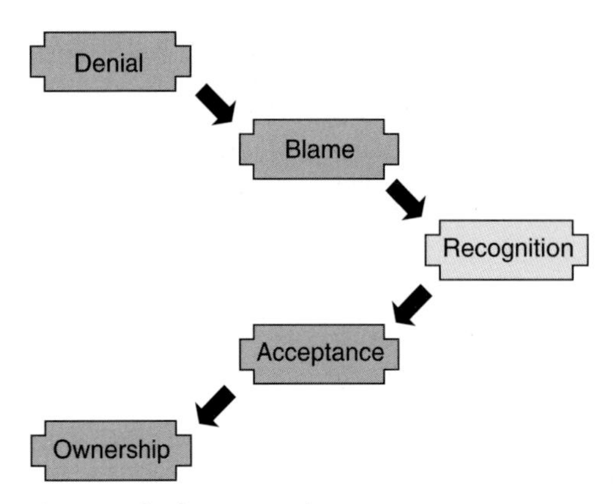

Fig. 9.5. The five stages of response to issues in natural resource management.

drain discharges from areas approved for agriculture and drained with continued encouragement from governments.

Despite their misgivings, farmers in McLeods Creek gave permission for researchers to work on their lands. The support of Tweed Shire Council for continued research and monitoring was a crucial element in persuading farmers to tolerate research directed at understanding the processes involved in the generation and export of acidity, its environmental impacts and the identification of amelioration options

Cooperative learning

In the early days of the research on the Tweed, there was some mutual mistrust between researchers and farmers. The culture and reward systems for both groups differed fundamentally. It took nearly 3 years for mutual understanding and trust to develop. This gradual evolution was assisted by the continued interest and support of local government staff and state-level initiatives in acid sulphate soils.

Researchers initially had a narrow focus on soil (Melville *et al.*, 1993), hydrological (White *et al.*, 1993) and environmental issues (Callinan *et al.*, 1993). Work had also expanded into the more biophysically and socially complex and larger Richmond River flood plain south of the Tweed, where there were also major acid discharges (see Fig. 9.6, Sammut *et al.*, 1996). Researchers came round to a growing appreciation that acid sulphate soil management was but one of a broad range of issues that had to be faced and dealt with by farmers, fishers and local governments. They recognized that farmers, fishers and regulators have to deal with and integrate a bewildering range of soil, climate, crop, nutrient, disease, pest, social, regulatory, financial, political and institutional issues in their daily tasks. It became increasingly apparent to researchers that they needed to understand these broader issues in order to transfer their research findings effectively.

Frequent informal discussions were held in the field, where both research results and information on the current issues faced by farmers, fishers and local government were exchanged. These were interspersed with more formal meetings with the NSW Sugar Industry and Tweed Shire Council representatives and various committees such as the Tweed River Management Plan Advisory Committee. It became increasingly clear to researchers that their task was to provide a range of practical management options for farmers, fishers and regulators, not single prescriptive solutions.

The training of graduate students and postdoctoral fellows in field acid sulphate soil and related research was a key component in the partnership. This has been an invaluable opportunity for cooperative learning. Graduate students and postdoctoral fellows are highly motivated and can focus on key issues exclusively. They are also less threatening than senior researchers. The opportunity for them to do research in partnership with experienced farmers is an unparalleled, two-way learning opportunity. Additionally, there remains a real dearth of trained people with field experience in coastal soils, hydrology and environmental impacts in Australia.

Farmers began to appreciate that researchers were not there to apportion blame. Indeed, researchers' message that acid sulphate soils were part of the global sulphur cycle, a naturally occurring phenomenon that has existed for hundreds of millions of years, brought about a shift in attitude. Farmers started to talk about *their*

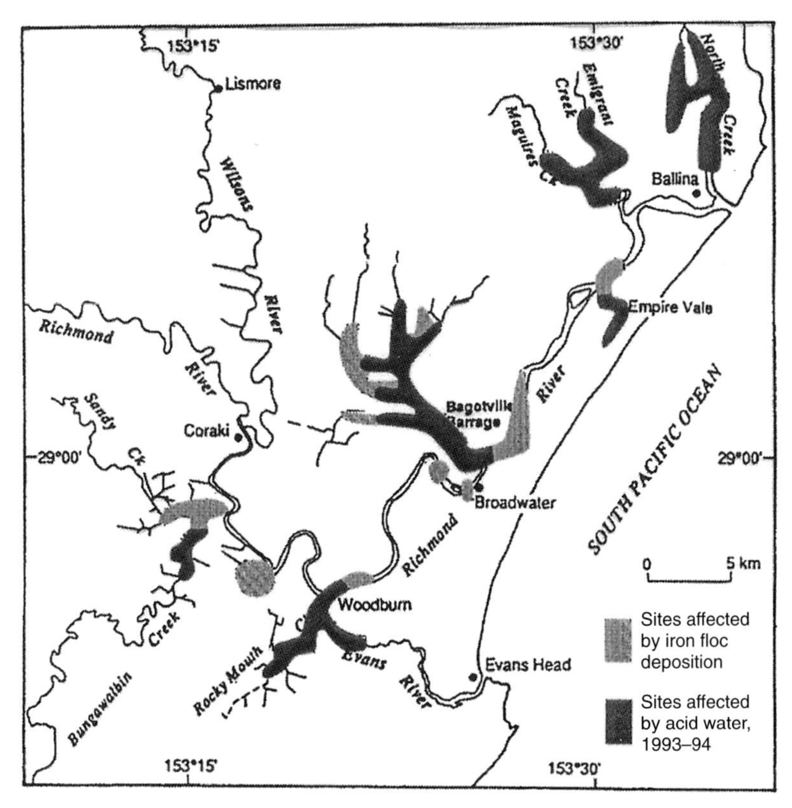

Fig. 9.6. Extent of acid-impacted areas in the Richmond River estuary, 1993–1994, from drainage of acid sulphate soils (from Sammut *et al.*, 1996).

acid sulphate soils and to carry pH meters. Once that happened, farmers became initiators of the research.

Research Findings

Some of the research findings in the Tweed were counter to the prevailing wisdom and resulted in significant land management changes. During dry periods, it was found that sugarcane survives on shallow groundwater (White *et al.*, 1993). The industry's quest to further lower water tables by increasing drain depths would result in decreased crop production. In addition, the shrink–swell nature of the gel sublayer dictates that lowering the water table also lowers the elevation of the soil surface (White *et al.*, 2001).

Because of the exceedingly small hydraulic conductivity of the oxidized clay gels making up the potential acid sulphate horizon (White, 2002; White *et al.*, 2003), crop evapotranspiration was the main determinant of water table level, not drain water level. Drains were essentially disconnected from the water table in the interior of cane blocks so that the prime function of drains was to remove surface water following floods and heavy rains, not to control water table elevation (White *et al.*, 1997). By using laser levelling to improve the shedding of surface water, waterlogging could be reduced and drain depth and density decreased (White *et al.*, 1997).

The research also shed light on a major source of acidity exported from acid sulphate soils. Vast quantities of existing acid products, equivalent to about 50 t/ha of sulphuric acid, from previous oxidation of acid sulphate soils and subsequent reactions with the soil, were found stored within the subsoil above the sulphidic layer (Smith, 2000; Donner and Melville, 2002). Sugarcane evolved in such environments and appears acid-tolerant, although responses to lime additions have been noted. The highest levels of stored acidity in the flood plain follow former natural drainage lines (Smith, 2000). This is consistent with the observation that most exported acidity is sourced from a few metres around drains (White *et al.*, 1997) and is readily released during the recession phase of floods, independent of any further oxidation of sulphides (Wilson *et al.*, 1999). Some of the exported, metal-rich acidic species were found to be stored as iron monosulphide in the drain-bottom sediments (Sammut *et al.*, 1996), others were sequestered in shallow sediments in receiving water bodies (Macdonald *et al.*, 2004b). The average annual flux of acidic products from drained flood plains (as sulphuric acid) was estimated to be from 0.3 to 0.5 t/ha/year (Sammut *et al.*, 1996; Wilson *et al.*, 1999) although some oxidation products are exported in the atmosphere as sulphur dioxide (SO_2) (Macdonald *et al.*, 2004a).

The store of acidic products in the soil has accumulated from more than just the influence of constructed drainage. Some are the result of prolonged dry periods (Lin and Melville, 1993) linked to the Pacific Decadal Oscillation, and some appear to be due to a small isostatic uplift of the east coast of Australia (Kinsela and Melville, 2004).

These findings changed the focus of management from the prevention of sulphide oxidation to the retention of as much of the stored acidity as possible within the soil-drain system by manipulating the volumes of surface run-off and flood recession flow. In addition, decreasing drainage density profoundly decreases acid export.

The collaborative research led to the development by farmers of new goals for drainage management. These were to leave as much of the stored soil acidity in the flood plain as possible and to prevent the development of any additional acidity.

Changed Land Management

Because research was being carried out in a cooperative learning process long before research was published, the farmers in the catchment were able to adapt land management strategies where appropriate. Researchers were amazed by the rapid implementation by farmers of even preliminary research findings, particularly since implementation often meant the expenditure of significant farmer funds. Farmers later explained that they weighed information very carefully before acting.

An early application of the research was that farmers abandoned plans for increased drainage. Instead, they focused on control and rapid removal of surface water. Progressive laser levelling of cane fields was undertaken. This removed waterlogged areas and enabled the elimination of some field drains. In some cases, drain density decreased by a factor of two and drain depth also decreased. The decrease in drain density made more land available for cane planting and eliminated the need for new drains. Lime was added to drainage lines and to cane fields at a rate that exceeded the mean annual acid export rate.

The rapid removal of surface water resulted in better-quality water during the rising limb of the hydrograph that could be discharged safely into streams. Farmers were then able to store as much of the acidic discharge as possible during the recession phase within drains. A range of options to treat acidic drainwater before discharge was also tested by farmers. Techniques included lime dosing of drain discharge, discharge of drainwater through closed-tank lime-bed reactors and discharge into constructed wetlands to precipitate iron monosulphides. As well, farmers modified tidal floodgates to permit their opening in dry periods. This allowed both tidal exchange with drainwater and increased fish passage.

Farmers recognized that the availability of organic matter was a key to driving the oxidation reaction backward (equation 9.1). They tested green cane harvesting with the retention of surface cane trash to increase organic matter availability. This has been coupled with the use of 'raised beds' for cane production, which, with trash retention, has greatly increased soil fauna.

Changes in land management came at a

considerable cost to landowners and were only possible because of the relatively high per-hectare returns of sugar production. Advisers in the industry had initially opposed the general use of lime, claiming it had no agronomic value. Farmer-initiated trials, however, have found about a 10% increase in yield with lime use in some soils. As well, removal of drains has increased the area for cane planting. The net result of these changes has been up to a 30% increase in production. The episodic nature of acid discharges makes it difficult to estimate precisely changes in acidic exports. Recent results suggest that these strategies have decreased the flux during alternate-year floods by 90%.

Development of guidelines for best practice

Most of the above changes in land management in the Tweed flood plain had the support of the Tweed Shire Council. There was limited interest in other cane-growing catchments in northern NSW and it was recognized that some form of regulation was necessary to prevent further acidification of coastal flood plains. The statutory instrument chosen was Local Environment Plans (Williams, 2002) administered by local governments.

It was recognized that the introduction of LEPs would result in development controls on even the most trivial of farming activities. The solution proposed by the NSW Cane Industry was self-regulation and the industry commenced developing the NSW Sugar Industry Best Practice Guidelines for Acid Sulphate Soils. Planning NSW, the state consent authority, authorized the NSW Sugar Milling Cooperative in conjunction with local governments to be the consent authorities and provided for strict audits and reviews of performance by local government and relevant state authorities. The research work on the Tweed and Richmond provided the basis for the guidelines, which were developed by Tweed Shire Council, the NSW Sugar Industry, cane farmers, NSW Agriculture and researchers. An acid sulphate soil survey by the industry of all farms underpinned the guidelines and helped raise awareness of acid sulphate soils among cane farmers outside the Tweed.

The intended purpose of Best Practice Guidelines is to provide guidelines based on the best available information for cane farmers with acid sulphate soils in order that they:

- minimize the export of acidity from their farms,
- minimize any downstream environmental impacts caused by acid export,
- maximize production from their land, and
- adhere to the intent of Local Environment Plans on acid sulphate soils.

Unlike wise practice agreements (UNESCO, 2002a,b), the NSW Sugar Industry Best Practice Guidelines for Acid Sulphate Soils are not voluntary. The Sugar Milling Cooperative is a cane farmers'-owned cooperative that can refuse to process cane from farmers who ignore the guidelines. In essence, the guidelines are mandatory.

Institutional and Policy Developments

Parallel to the developments on the Tweed, progress was occurring simultaneously at the state level. Concern over fish kills on the Tweed and in other coastal rivers spread rapidly throughout NSW and into neighbouring Queensland. The commercial fishing industry seized on environmental problems generated by acid sulphate soil drainage, especially fish kills, as an indicator of general malaise in coastal and estuarine management in eastern Australia. As late as 2002, the Healthy Rivers Commission (2000, 2002) concluded that, because so many government agencies had responsibilities for coastal lake and estuary management, ultimately no one had an abiding responsibility. The fishing industry made effective use of the media, which both raised public awareness of the problems of acid sulphate soils and spread the conflict to other regions.

In 1993, the NSW Department of Agriculture, the lead agency in the state, sponsored the first Australian National Conference on acid sulphate soils in

Coolangatta on the NSW/Queensland border. This conference served as a venue to present what was known of the properties and distribution of acid sulphate soils in Australia and to learn of overseas experience. It did nothing, however, to abate the conflicts over the use and management of acid sulphate soils. Many in the farming industry, local government and indeed in other states were in the denial phase (see Fig. 9.5). Queensland, the neighbouring state to the north of NSW, continued officially to deny the existence of acid sulphate soils until 1995.

Acid Sulphate Soil Management Advisory Committee

Faced with escalating conflicts between two industries within his portfolio, the then NSW Minister for Agriculture and Fisheries, Mr Ian Causely, in 1994 formed the whole-of-government Acid Sulphate Soil Management Advisory Committee (ASSMAC) to provide him with advice on the use and management of acid sulphate soils. ASSMAC initially consisted of representatives of the then five government agencies with significant responsibilities in coastal and estuarine matters, together with representatives from the NSW Farmers' Association, the seafood industry and the research community. Membership later broadened to include all the significant coastal industries, including the oyster, sugarcane, dairying, tea tree and urban development industries. The predominant focus of ASSMAC, however, remained on agricultural solutions and strategies, since the predominant developments on acid sulphate soil were agricultural. At farmers' insistence, ASSMAC was chaired by NSW Agriculture, which also provided technical support. ASSMAC was supported by a technical committee to provide it with specialist advice on acid sulphate soils.

Tensions between the farming and fishing industry representatives were apparent, with one in denial mode and the other insisting on the immediate reflooding of coastal lowlands. Additional differences between state management and regulatory agencies surfaced. Some advocated the introduction of Environmental Impact Statements for farming activities. Others suggested a specific acid sulphate soil State Environment Protection Plan (SEPP), a NSW instrument loathed by landowners, and some suggested use of the prosecuting powers of the Clean Waters Act for discharge of acidic drainage. More moderate opinions recognized that the Government was ultimately responsible for the developments on acid sulphate soils within the state.

Faced with these tensions, ASSMAC developed a strategic approach to its task and identified the prevention of further disturbance of acid sulphate soils and the remediation of problem areas as its main functions. ASSMAC's approach was based on raising awareness of acid sulphate soils, on promoting education and training on their properties and use, on sponsoring appropriate research and on exploring the adequacy of existing policy and legislation. One essential task in this strategic approach was the military-like production and promotion of acid sulphate risk maps for the entire state coastal zone in 1995 (see Fig. 9.3, Naylor *et al.*, 1995). These showed that the state had 4000 km^2 of high-risk acid sulphate soils.

Other early tasks included the publication of acid sulphate soil management guidelines in 1995 (Blunden and Naylor, 1995) and the formation of acid sulphate soil local action groups, ASSLAGs, modelled on the committee set up by the Tweed Shire Council in 1990. ASSLAGs had varying successes depending on the support and enthusiasm of local governments and the involvement of researchers and facilitators on the committees. Additional educational and publicity materials and a national newsletter (ASSAY) were also produced, and the second national conference on acid sulphate soils was sponsored by ASSMAC and held in 1996.

The review of legislation revealed that there were already adequate laws in NSW to prevent increasing acidic discharge. However, it was reasoned that, without a specific instrument, continued disturbance was possible. Since local government approves most flood plain developments, it was concluded that state government–mandated LEPs, that forced local governments to develop an LEP covering the disturbance of acid sulphate soils and approved by the relevant state agency, were the preferred statutory instrument. A model LEP, based on the published acid sulphate soil risk maps for each estuary, was developed in consultation with coastal councils, government agencies and stakeholders (Williams, 2002).

ASSMAC received a boost to its meagre resources through the Acid Soil Action Programme in 1997, which provided AUS$ 2.1 million funding over 3 years. This funded the Acid Sulphate Soils Programme (ASSPRO) initiative to support research and trials of remediation strategies for problematic acid sulphate areas in the state. The Tweed region benefited significantly from the ASSPRO initiative. ASSPRO also commissioned a survey identifying acid sulphate soil 'hot spots' in estuaries, those areas likely to be contributing disproportionately to acid fluxes. This formed the basis of a 1999 Hot Spot Programme designed to remediate problem areas. Only a small fraction of the state government's promised AUS$ 13.4 million was allocated to the production of remediation plans. Major structural rearrangements in natural resource management in NSW, at the encouragement of the federal government, through the establishment of Catchment Management Authorities and the Natural Resources Commission, led to the abolition of the Acid Soil Action Programme, the demise of ASSMAC in 2004 and reallocation of Hot Spot funds. The impetus for change and acid sulphate soil management have been set back by this shift in the government natural resource management service delivery model. Most other Australian state jurisdictions that are making progress are managing acid sulphate soils as a specific issue and seeking support from regional service delivery organizations.

National strategy for managing coastal acid sulphate soils

As part of its strategic approach, ASSMAC encouraged the establishment of a similar body in Queensland, QASSMAC, which by 1995 recognized that it also had acid sulphate soils. Major urban developments have continued to occur in coastal southern Queensland over the past decade, and the focus of QASSMAC, unlike the agricultural thrust of ASSMAC, has been largely on urban development on acid sulphate soils.

ASSMAC also worked towards the development of an Australia-wide strategy on acid sulphate soils. Eventually, the Commonwealth's Standing Committee on Agricultural and Resource Management formed a national working party in 1998 to develop a national strategy for managing coastal acid sulphate soils. Both ASSMAC and QASSMAC were represented on this working party. The national strategy was accepted and released by the Agricultural and Resource Management Council of Australia and New Zealand, the Australian and New Zealand Environment and Conservation Council, and the Ministerial Council for Fisheries, Forestry and Aquaculture in 2000.

The aims of this national strategy are as follows:

1. Improve the management and use of coastal acid sulphate soils in Australia to protect and improve water quality in coastal flood plains and embayments.
2. Assist governments, industry and the community in identifying and playing their roles in managing coastal acid sulphate soils.

To achieve these overriding aims, the strategy has the following four principal objectives:

1. Identify and define coastal acid sulphate soils in Australia.
2. Avoid disturbance of coastal acid sulphate soils.
3. Mitigate impacts when acid sulphate soil disturbance is unavoidable.
4. Rehabilitate disturbed acid sulphate soil and acid drainage.

To date, the national strategy has enabled the rapid transfer of techniques and information and has catalysed policy development and the mapping of acid sulphate soils in other states and territories in Australia.

Concluding Comments

Acidic discharge from acid sulphate soils has had significant environmental impacts throughout the world. We have examined here a particular case study where fish kills in eastern Australia due to acid sulphate drainage eventually culminated in publication of the world's first national strategy on acid sulphate soil management. These fish kills led to the evolution of a participatory approach to agricultural land management in a coastal flood plain that involved farmers, local government and researchers as equal partners. As mutual trust grew, the process evolved into a cooperative learning opportunity. This participatory process generated new information on acid sulphate soils, which was almost immediately translated into a range of management options that both significantly improved cane production and decreased acid discharge, and formed the basis for the establishment of enforceable and mandatory best management guidelines for the NSW sugarcane industry.

Some of the outcomes here are dependent on the particular institutions and governance arrangements for natural resources in Australia. However, we believe that some lessons are applicable elsewhere. We have attempted to show that local government played a vital role in identifying the issues, in initiating the participation process and in providing support for the developing collaboration. Farmers belonged to an industry that was a well-structured and profitable cooperative that fosters innovation. This ensured the rapid translation of collaborative research results into practice. Researchers and students had a broad range of skills in soil science, hydrology, geomorphology, geochemistry, agriculture and environmental management. There were sources of funding to support the work and sufficient time for the partnership to develop. Regulation and policy frameworks existed that were directed at positive environmental outcomes. Lastly, the 1987 fish kills were a highly visible and widely publicized environmental impact that galvanized attention at all levels. Many of the fundamental principles of participatory resource management (Ashby, 2003) are evident in this case study, such as the importance of including stakeholders in joint enquiry, co-development of new resource management regimes and the need to combine local and scientific knowledge and expertise.

Other lessons emerging from the statewide response are that a whole-of-government and industry approach, embodied in ASSMAC, was more effective in addressing problems caused by acid sulphate soils than approaches by individual agencies. There, a strategic approach based on consensus and cooperation, rather than heavy-handed application of existing regulations, brought about quantifiable behavioural change. Focusing a whole-of-government effort on a single issue, such as acid sulphate soils, at a strategic state level and fostering action at the local level appears to have been a far more effective approach than natural resource management reforms based on broadscale integrated catchment management. Finally, the parallel development of a national strategy on acid sulphate soils has been very effective in coordinating a national focus on the problem. It has enabled the rapid and effective transfer of knowledge to states and territories in Australia where the problems of acid sulphate soils are only just being recognized.

Much has been written about integrated natural resource management and participatory processes. In the case study examined here, farmers and local government provided both the integration and the participation. Farmers integrate a broad range of soil, climate, crop, disease, economic, social, regulatory and institutional issues in running their farms. Local, democratically elected governments, responsive to their electorates, are aware of the social, environmental and economic aspects of improved coastal management. The researchers' understanding of these issues greatly expanded in the process. Mutual understanding of the different cultures under which the partners operated took at least three years to develop. Once that occurred, the dynamics of the process changed and farmers started to initiate and drive the research. An important aspect of this case study was that farmers had sufficient resources to initiate changes in land management.

Acknowledgements

This work has been supported by the Australian Research Council, ASSPRO, NSW Canegrowers, NSW Sugar Milling Cooperative, the Natural Heritage Trust, the Tweed River Management and Planning Committee, the Tweed Shire Council and Environment Australia's Coastal Acid Sulphate Soil Programme.

References

Ashby, J. (2003) Introduction: uniting science and participation in the process of innovation – research for development. In: Pound, B., Snapp, S., McDougall, C. and Braun, A. (eds) *Managing Natural Resources for Sustainable Livelihoods: Uniting Science and Participation*. Earthscan Publications, London.

Blunden, B. and Naylor, S.D. (1995) *Assessing and Managing Acid Sulphate Soils: Guidelines for Land Management in NSW Coastal Areas.* NSW EPA, Chatswood, NSW, Australia.

Bousquet, F., Barreteau, O., d'Aquino, P., Etienne, M., Boissau, S., Aubert, S., Le Page, C., Babin, D. and Castella, J.-C. (2002) Multi-agent systems and role games: collective learning processes for ecosystem management. In: Janssen, M.A. (ed.) *Complexity and Ecosystem Management: The Theory and Practice of Multi-Agent Systems*. Edward Elgar Publishers, Cheltenham, UK, pp. 248–285.

Brown, T.E., Morley, A.W., Sanderson, N.T. and Tait, R.D. (1983) Report on a large fish kill resulting from natural acid water conditions in Australia. *Journal of Fish Biology* 22, 333–350.

Buddemeier, R.W., Smith, S.V., Swaney, D.P. and Crossland, C.J. (eds) (2002) *The Role of the Coastal Ocean in the Disturbed and Undisturbed Nutrient and Carbon Cycles.* LOICZ Reports and Studies No. 24, LOICZ, Texel, The Netherlands, 83 pp.

Bush, R.T., McGrath, R. and Sullivan, L.A. (2004) Occurrence of marcasite in an organic-rich Holocene estuarine mud. *Australian Journal of Soil Research* 42, 617–621.

Callinan, R.B., Fraser, G.C. and Melville, M.D. (1993) Seasonally recurrent fish mortalities and ulcerative disease outbreaks associated with acid sulfate soils in Australian estuaries. In: Dent, D.L. and van Mensvoort, M.E.F. (eds) *Selected Papers from the Ho Chi Minh City Symposium on Acid Sulphate Soils,* March 1992. IILRI Pub. No. 53, International Institute for Land Reclamation and Improvement, Wageningen, Netherlands, pp. 403–410.

Chambers, R. (1983) *Rural Development: Putting the First Last.* Pearson Education, Harlow, UK.

Costanza, R., D'Arge, R., de Groot, R., Faber, S., Grasso, M., Hannon, B., Limburg, K., Naeem, S., O'Neill, R.V., Paruello, J., Raskin, R.G., Sutton, P. and Van Den Belt, M. (1997) The value of the world's ecosystem services and natural capital. *Nature* 387, 253–260.

Curtis, A. (1998) The agency/community partnership in Landcare: lessons from state-sponsored citizen resource management. *Environmental Management* 22, 656–674.

Dennison, W.C., O'Neil, J.M., Duffy, E., Oliver, P. and Shaw, G. (1997) Blooms of the cyanobacterium *Lyngbia majuscula* in coastal waters of Queensland. In: Charpy, L. and Larkum, A.W.D. (eds) *Proceedings of the International Symposium on Marine Cyanobacteria,* November 1997. Institut Oceanographique, Paris.

Dent, D.L. (1986). *Acid Sulphate Soils: A Baseline for Research and Development.* IILRI Publication No. 39, International Institute for Land Reclamation and Improvement, Wageningen, Netherlands.

Donner, E.N.S. and Melville, M.D. (2002) The effects of flood plain land use on the chemistry and hydrology of coastal acid sulfate soils. In: *5th International Acid Sulfate Soils Conference,* Tweed Heads, Australia, 26–30 August 2002. ASSMAC, Wollongbar, NSW, Australia, pp. 1–2.

Dovers, S.R. and Mobbs, C. (1999) An alluring prospect? Ecology and the requirements of adaptive management. In: Bossemann, K. and Richardson, B. (eds) *Environmental Justice and Market Place Mechanisms.* Kluwer International, London.

Easton, C. (1989) The trouble with the Tweed. *Fishing World,* March 1989, pp. 58–59.

Esman, M.J. and Uphoff, N.T. (1984) *Local Organisations: Intermediaries in Rural Development.* Cornell University Press, New York.

GESAMP (2001) *A Sea of Troubles.* GESAMP (IMO/FAO/UNESCO-IOC/WMO/IAEA/UN/UNEP Joint

Group of Experts on the Scientific Aspects of Marine Environmental Protection) Report Study GESAMP No. 70, p. 35 and No. 71, p. 162.

Harris, G.P. (1999) Comparison of the biogeochemistry of lakes and estuaries: ecosystem processes, functional groups, hysteresis effects and interaction between macro- and microbiology. *Marine and Freshwater Research* 50, 791–811.

Hart, B.T., Ottoway, E.M. and Noller, B.N. (1987) Magela Creek System, Northern Australia. I. 1982–83 wet season water quality. *Australian Journal of Marine and Freshwater Research* 38, 261–288.

Healthy Rivers Commission (2000) *Securing Healthy Coastal Rivers: a Strategic Perspective.* Healthy Rivers Commission of New South Wales, Sydney, Australia.

Healthy Rivers Commission (2002) *Independent Inquiry into Coastal Lakes.* Final Report, April 2002. Healthy Rivers Commission of New South Wales, Sydney, Australia, p. 74.

Jiggins, J. (2002) Interagency learning process. In: *OECD Workshop on an Interdisciplinary Dialogue: Agricultural Production and Integrated Ecosystem Management of Soil and Water,* Balina, Australia, 10–16 November 2002. OECD, NSW Agriculture, University of Western Sydney, Australia, pp. 86–94.

King, C.J. (1948) The first fifty years of agriculture in New South Wales. *Review of Marketing and Agricultural Economics, Sydney* 16, 362–386.

Kinsela, A.S. and Melville, M.D. (2004) Mechanisms of acid sulfate soil oxidation and leaching under sugarcane cropping. *Australian Journal of Soil Research* 42, 569–578.

Kremer, H.H. and Crossland, C.J. (2002) Coastal change and the 'Anthropocene'. Past and future directions of IGBP-LOICZ project. In: *Low-lying Coastal Areas – Hydrology and Integrated Coastal Zone Management.* International Symposium, Bremerhaven, Germany, 9–12 September 2002. Deutches IHP/OHP-Nationalkomittee, Koblenz, Germany, pp. 3–19.

Lin, C. and Melville, M.D. (1993) Control of soil acidification by fluvial sedimentation in an estuarine flood plain, eastern Australia. *Sedimentary Geology* 85, 1–13.

Macdonald, B.C.T., Denmead, O.T., White, I. and Melville, M.D. (2004a) Natural sulfur dioxide emissions from sulfuric soils. *Atmospheric Environment* 38, 1473–1480.

Macdonald, B.C.T., Smith, J., Keene, A.F., Tunks, M., Kinsela, A. and White, I. (2004b) Impact of run-off from sulfuric soils on sediment chemistry in an estuarine lake. *Science of the Total Environment* 329, 115–130.

Melville, M.D., White, I. and Willett, I.R. (1990) Problems of acid sulphate soils and water degradation in Holocene pyritic systems. In: *Proceedings of the Workshop on Applied Quarternary Studies,* Australian National University, 2–3 July 1990, Canberra, Australia, pp. 89–95.

Melville, M.D., White, I. and Lin, C. 1993. The origins of acid sulphate soils. In: *Proceedings of the National Conference on Acid Sulphate Soils.* Coolangatta, 24–25 June 1993, NSW Agriculture, Wollongbar, Australia, pp. 19–25.

Naylor, S.D., Chapman, G.A., Atkinson, G., Murphy, C.L., Tulau, M.J., Flewin, T.C., Milford, H.B. and Morand, D.T. (1995) *Guidelines for the Use of Acid Sulphate Soils Risk Maps.* NSW Soil Conservation Service, Department of Land and Water Conservation, Sydney, Australia.

Noller, B.N. and Cusbert, P.J. (1985) Mobilization of aluminium from a tropical flood plain and its role in natural fish kills: a conceptual model. In: Lekkas, T.D. (ed.) *Proceedings of the 5th Conference on Heavy Metals in the Environment,* September 1985, Athens. CEP Consultants Ltd., Edinburgh, UK, pp. 700–702.

Perez, P., Dray, A., White, I., LePage, C., and Falkland, A. (2003). AtollScape: simulating freshwater management in Pacific atolls. Spatial processes and time dependence issues. In: *Proceedings MODSIM, International Congress on Modelling and Simulation, Integrative Modelling of Biophysical, Social and Economic Systems for Resource Management Solutions* (14–17 July 2003, Townsville, Australia). Modelling and Simulation Society of Australia and New Zealand, Inc., Canberra, Australia, pp. 514–518.

Pons, L.J. (1973). Outline of the genesis, characteristics, classification and improvement of acid sulphate soils. In: Dost, H. (ed.) *Proceedings of the International Symposium on Acid Sulphate Soils,* 13–29 August 1972, Wageningen. IILRI Publication No. 18, Vol. 1, International Institute for Land Reclamation and Improvement, Wageningen, Netherlands, pp. 3–27.

Roling, N. (1988) *Extension Science: Information Systems in Agricultural Development.* Cambridge University Press, Cambridge, UK.

Sammut, J., Melville, M.D., Callinan, R.D. and Fraser, G.C. (1995) Estuarine acidification: impacts on aquatic biota of draining acid sulphate soils. *Australian Geographical Studies* 33, 89–100.

Sammut, J., White, I. and Melville, M.D. (1996) Acifidication of an estuary due to drainage from acid sulfate soils. *Marine and Freshwater Research* 4, 669–684.

Simpson, H.J. and Pedini, M. (1985) *Brackish Water Aquaculture in the Tropics: the Problem of Acid Sulphate Soils*. FAO Fish Circular No. 791. FAO, Rome.

Smith, J. (2000) An assessment of the spatial variations in actual acidity at McLeods Creek. BSc Honours thesis, School of Geography, the University of New South Wales, Sydney, Australia.

Teakle, L.J.H. and Southern, B.L. (1937) The peat soils and related soils of Western Australia, II. A soil survey of Herdsman Lake. *Journal of Agriculture Western Australia* 14, 404–424.

Thom, B.G. (2002) *Environmental History and Decision Making*. Keynote address, 5th International Acid Sulfate Soils Conference, Tweed Heads, NSW, Australia, 26–30 August 2002. ASSMAC, Wollongbar, NSW, Australia.

Thom, B.G. and Harvey, N. (2000) Triggers for late twentieth century reform of Australian coastal management. *Australian Geographical Studies* 38, 275–290.

UNESCO (2002a) Wise practices for managing conflict situations. In: *Continental Coastal Regions. Results of a workshop on 'Wise Practices for Coastal Conflict Prevention and Resolution'*, Maputo, Mozambique, 19–23 November 2001. Coastal region and small island papers, UNESCO, Paris.

UNESCO (2002b) Wise practises for conflict prevention and resolution in small islands. In: *Results of a Workshop on 'Furthering Coastal Stewardship in Small Islands'*, Dominica, 4–6 July 2001. Coastal region and small island papers 11, UNESCO, Paris, 70 pp.

Walker, P.H. (1972) Seasonal and stratigraphic controls in coastal flood plain soils. *Australian Journal of Soil Research* 10, 127–142.

White, I. (2002) Swelling and the hydraulic properties and management of acid sulfate soils. In: Lin, C., Melville, M.D. and Sullivan, L.A. (eds) *Acid Sulfate Soils in Australia and China*. Science Press, Beijing, China, pp. 38–62.

White, I. and Melville, M.D. (1993) *Treatment and containment of potential acid sulphate soils: formation, properties and management of potential acid sulphate soils*. Report for Roads and Traffic Authority, NSW. CSIRO Centre for Environmental Mechanics, Canberra, Australia, Technical Report T53.

White, I., Melville, M.D., Wilson, B.P., Price, C.B. and Willett, I.R. (1993) Understanding acid sulphate soils in canelands. In: *Proceedings of the National Conference on Acid Sulphate Soils*, Coolangatta, Australia, 24–25 June 1993. NSW Agriculture, Wollongbar, Australia, pp. 130–148.

White, I., Melville, M.D., Sammut, J., Wilson, B.P. and Bowman, G.M. (1996) Downstream impacts from acid sulfate soils. In: Hunter, H., Eyles, A. and Rayment, G. (eds) *Downstream Impacts of Land Use*. Department of Natural Resources, Brisbane, Australia, pp. 165–172.

White, I., Melville, M.D., Wilson, B.P. and Sammut, J. (1997). Reducing acid discharge from estuarine wetlands in eastern Australia. *Wetlands Ecology and Management* 5, 55–72.

White, I., Macdonald, B.C.T. and Melville, M.D. (2001) Modelling groundwater in soft, sulfidic, coastal sediments. In: Ghassemi, F., Whetton, P., Little, R. and Littleboy, M. (eds) *Proceedings of MODSIM, International Congress on Modelling and Simulation*, 10–13 December 2001, Australian National University, Canberra, Australia. Vol. 2, Natural Systems (Part 2), Modelling and Simulation Society of Australia and New Zealand, Inc., Canberra, Australia, pp. 567–572.

White, I., Smiles, D.E., Santomartino, S., van Oploo, P., Macdonald, B.C.T. and Waite, T.D. (2003) Dewatering and the hydraulic properties of soft sulfidic soils. *Water Resources Research* 39, 1295, doi:10.1029/2002WR001324.

Willett, I.R. and Walker, P.H. (1982) Soil morphology and distribution of iron and sulphur fractions in a coastal flood plain toposequence. *Australian Journal of Soil Research* 20, 283–294.

Williams, J. (2002) Acid sulfate soils. In: *OECD Workshop on An Interdisciplinary Dialogue: agricultural production and integrated ecosystem management of soil and water*, Balina, Australia, 10–16 November 2002, OECD, NSW Agriculture, University of New South Wales, Sydney, pp. 114–122.

Wilson, B.P., White, I. and Melville, M.D. (1999) Floodplain hydrology, acid discharge and water quality associated with a drained acid sulfate soil. *Marine and Freshwater Research* 50, 149–157.

Woodward, H.P. (1917) *Investigation into the cause of mineralisation of the 'seven-mile' swamp at Grassmere, Southwest Division*. Western Australia Department of Mines Annual Report, pp. 49–57.

WRI (World Resources Institute) (2000) *World Resources 2000–2001. People and Ecosystems: The Fraying Web of Life*. World Resources Institute, Washington, D.C., 389 pp.

10 Mangrove Dependency and the Livelihoods of Coastal Communities in Thailand

E.B. Barbier
Department of Economics and Finance, University of Wyoming, Laramie, Wyoming, USA, e-mail: ebarbier@uwyo.edu

Abstract

From 1961 to 1996, Thailand lost 50–60% of its mangrove forests, mainly because of conversion to shrimp aquaculture. The speed and scale of deforestation have affected many coastal communities. This chapter highlights the importance of mangroves to four case study villages. Households depend directly on mangrove forests for fish and wood collection and/or benefit indirectly from the mangroves' support to coastal fisheries. Mangrove loss therefore affects the decision of households to look for outside employment. In response to deforestation, female household members allocate more hours to employment relative to mangrove-dependent activities, whereas males allocate fewer hours to outside work. Awareness of community conservation efforts and of the environmental damage imposed by shrimp farms also motivates households to participate in replanting activities. Efforts to control mangrove deforestation and promote community-based management of remaining mangrove forests, as well as replanting, would help to mitigate some of the worst impacts on coastal villages. By developing institutions to support local community management, the government of Thailand could help avoid excessive mangrove deforestation and conflicts over uses. Such a framework could also provide important lessons in coastal resource management for other countries in South-east Asia and elsewhere.

Introduction: Shrimp Farm Expansion and Mangrove Loss in Thailand

The issue of coastal land conversion for commercial shrimp farming is a highly controversial topic in Thailand. Frozen shrimp are a major export product of Thailand, earning more than US$1.6 billion each year, and the government has been keen to expand these exports (Tokrisna, 1998; Barbier and Sathirathai, 2004). Yet, the expansion of shrimp exports has caused much devastation to Thailand's coastline and has affected other commercial sectors, such as fisheries.

Thailand's coastline is vast, stretching for 2815 km, of which 1878 km are on the Gulf of Thailand and 937 km on the Andaman Sea (Indian Ocean) (Kaosa-ard and Pednekar, 1998). In recent decades, the expansion of intensive shrimp farming in the coastal areas of southern Thailand has led to rapid conversion of mangroves (Barbier and Sathirathai, 2004). Between 1961 and 1996, Thailand lost around 20,500 km^2 of mangrove forests, or about 56% of the original area, mainly because of shrimp aquaculture and other coastal developments (Charuppat and Charuppat, 1997). Estimates of the

 Environment and Livelihoods in Tropical Coastal Zones (eds C.T. Hoanh, T.P. Tuong, J.W. Gowing and B. Hardy)

amount of mangrove conversion caused by shrimp farming vary, but recent studies suggest that up to 65% of Thailand's mangroves have been lost to shrimp farm conversion since 1975 (Dierberg and Kiattisimkul, 1996; Charuppat and Charuppat, 1997; Aksornkoae and Tokrisna, 2004). The rate of mangrove deforestation slowed in the 1990s, but in the mid-1990s the annual loss was estimated to be around 3000 ha/year (Sathirathai, 1998).

Although mangrove conversion for aquaculture began in Thailand as early as 1974, the boom in intensive shrimp farming through mangrove clearing took off in 1985 when the increasing demand for shrimp in Japan pushed up the border-equivalent price to $100/kg (Barbier and Sathirathai, 2004). For example, from 1981 to 1985 in Thailand, annual shrimp production through aquaculture was around 15,000 t, but by 1991 it had risen to over 162,000 t and by 1994 to over 264,000 t (Kaosa-ard and Pednekar, 1998).

Shrimp farm area expanded from 31,906 to 66,027 ha from 1983 to 1996. However, much of the semi-intensive and intensive shrimp farming in Thailand is short-term and 'unsustainable', that is, poor water quality and disease problems mean that yields decline rapidly and farms are routinely abandoned after 5–6 years of production (Flaherty and Karnjanakesorn, 1995; Dierberg and Kiattisimkul, 1996; Tokrisna, 1998; Vandergeest *et al.*, 1999).

Although shrimp farm expansion has slowed in recent years, unsustainable production methods and lack of know-how have meant that more expansion still takes place every year simply to replace unproductive and abandoned farms. In provinces close to Bangkok, such as Chanthaburi, mangrove areas have been devastated by shrimp farm developments (Raine, 1994). More recently, Thailand's shrimp output has been maintained by the expansion of shrimp-farming activities to the southern and eastern parts of the Gulf of Thailand, and across to the Andaman Sea coast (Flaherty and Karnjanakesorn, 1995; Sathirathai, 1998; Vandergeest *et al.*, 1999).

Moreover, conversion of mangroves to shrimp farms is irreversible. Without careful ecosystem restoration and manual replanting efforts, mangroves do not regenerate even in abandoned shrimp farm areas. In Thailand, most of the estimated 11,000 ha or more of replanted areas between 1991 and 1995 have been on previously unvegetated tidal mudflats (Lewis *et al.*, 2000). Currently, in Thailand, there is no legal requirement that shrimp farm owners invest in replanting and restoring mangroves once farming operations have ceased and the ponds are abandoned.

Much of the financial investment in coastal shrimp farms is from wealthy individual investors and business enterprises from outside of the local community (Flaherty and Karnjanakesorn, 1995; Goss *et al.*, 2000, 2001). Although some hiring of local labour occurs, it is reported that many shrimp farm owners in coastal areas have hired Burmese workers, as their wage rates are much lower.

Ill-defined property rights have accelerated the rapid conversion of mangroves to shrimp farms in Thailand. Historically, this has been a common problem for all forested areas in Thailand (Feder *et al.*, 1988; Thomson *et al.*, 1992; Feeny, 2002). Although the state, through the Royal Forestry Department, ostensibly owns and controls mangrove areas, in practice they are *de facto* open-access areas on to which anyone can encroach. This has had three effects on mangrove deforestation attributable to shrimp farms. First, the open-access conditions have allowed illegal occupation of mangrove areas for establishing shrimp farms in response to the rising prices and profits from shrimp aquaculture (Barbier and Sathirathai, 2004). Second, insecure property rights in cleared forest areas have been associated with underinvestment in land quality and farm productivity (Feder and Onchan, 1987; Feder *et al.*, 1988; Thomson *et al.*, 1992). The lack of tenure security for shrimp farms in southern Thailand also appears to be a major factor in the lack of investment in improving productivity and adopting better aquaculture methods, leading to more mangrove areas being cleared than necessary (Barbier and Sathirathai, 2004). Third, several studies have pointed out how open-access forest lands are more vulnerable to rapid deforestation and conversion to agricultural and other commercial uses as the development of roads and the highway network makes these lands more 'accessible' (Cropper *et al.*, 1999; Feeny, 2002).

Despite the lack of secure property rights and frequently illegal occupation of mangrove areas, owners have an incentive to register their shrimp farms with the Department of Fisheries. In doing so, they become eligible for the preferential subsidies for key production inputs, such as shrimp larvae, chemicals and machinery, and for preferential commercial loans for land clearing and pond establishment (Tokrisna, 1998; Barbier and Sathirathai, 2004). Such subsidies inflate artificially the commercial profitability of shrimp farming, thus leading to more mangrove conversion, even though estimates of the economic returns to shrimp aquaculture in Thailand suggest that such conversion is not always justified (Sathirathai and Barbier, 2001). Combined with insecure property rights, the subsidies also put further emphasis on shrimp aquaculture as a commercial activity for short-term exploitative financial gains rather than as a long-term sustainable activity.

Case Study in Four Coastal Villages

A case study of the labour allocation decisions of rural households from four representative villages illustrates the importance of mangroves to the livelihoods of coastal communities in Thailand. The four case study villages are Ban Khlong Khut and Ban Gong Khong in Nakhon Si Thammarat Province on the Gulf of Thailand, and Ban Sam Chong Tai and Ban Bang Pat in Phang-nga Province on the Andaman Sea (see Fig. 10.1). Further background details on these case study villages can be found in Aksornkoae *et al.* (2004).

These four villages have experienced rates of mangrove loss similar to those which have occurred nationally in Thailand. Such mangrove deforestation has had important, albeit varying, impacts on the livelihoods of villagers. Some households in these four communities derive their income and subsistence directly from mangrove forests, in terms of fish collection, wood products and firewood. Other households benefit indirectly from the protection and support the mangroves give to coastal fisheries. A few engage in aquaculture.

A randomly stratified survey at the four village study sites was conducted during April and July 2000. Interviews with the heads of the households were conducted by trained enumerators speaking the local language under the supervision of a team of Thailand-based researchers, using a pretested survey designed by the author. Pretesting of the questionnaires was conducted in February 2000. The first stage of the survey was conducted in Phang-nga from 17 to 23 April 2000. The second stage of the survey, in Nakhon Si Thammarat, was carried out from 2 to 8 July 2000.

The survey gathered information on household involvement in outside employment and important household characteristics such as age, education, household composition, number of children, debt and size of landholding, and various production/income characteristics. The survey also collected detailed information on the mangrove-based activities of households, including the area of mangrove used by the household for such activities. Details on household labour allocation were also obtained to establish whether the household was undertaking other activities that were not dependent on mangroves.

Ban Sam Chong Tai and Ban Bang Pat are located on Phang-nga Bay, the former having only poor road access and consisting mainly of traditional fishing households that also collect many products from the mangroves. Ban Bang Pat is quite different. This village is located on the main highway and is highly commercialized and relatively modern. Although the villagers here still engage in coastal fishing, they generally do less traditional collection from the mangrove areas. Some female villagers also conduct various agricultural activities, including tending any rubber plantations owned by the household. The Nakhon Si Thammarat villages, Ban

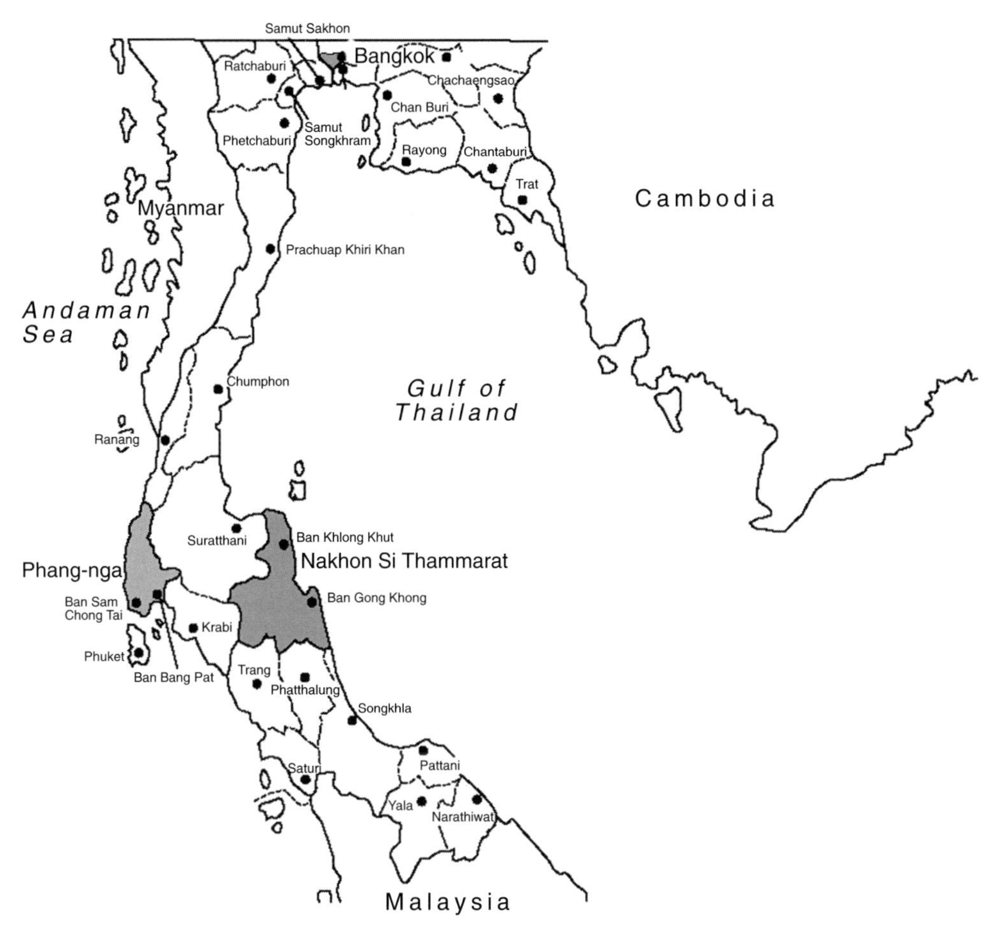

Fig. 10.1. Case study villages, Thailand (from Barbier and Sathirathai, 2004).

Gong Khong and Ban Khlong Khut, have relatively high levels of urbanization and commercialization. As both villages are located on the coast, fishing is still a major activity for many households. Villagers in Ban Gong Khong still engage in traditional collection activities from the mangrove areas, but households in Ban Khlong Khut do much less collection. In Ban Khlong Khut, some households have their own shrimp ponds, which occupy much of the female labour of the household. A large percentage of household members in both villages in Nakhon Si Thammarat turn to outside employment. The main source of employment for female villagers is at nearby factories, whereas male villagers often work on commercial shrimp farms owned by outsiders.

The mangrove areas in Ban Sam Chong Tai have been degraded mainly because of forest concessions. According to Thai law, forest concessionaires are required to replant but, in reality, reforestation has never taken place. Although the forests have not been completely cleared, extensive damage has occurred in many of the forest areas. In Ban Bang Pat, the mangrove forests were first cleared by tin-mining concessions. These activities not only destroyed the forests but also created extensive water pollution in the area. After the prices of tin fell drastically, coupled with the severe decline of mangroves, a Cabinet resolution on 23 July 1991 abolished tin mining in mangrove forests throughout Thailand. However, the unintended consequence was that the forests became open-access areas and became susceptible to conversion to shrimp farming, which is the current threat to the mangroves near Ban Bang Pat.

Mangrove areas in Nakhon Si Thammarat have decreased by as much as 53,811 ha, or 87.93%, from 1961 to 1996 (Charuppat and Charuppat, 1997). This mangrove loss was much higher than the deforestation level of 19,742 ha, or 33.56%, in Phang-nga over the same period. At present in Phang-nga, 38,138 ha of mangrove area still remain compared with only 7,389 ha in Nakhon Si Thammarat.

In Ban Sam Chong Tai village, the local community is very active in the conservation of mangroves. They consider an area of around 60 ha, which is legally owned by the state, as their own community forest. These villagers are small-scale fishermen, who, when questioned during our survey, expressed knowledge that their local mangrove areas serve as breeding grounds and fry nurseries for coastal fisheries. In Ban Bang Pat, the local community also participates actively in the replanting of mangroves, but less so than in Ban Sam Chong Tai. The replanting projects in Ban Bang Pat were not initiated originally by the community but by outside non-governmental organizations (NGOs). The situation is similar in the two villages surveyed in Nakhon Si Thammarat, where local replanting schemes were initiated by NGOs or by the Royal Forestry Department.

The survey of the four villages elicited from households their allocation of male and female labour to their main income-producing and other activities, as well as key socio-economic characteristics. Information on the employment of male and female labour in work outside of the household also included wage rates and detailed time allocations. In total, 201 households were surveyed, although two households reported no direct or indirect income-producing activities that depended on mangroves and were excluded from the sample, leaving a total of 199 households. Of the latter households, 61 reported having at least one male member undertaking outside employment, and 33 have at least one female member participating in paid outside work.

Table 10.1 provides a brief set of summary statistics for the entire sample and shows that 32 households were involved in offshore fishing but did not collect mangrove products, 61 households were involved in direct-use collection activities only and 104 households did a combination of both. From those three groups, the collect-only group had the greatest percentage of households devoting time to outside employment (49% of males and 32% of females). The households involved in both activities had the lowest percentages participating in outside employment (males 17%, females 7%). As might be expected, the need to supplement income in households involved only in collection activities is paramount. This result is reinforced by the figures for percentage of income that is mangrove-based. The households that are solely involved in collection activities have the lowest proportion of mangrove-dependent income. At the other end of the scale, for those households involved in both direct and indirect mangrove-based activities, almost all of their income comes from these activities.

Table 10.2 reports similar data by village. An interesting pattern also emerges here, in that the villages in Nakhon Si Thammarat have a lower proportion of income coming from mangrove-based activities and a higher percentage of households working in outside employment. Households in Phang-nga, on the other hand, obtain a much higher percentage of their income from mangrove-related activities and engage in less outside employment, with the majority of households choosing to devote their time to both direct collection and indirect mangrove production activities.

Table 10.3 summarizes by village the household male and female labour allocation, in terms of average hours per year, for mangrove-dependent activities, agriculture, replanting and outside employment. For all villages surveyed, collection of fish (mainly shellfish and crabs) from the mangrove swamps and coastal fishing are the principal sources of mangrove-dependent employment for male and female labour. Across the entire 199 surveyed households, both male

Table 10.1. Summary statistics for outside employment by household type.

	Fish only (*n* = 32)		Collect only (*n* = 63)		Fish and collect (*n* = 104)	
	Male	Female	Male	Female	Male	Female
Number (*n*)	12	6	31	20	18	7
(% of total)	(38)	(19)	(49)	(32)	(17)	(7)
Mangrove-dependent income share of total Income (%)	82		75		93	

Table 10.2. Summary statistics for outside employment by village.

	Phang-nga				Nakhon Si Thammarat			
	Ban Sam Chong Tai (*n* = 55)		Ban Bang Pat (*n* = 41)		Ban Gong Khong (*n* = 52)		Ban Khlong Khut (*n* = 51)	
	M	F	M	F	M	F	M	F
Number (*n*)	8	4	7	7	31	12	15	10
(% of total)	(15)	(7)	(17)	(17)	(60)	(23)	(29)	(20)
Mangrove-dependent income share of total income (%)	95		89		66		83	
Fish only	1		5		3		23	
Collect only	19		3		34		7	
Fish and collect	35		33		15		21	

M, male; F, female.

and female household members devote a substantial number of hours each year to mangrove-dependent activities. This is not surprising, given that, in the sample, mangrove-based income accounts on average for 83% of all household income, with a relatively small deviation across households.

Mangrove-based activities appear to require more male than female household labour. The exceptions are that females spend more time making shrimp paste in Ban Sam Chong Tai and in producing dried fish in Ban Gong Khong. However, as Table 10.3 indicates, these two activities do not require a considerable amount of labour compared with the other mangrove-dependent activities conducted by the households. In the two Phang-nga villages, households allocate on average almost three times as many male hours as female hours per year to all mangrove-dependent fishing and collection activities. In the Nakhon Si Thammarat villages, the ratio of total male to female hours spent per year on these activities is around 3.7 for Ban Gong Khong and 4.3 for Ban Khlong Khut. On average across all four villages, males spend over three times as many hours on mangrove-dependent activities as females.

In contrast, females spend proportionately much more of their time in outside employment relative to mangrove-based activities. Across all households, the ratio of the average hours in outside employment to hours in all mangrove-based activities ranges from 41% to 74% for females, whereas the ratio for males ranges from 11% to 28%. The difference between males and females is even more striking when comparing average labour allocation rates for only those households whose members participate in outside work. In all four villages, for those house-

Table 10.3. Summary statistics for labour allocation, by village (average hours per year).

Activity	Phang-nga				Nakhon Si Thammarat			
	Ban Sam Chong Tai ($n = 55$)		Ban Bang Pat ($n = 41$)		Ban Gong Khong ($n = 52$)		Ban Khlong Khut ($n = 51$)	
	Male	Female	Male	Female	Male	Female	Male	Female
Wood collection	5.18	0.84	3.76	0.05	10.67	4.41	10.57	0.76
Fuelwood and charcoal	5.82	0	0	0	3.96	2.56	0	0.47
Fish collection	610.85	218.89	125.73	110.41	1367.73	386.88	324.02	35.22
Shrimp paste	28.87	123.35	23.27	2.93	4.35	4.73	0	0
Dried fish	0	0	0	0	9.81	12.23	1.73	1.12
All collection	650.73	343.07	152.76	113.39	1396.51	410.82	336.31	37.57
Coastal fishing	783.20	194.09	965.20	279.73	857.50	195.69	2231.98	520.27
Aquaculture	37.07	12.75	19.46	14.66	0	4.65	19.96	47.65
All fishing	820.27	206.84	984.66	294.39	857.50	200.35	2251.94	567.92
All mangrove-based activities	1471.00	549.91	1137.41	407.78	2254.01	611.16	2588.25	605.49
Agriculture	0	0	0	35.12	452.69	253.15	82.37	33.02
Replanting mangroves	22.69	14.07	18.94	9.43	61.69	18.77	2.37	0.31
Outside work[a]	157.56	225.64	176.66	301.68	621.67	361.52	541.67	393.41
(% of mangrove-based hours)[b]	(11)	(41)	(16)	(74)	(28)	(59)	(21)	(65)
Adjusted outside work[c]	1083.25	3102.50	1034.71	1767.00	1042.81	1566.58	1841.67	2006.40
(% of mangrove-based hours)[b]	(130)	–	(86)	(542)	(61)	(155)	(59)	(490)

[a] Hours in outside employment averaged across all households.
[b] Ratio of average hours in outside employment to average hours in all mangrove-based activities.
[c] Hours in outside employment averaged across households whose members participate in such work.

holds reporting individuals engaged in outside work, the total number of average hours per year spent in outside employment by females exceeds that of males.

However, males clearly receive higher wages for outside work than females. For the 32 households whose female members participated in outside employment, the average hourly wage received was 22.8 baht ($0.57/h).[1] For the 60 households whose male members participated in outside work, the average hourly wage received was 44.5 baht/h ($1.11/h).

Mangrove Loss and Labour Allocation in the Case Study Villages

The above case study survey of four coastal villages is ideal for analysing the impacts of mangrove loss on labour allocation decisions in several respects. First, the livelihoods of the surveyed households from these villages clearly depend on the surrounding mangrove ecosystems (Aksornkoae *et al.*, 2004). Second, although a few households in these four villages also engage in agriculture, the main alternative to mangrove-dependent activities is employment as wage earners outside of the household. Thus, any depletion or degradation of local mangrove forests will affect the income earned by villagers from mangrove-dependent activities and influence their decision to seek outside employment.

Using a three-step Heckman selection model, Barbier (2004) estimates the total effect of a change in mangrove area on the supply of labour to outside employment by mangrove-dependent households in the survey.[2] The results are reported in Table 10.4 in terms of both marginal effects (a 1 ha change in mangrove area) and elasticities (a 1% change in mangrove area). Two interesting findings emerge from the analysis.

First, both males and females appear to have 'backward-bending' supply curves with respect to the number of hours spent in outside employment, implying that higher wages lead to income effects that are greater than the substitution effects. The result is that, as males and females receive higher wages for outside employment, the total number of hours that they spend engaged in such work actually declines. Such a negative 'own-wage effect' is also found in other 'household outside employment' studies in developing countries (Rosenzweig, 1980; Hernández-Licona, 1997), and is consistent with the situation where households receive low market wages yet their minimum subjective requirement of income for subsistence cannot be achieved without outside employment. It is very likely that these conditions hold for the mangrove-dependent households surveyed in coastal Thailand.

Second, a change in mangrove area may affect the amount of labour supplied to outside employment in two ways: through a direct effect on hours worked and through an indirect impact on hours worked via the wage rate. Table 10.4 indicates that there is a direct effect of a change in mangrove area on the number of hours worked in outside employment for females, but not for males. Instead, mangrove changes influence the labour supplied by males for outside work indirectly through influencing the 'own-wage effect' described above and a 'cross-wage effect' via female wages. The latter effect indicates the extent to which males adjust their hours devoted to outside

[1] The exchange rate at the time of the survey (July 2000) was 40 baht = US$1.

[2] Applying standard ordinary least squares (OLS) regression analysis to estimate this relationship would yield biased parameter estimates, since an OLS regression cannot take into account the censored nature of the labour allocation decision of the mangrove-dependent household. Although the household always engages in some form of mangrove-based activity, it may not participate in outside employment. This means in turn that the market wage rate and the amount of hours in paid work will be observed only if the household decides to participate in outside employment; if the household decides not to undertake outside work, no wages or hours worked will be observed. To avoid sample selection bias arising from this participation decision, a standard approach adopted in the off-farm labour supply literature is to use a three-step Heckman procedure for conditioning the estimations of wages and hours supplied (for further discussion, see Abdulai and Delgado, 1999; Barbier, 2004).

Table 10.4. The effect of a change in mangrove area on the supply of outside labour (from Barbier, 2004).

	Direct effect	Indirect effect (via male wages)	Indirect effect (via female wages)	Total effect
Males				
Marginal effects	–	0.13	−0.09	0.04
Elasticities (%)	–	2.30	−1.60	0.70
Females				
Marginal effects	0.26	–	−0.35	−0.09
Elasticities (%)	5.36	–	−7.25	−1.88

employment as the wage paid to female household members for outside work changes. Both of these indirect wage effects of changes in mangrove area are therefore shown in Table 10.4, and the sum of these two effects equals the total effect of a change in mangrove area on labour supply by males. In contrast, only the own-wage indirect effect is significant in affecting the hours worked in outside jobs by females. As shown in Table 10.4, the latter indirect effect plus the direct impact of a change in mangrove area equal the total effect of a change in natural capital on female labour supply to outside employment.

The results reported in Table 10.4 suggest that, for the surveyed mangrove-dependent households, the dominant impacts of a loss of natural capital on the supply of both male and female labour to outside employment arise through indirect own-wage effects. Because mangrove loss leads to a reduction in the wages that females will receive from outside employment, the result is that females will increase the hours they work. In contrast, mangrove deforestation increases the wages that males receive from casual work and, as a result, they will work fewer hours in such employment.

Thus, the total effect of a loss in mangrove area is to reduce the supply of male labour to outside employment but to increase the supply of female members. Across the 199 surveyed households, a 1% decline in the local mangrove forests will cause the number of hours that males work in outside employment to decline by 0.7% while increasing the number of hours worked by females by 1.88%. Given the large losses in mangrove forests that have occurred at the two case study sites, such deforestation clearly has had a significant impact on the allocation of household labour in these coastal communities.

Mangrove Dependency and Participation in Conservation Efforts

Barbier *et al.* (2004) test the hypothesis that the degree of mangrove dependency is a major causative factor in the active participation of households from the four case study villages in conservation efforts. The hypothesis is that, once households realize that as mangrove area declines they will experience impacts on their livelihoods leading to income losses, they will participate in the replanting of mangroves. Whether households choose to be involved in mangrove conservation is also likely to vary with their characteristics and location, land ownership and tenure considerations, awareness of and attitudes towards community conservation efforts – including the replanting programmes sponsored by non-governmental organizations (NGOs) and some international organizations, and with concerns over the threat of the environmental effects of shrimp farms. In addition, the decision to participate in mangrove conservation may vary between male and female members of the household.

As indicated in Table 10.3, all mangrove-dependent households in the four case study villages allocate some time to replanting activities. However, the average hours per year spent replanting vary considerably across the villages. Males generally spend more time replanting than females. Barbier *et al.* (2004) depict a mangrove-dependent household's choice of whether or not to participate in mangrove conservation as a binary decision, which can be empirically estimated through a bivariate probit estimation for household males and females. The regression results are depicted in Table 10.5.

The results show that the male decision to participate is mostly influenced by household awareness of community conservation efforts and use rules, as shown by the positive coefficient and highest marginal probability. The degree of mangrove-dependent income is the second-most important positive influence, with a marginal probability of 0.28. The household's awareness of the environmental impact of shrimp farming is the other significant variable in the male equation. The positive coefficient value and marginal probability of 0.13 suggest that males from households that are aware of the negative environmental impact of shrimp farms are more likely to participate in replanting.

For females, the degree of dependence of the household on mangrove-based income is significant at the 10% level and is the most important variable influencing its participation. Distance to the mangroves from the household is the next most important influence. The negative coefficient suggests that females from households that live increasing distances from the mangroves are less likely to participate in replanting. The area of mangrove used by the household is also important in the female decision. The result suggests that females from households that collect and fish in larger mangrove areas are less likely to participate in conservation. This might reflect that the household recognizes that smaller mangrove areas require more replanting effort. The number of children under 6 years of age, as might be expected, also influences the female decision to participate. The household awareness of community conservation and use rules positively affects female participation in replanting.

Finally, the variable ρ (1,2) measures the degree to which a household determines simultaneously, or jointly, whether males and females should participate in mangrove conservation. This variable is positive and significant, suggesting that the participation decision is determined jointly.

Conclusions and Policy Implications

Drawing on a case study of four coastal villages surveyed in Thailand, this chapter has shown that continuing mangrove deforestation not only has a significant impact on the allocation of household labour in Thai coastal villages that traditionally exploit these forests but also affects the intra-household division of labour. In response to such deforestation, for those households whose members participate in some outside work, females will continue to allocate more hours to such employment relative to mangrove-dependent activities, whereas males will allocate fewer hours to such work. One might also expect other mangrove-dependent households to send their non-working females out to look for outside employment.

Two concerns arise from this intra-household allocation of labour in response to mangrove deforestation. First, for the households in the case study survey, the average hourly wage received by females ($0.57) is barely half that received by males ($1.11). If the households require income from outside employment to meet overall needs, then they may fall short of their outside income target if the households increasingly rely on female members to participate in such employment. Even if the households do achieve their target by supplying more female labour to outside work, there may be an impact on other non-income activities important to the welfare of the household that are traditionally undertaken by females, such as child rearing, food preparation, care of the elderly and house cleaning. Secondly, the decline in the number of hours spent by males in outside employment accompanying deforestation presumably means that the males will be more productively employed at the margin in mangrove-based activities. If this is the case, household income from these activities

Table 10.5. Male and female participation in mangrove replanting efforts (from Barbier *et al.*, 2004).

Variable	Males			Females		
	Coeff.	t-ratio	Marginal prob.	Coeff.	t-ratio	Marginal prob.
Constant	−2.4374	−2.4800	−0.6146	−1.9948	−1.6098	0.2741
Mangrove-dependent income as a proportion of total income	1.1070	1.9772	0.2792	1.5465	1.7843	−0.2125
Area of mangrove used by household (ha)	−0.0002	−0.1428	0.0000	−0.0057	−2.2296	0.0008
If household is aware of community conservation efforts and use rules (AWARE, 1; otherwise, 0)	1.1357	4.3000	0.2864	0.8765	2.6493	−0.1204
If household believes shrimp farming has a negative environmental impact (ATSFARM, 1; otherwise, 0)	0.5012	1.6316	0.1264	0.8173	1.4319	−0.1123
Average age of household members	0.0131	0.7055	0.0033	−0.0093	−0.6093	0.0013
Number of children < 6	−0.0308	−0.1129	−0.0078	−0.6560	−2.5976	0.0901
Number of children 6–12	0.1048	0.5644	0.0264	−0.1447	−0.7812	0.0199
Distance of household to mangroves	−0.0307	−1.4999	−0.0077	−0.0484	−2.1469	−0.0066
Average years of male education	0.0043	0.0982	0.0011			
Number of adult males in household	0.2142	1.3659	0.0540			
If any household males participate in outside employment (DM, 1; otherwise, 0)	0.1334	0.4194	0.0336			
Average years of female education				0.0412	0.5620	−0.0057
Number of adult females in household				0.0369	0.2810	−0.0051
If any household females participate in outside employment (DF, 1; otherwise, 0)				0.2451	0.4069	−0.0337
ρ (1,2)	0.6817	4.2975				

McFadden R^2 = 0.40
Log-likelihood ratio statistic = 194.54
Log-likelihood ratio test for homoskedasticity = −21.46
Note: The McFadden R^2 is calculated as $R^2 = 1 - L_{UR}/L_R$, where L_{UR} is the unrestricted maximum likelihood and L_R is the restricted maximum likelihood with all slope coefficients set equal to zero. The log-likelihood ratio statistic is given by $2(L_{UR} - L_R)$ and is asymptotically distributed as an χ^2 random variable. The log-likelihood ratio test for homoskedasticity was computed by $\chi^2 = -2(LR_{HOMO} - LR_{HETO})$, where LR_{HOMO} is the maximum likelihood in the homoskedastic regression and LR_{HETO} is the maximum likelihood in a regression corrected for heteroskedasticity.

should increase. However, as noted above, the loss of mangrove area at the four case study sites has been far from marginal. The large-scale land-use changes that have occurred have already led to substantial losses to the local mangrove forests. Any large, and decidedly non-marginal, losses in the remaining mangrove areas, such as the current threat posed by conversion to commercial shrimp farms, would have devastating consequences for the livelihoods of the mangrove-dependent households. The current mangrove-based collection and fishing activities conducted by these households would be in danger of collapsing, and the amount of time that males spend in such activities would not increase but would decline drastically.

The analysis of the decision by male and female members of mangrove-dependent households to participate in replanting activities suggests that awareness of community conservation efforts and of the environmental damage imposed by shrimp farms is a powerful motivating force. The degree of dependence of the household on mangrove-based income is also an important factor. However, participation in replanting by females appears to face additional considerations, such as the distance of the household to mangroves, the number of children under 6 in the household and the size of the mangrove area.

The insights from the case study analysis of mangrove-dependent households in Thailand suggest two main policy implications. First, there is an urgent need to address the main institutional failure concerning management of local mangrove resources in coastal areas of Thailand. The present law and formal institutional structures of resource management in Thailand do not allow coastal communities to establish and enforce their local rules effectively. This has an important impact on the ability and willingness of these communities to conserve and protect their local mangrove forests. For example, in Ban Sam Chong Tai village in Phang-nga, the local community is very active in the conservation of mangroves. The community considers an area of around 60 ha as its own community forest, even though it is legally owned by the state and still faces a threat from possible conversion to shrimp farming by outside investors. In the other three surveyed villages, replanting projects were not initiated by the community but by outside NGOs or the Royal Forestry Department. These villagers are less motivated to participate in the replanting schemes and also have less say in the management of the remaining mangrove forests.

A new institutional framework for coastal mangrove management in Thailand that could make a difference to these and other coastal communities might contain the following features (Barbier and Sathirathai, 2004). First, remaining mangrove areas should be designated as conservation (i.e. preservation) and economic zones. Shrimp farming and other extractive commercial uses (e.g. wood concessions) should be restricted to the economic zones only. However, local communities that depend on the collection of forest and fishery products from mangrove forests should be allowed access to both zones, as long as such harvesting activities are conducted on a sustainable basis. Second, the establishment of community mangrove forests should also occur in both the economic and conservation zones. However, the decision to allow such local management efforts should be based on the capability of communities to effectively enforce their local rules and manage the forest sustainably. Moreover, such community rights should not involve full ownership of the forest but be in the form of user rights. Third, the community mangrove forests should be co-managed by the government and local communities. Such effective co-management will require the active participation of existing coastal community organizations, and will allow the representatives of such organizations to have the right to express opinions and make decisions regarding the management plan and regulations related to the use of mangrove resources. Finally, the government must provide technical, educational and financial support to the local community organizations participating in managing the mangrove forests. For example, if only user rights (but not full ownership rights) are granted to local communities, the latter's access to formal credit markets for initiatives such as investment in mangrove conservation and replanting may be restricted. The government may need to provide special lines of credit to support such community-based activities.

A second policy initiative would be to focus on improvements in education and skills training, especially for females. Of the surveyed households, over two-thirds of the households with female members employed in outside work are from the two villages in Nakhon Si Thammarat (see Table 10.2), where the main source of employment is at nearby factories hiring relatively unskilled and young female workers in textiles and other light manufacturing occupations. The very low average female wage rate across all households suggests that outside employment for all females involves few or no skills.

Given the current reliance of mangrove-dependent households on their female members participating in outside employment, and that this reliance will only increase as mangrove deforestation continues, improved education and skills training for young females in the households may be increasingly important for the future income-earning potential and welfare of these households.

References

Abdulai, A. and Delgado, C. (1999) Determinants of non-farm earnings of farm-based husbands and wives in northern Ghana. *American Journal of Agricultural Economics* 81, 117–130.

Aksornkoae, S., Sugunnasil, W. and Sathirathai, S. (2004) Analytical background of the case studies and research sites: ecological, historical and social perspectives. In: Barbier, E.B. and Sathirathai, S. (eds) *Shrimp Farming and Mangrove Loss in Thailand*. Edward Elgar, London, pp. 73–95.

Aksornkoae, S. and Tokrisna, R. (2004) Overview of shrimp farming and mangrove loss in Thailand. In: Barbier, E.B. and Sathirathai, S. (eds) *Shrimp Farming and Mangrove Loss in Thailand*. Edward Elgar, London, pp. 37–51.

Barbier, E.B. (2004) *Natural Capital and Labor Allocation: Mangrove-Dependent Households in Thailand*. Department of Economics and Finance, University of Wyoming, Laramie, Wyoming.

Barbier, E.B., Sathirathai, S. (eds) (2004) *Shrimp Farming and Mangrove Loss in Thailand*. Edward Elgar, London.

Barbier, E.B., Cox, M. and Sarntisart, I. (2004) Household use of mangrove and mangrove conservation decisions. In: Barbier, E.B. and Sathirathai, S. (eds) *Shrimp Farming and Mangrove Loss in Thailand*. Edward Elgar, London, pp. 115–130.

Charuppat, T. and Charuppat, J. (1997) *The Use of Landsat-5 (TM) Satellite Images for Tracing the Changes of Mangrove Forest Areas of Thailand*. Royal Forestry Department, Bangkok, Thailand.

Cropper, M., Griffiths, C. and Mani M. (1999) Roads, population pressures, and deforestation in Thailand, 1976–1989. *Land Economics* 75(1), 58–73.

Dierberg, F.E. and Kiattisimkul, W. (1996) Issues, impacts and implications of shrimp aquaculture in Thailand. *Environmental Management* 20(5), 649–666.

Feder, G. and Onchan, T. (1987) Land ownership security and farm investment in Thailand. *American Journal of Agricultural Economics* 69, 311–320.

Feder, G., Onchan, T., Chalamwong, Y. and Hongladarom, C. (1988) Land policies and farm performance in Thailand's forest reserve areas. *Economic Development and Cultural Change* 36(3), 483–501.

Feeny, D. (2002) The co-evolution of property rights regimes for man, land, and forests in Thailand, 1790–1990. In: Richards, J.F. (ed.) *Land, Property and the Environment*. Institute for Contemporary Studies Press, San Francisco, California, pp. 179–221.

Flaherty, M. and Karnjanakesorn, C. (1995) Marine shrimp aquaculture and natural resource degradation in Thailand. *Environmental Management* 19(1), 27–37.

Goss, J., Burch, D. and Rickson, R.E. (2000) Agri-food restructuring and third world transnationals: Thailand, the CP Group and the Global Shrimp Industry. *World Development* 28(3), 513–530.

Goss, J., Skladany, M. and Middendorf, G. (2001) Dialogue: shrimp aquaculture in Thailand: a response to Vandergeest, Flaherty and Miller. *Rural Sociology* 66(3), 451–460.

Hernández-Licona, G. (1997). Oferta laboral familiar y desempleo en México. *Trimestre Económico* (64), 531–568.

Kaosa-ard, M. and Pednekar, S.S. (1998) Background report for the Thai Marine Rehabilitation Plan 1997–2001. Report submitted to the Joint Research Centre of the Commission of the European Community and the Department of Fisheries, Ministry of Agriculture and Cooperatives, Thailand Development Research Institute, Bangkok, Thailand.

Lewis, R.R. III, Erftemeijer, P.L.A., Sayaka, A. and Kethkaew, P. (2000) *Mangrove Rehabilitation after Shrimp Aquaculture: a Case Study in Progress at the Don Sak National Forest Reserves, Surat Thani, Southern Thailand*. Mangrove Forest Management Unit, Surat Thani Regional Forest Office, Royal Forest Department, Surat Thani, Thailand.

Raine, R.M. (1994) Current land use and changes in land use over time in the coastal zone of Chanthaburi Province, Thailand. *Biological Conservation* 67, 201–204.

Rosenzweig, M.R. (1980) Neoclassical theory and the optimizing peasant: an econometric analysis of market family labor supply in a developing country. *Quarterly Journal of Economics* 94, 31–55.
Sathirathai, S. (1998) *Economic Valuation of Mangroves and the Roles of Local Communities in the Conservation of the Resources: Case Study of Surat Thani, South of Thailand*. Final report submitted to the Economy and Environment Program for South-east Asia (EEPSEA), Singapore.
Sathirathai, S. and Barbier, E.B. (2001) Valuing mangrove conservation, southern Thailand. *Contemporary Economic Policy* 19(2), 109–122.
Thomson, J.T., Feeny, D.H. and Oakerson, R.J. (1992) Institutional dynamics: the evolution and dissolution of common property resource management. In: Bromley, D.W. (ed.) *Making the Commons Work: Theory, Practice, and Policy*. Institute for Contemporary Studies Press, San Francisco, California, pp. 129–160.
Tokrisna, R. (1998) The use of economic analysis in support of development and investment decision in Thai aquaculture: with particular reference to marine shrimp culture. A paper submitted to the Food and Agriculture Organization of the United Nations.
Vandergeest, P., Flaherty, M. and Miller, P. (1999) A political ecology of shrimp aquaculture in Thailand. *Rural Sociology* 64(4), 573–596.

11 Mangroves, People and Cockles: Impacts of the Shrimp-Farming Industry on Mangrove Communities in Esmeraldas Province, Ecuador

P. Ocampo-Thomason
School of Geography, Politics and Sociology, University of Newcastle, Newcastle upon Tyne, United Kingdom, e-mail: p.o.thomason@ncl.ac.uk

Abstract

The Ecological Mangrove Reserve Cayapas-Mataje is located in the delta formed by the estuary of the Cayapas–Santiago–Mataje rivers in Esmeraldas Province, Ecuador, on the border with Colombia. This area harbours the most pristine mangrove ecosystem of Ecuador and is one of the last sites where traditional mangrove resource exploitation activities have not yet been displaced by other uses. Some 6000 inhabitants rely on the mangrove forest for their livelihood; however, changes brought about by new developments such as African palm culture and commercial shrimp farming are having an impact on the mangrove ecosystem. This research examined how these effects on the mangrove ecosystem are affecting local communities. Research found that fishing and cockle gathering are the most important economic activities, with 85% of the households depending on them. In contrast, the 3000 ha of shrimp farms employ only 0.6% of the locals. Construction of shrimp farms has led to the destruction of cockle-gathering grounds and damage to agricultural land. Local people responded to these changes by creating new management strategies, from the creation of mangrove defence groups to the implementation of a novel stewardship practice called 'custodias'.

Introduction

The rapid development of shrimp farming has been accompanied by increasingly controversial debates over its environmental, social and economic impacts. There is considerable uncertainty about appropriate policy and management responses, not least because of the perception that shrimp culture generates substantial benefits in coastal regions and nationally. Recently, increasing publicity locally, nationally and internationally has been given to environmental and social issues related to shrimp farming, such as sustainable development, environmental interactions and the long-term sustainability of aquaculture (Chamberlain and Rosenthal, 1995; Reinertsen and Haaland, 1995; Paez-Osuna, 2001). Since 99% of shrimp farms are located in tropical areas, the impact of the industry on developing countries has received special attention (Pullin, 1993; Bagarinao and Flores, 1994; Parks and Bonifaz, 1994; FAO/NACA, 1995; Menasveta, 1997; Nambiar and Singh, 1997; Hein, 2000; EJF, 2003, 2004; Barbier and Cox, 2004).

In Ecuador, the first commercial shrimp pond was constructed in 1969, and by 1982 Ecuador had the world's largest area under shrimp production. By 1991, 132,000 ha of coastal land had been converted to shrimp ponds (Tobey *et al.*, 1998) and, according to the Ecuadorian Forestry and Natural Areas Wild Life Institute (cited in FUNDECOL, 2000a), this had increased to 208,714 ha by 1999.

One of the most significant impacts of the industry in Ecuador has been the cutting of mangroves for the construction of ponds. Mangrove cover was 362,727 ha in 1969 (MAG, 1987) and this had dropped to 154,087 ha in 1999: a loss of 57% in just 30 years. This rapid loss has been attributed mainly to the uncontrolled expansion of shrimp aquaculture (Bodero and Robadue, 1998). Figure 11.1 illustrates the increase in shrimp-farming construction and the decrease in mangrove cover in Ecuador since 1969.

Some of the factors that made this rapid expansion possible were the incentives given by the Ecuadorian government to the shrimp farmers, plus the absence of clear property rights and effective management regimes for mangroves. The objective of this research was to examine how the degradation and loss of the mangrove ecosystem are affecting local communities. This chapter describes effects of the shrimp-farming industry in Ecuador by focusing on the Ecological Mangrove Reserve (REMACAM), the last remaining fully functional mangrove ecosystem in Ecuador (Rosero, 1999). In particular, it examines changes in the use of natural resources, the responses from the local communities and the new management strategies created to protect the remaining mangroves and the livelihoods associated with them.

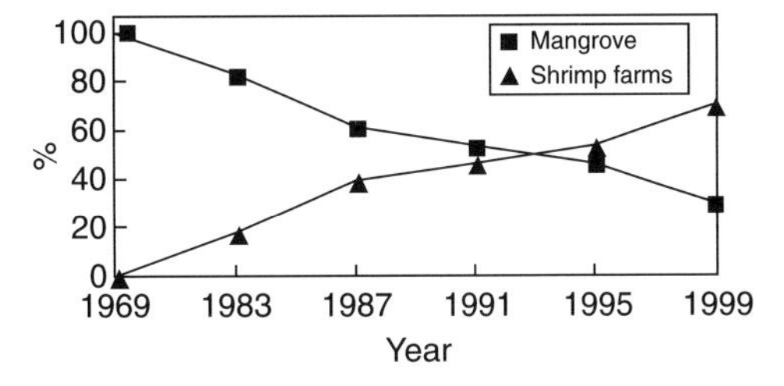

Fig. 11.1. Percentage change in mangrove cover and number of shrimp farms in the Ecuadorian coastal zone (adapted from FUNDECOL, 2000b).

Study Area

The Ecological Mangrove Reserve Cayapas-Mataje (REMACAM)

REMACAM is located in the delta formed by the estuary of the Cayapas–Santiago–Mataje rivers in Esmeraldas Province of Ecuador on the border with Colombia (Fig. 11.2). It is part of a continuous mangrove belt that commences in the central area of the Colombian Pacific coast (Cape Corrientes) and finishes in the south of Esmeraldas.

The reserve encompasses 53,200 ha, of which 32,250 ha are terrestrial habitats and 18,000 ha are mangroves. REMACAM is part of the Protected Areas National System of Ecuador and was established in January 1996 (Resolution 001 DE 052-A-DE). It is under the jurisdiction of the state, making the state the only legal owner. It is administered and protected through the Ministry of the Environment.

REMACAM is one of the last places in Ecuador where traditional mangrove resource exploitation activities have not yet been displaced by other uses. Its inhabitants, most of whom are Afro-Ecuadorians, rely directly on the mangroves and other local natural resources for their livelihood. They are grouped mainly in small communities along the rivers and on the mangrove islands. There are 31 rural communities in the reserve, with a total of about 5600 inhabitants. Geographically, the reserve includes the large urban towns of San Lorenzo and Limones (Fig. 11.2), with 13,000 and 7000 inhabitants, respectively. Administratively, they do not belong to the reserve, but in practice they use the reserve and its resources.

Although the reserve is considered the most pristine mangrove system of Ecuador, the shrimp-farming industry is already present. According to the latest survey (FEPP-Manglares, 2002), there are 45 shrimp farms in the reserve, occupying a total of 3114 ha, of which 90% are illegal. The big shrimp farms are located in the central and southern parts of REMACAM and thus they affect some communities (e.g. Tambillo and Olmedo) more than others.

Methodology

The fieldwork focused on the rural communities located in the mangrove islands where the majority (87%) of REMACAM inhabitants live. Socio-economic information was gathered using 170 socio-economic surveys (SES) and 100 semi-structured interviews (SSI). These were conducted in 12 different communities, ranging from small to large, dispersed throughout the reserve (Fig. 11.2). The SES included information on the different economic activities undertaken in the area, and the uses of mangroves and other resources associated with mangroves.

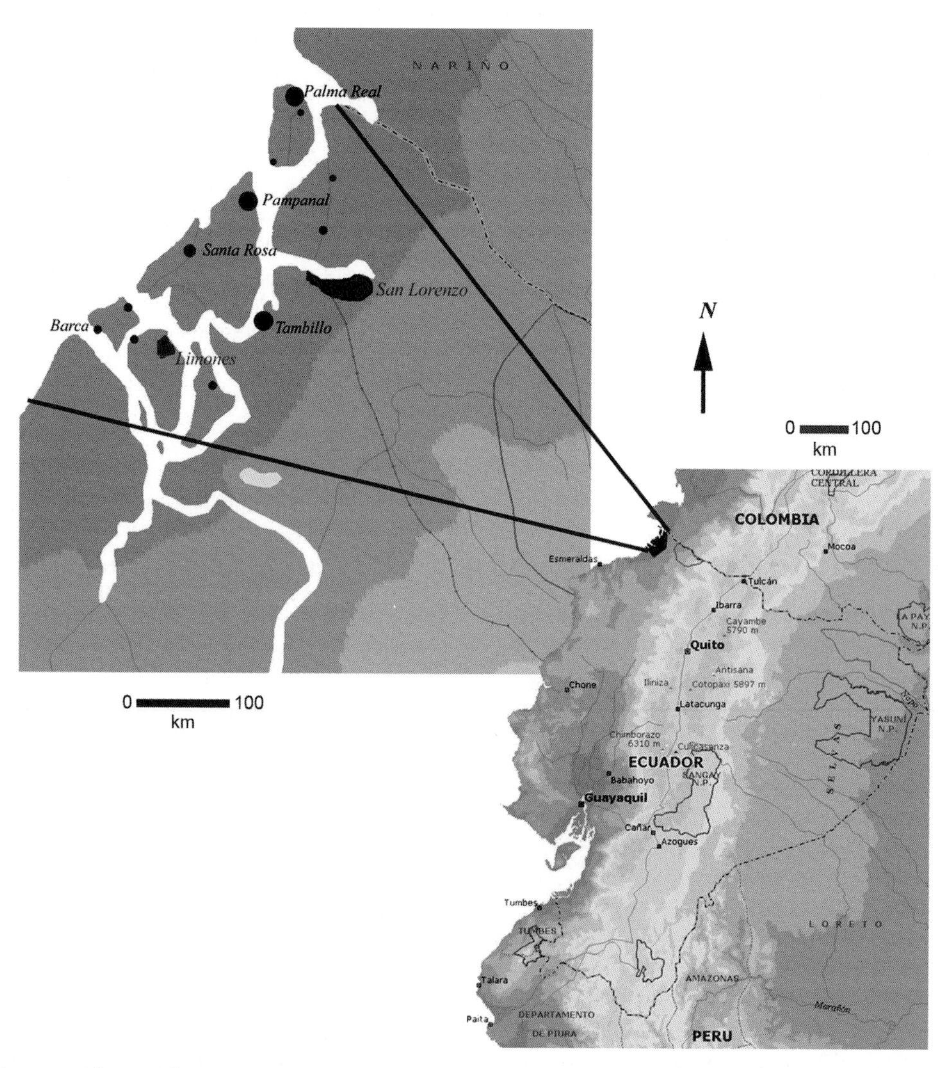

Fig. 11.2. The Ecuadorian coastal zone showing the location of REMACAM. The detailed inset shows the communities where the socio-economic survey and interviews were carried out. The size of the communities is represented by the size of the dots. Urban towns (San Lorenzo and Limones) were not part of the study and are depicted for information purposes only.

Information regarding education, social organizations and migration was also collected. The SSIs and additional informal questioning were used to understand people's attitudes and thoughts about mangrove defence, fishing and cockle gathering, and the new management strategies. To fully understand the history of the mangrove defence movements in Ecuador, the creation of the REMACAM reserve and the background to the custody process, in-depth interviews were conducted with social movement leaders working in the area, government authorities and technicians working with various NGOs.

One of the most important methodological components of the research was participant observation; the researcher lived full-time in the area as part of the Mangrove Project team (implemented by the Fondo Ecuatoriano Popularum Progressio, FEPP). Fieldwork was done between March 2002 and February 2003.

Results and Discussion

Socio-economic characteristics of the communities

The communities in REMACAM are well adapted to the mangrove ecosystem. They are always located in sheltered areas behind mangrove stands so they are protected from strong waves and winds. The houses are commonly built on stilts (Fig. 11.3). The communities are isolated from the continent and the only access to them is by sea. During high tides, the sheltered estuaries allow people to navigate to very distant places in the small open boats they call 'water horses' (*potros*).

REMACAM communities are normally small: 74% of them have from four to 28 households, only 13% of them have more than 130 households and the biggest (Pampanal) has 225 households. The average is 4.46 people per household. The houses are normally small, and 88.7% of them are built of wood, of which 55% is mangrove wood. Others are made from a mix of wood and other materials such as bricks and concrete. In almost all cases, the roof is made of corrugated iron.

Basic social infrastructure in the communities is scarce. There is a correlation between the size of the communities and the type of infrastructure: the smaller the community, the scarcer the infrastructure. For example, electricity and basic health care facilities (a small medical centre with a community nurse) are found only in the larger communities (Pampanal, Palma Real and Tambillo). The small communities have to rely on irregular health campaigns from San Lorenzo and Limones hospitals. A few communities, for example, Pichangal and El Viento, have a small solar plant that produces electricity for 4 h in the evening. Two communities, La Barca and Canchimalero, have no access to electricity. Palma Real has a basic piped water system, and a treated water system is being installed at Pampanal and Tambillo. In the smaller communities, rain is sometimes the only and, in all cases, the most important source of water. Some communities also have small wells for use during dry periods. Rainwater and well water are treated (boiled or bleached) by 23% of the households. There is no sewage system in any of the communities and household wastes are discharged directly into the estuaries. Only a very small percentage of households in the bigger communities have latrines (15.5%), and even these have direct discharges.

There is little use of mangrove wood for charcoal production, fuel-wood and house construction in REMACAM. Natural gas is the most important cooking resource, with 87% of the households having gas stoves. Only the poorer families use mangrove wood for cooking, though some households occasionally use mangrove wood for cooking when the bottled gas runs out. Nobody reported the use of mangrove charcoal and only one family had the production of charcoal as its main economic activity. These findings do not support local concerns about mangrove damage coming from excessive fuel-wood use and charcoal production by local communities.

Even though several communities (Bajito, Viento and Canchimalero) have more than 100 children, there was no school. Furthermore, in the communities that have schools, these are only very basic primary

Fig. 11.3. A typical community in REMACAM.

schools and they are understaffed. Palma Real has only one secondary school. Because of this poor educational infrastructure, parents are forced to send their youngsters to urban areas for continuing education.

Household income in REMACAM stems from multiple sources, and these activities are normally carried out by different members of the family. The husband, wife and children all normally contribute to the household income. Table 11.1 shows the occupational structure in REMACAM. Mangroves are clearly the most important source of income, with fishing and/or cockle gathering being carried out by more than 85% of the households.

The commercial dependence on the mangrove ecosystem varies among the communities (Table 11.2). Fishing and cockle gathering are the most important activities in all communities, but, in two communities (Campanita and Bajito), a high percentage of households depend on agriculture. Many households own small plantations that are normally worked during the fishing or cockle gathering off-season. In this case, agriculture is not perceived as an economic activity but as an insurance policy for bad times. In addition, many of those who make their living through agriculture or commercial activities collect cockles, fish, crabs, timber and firewood on a subsistence basis.

Two important aspects should be recognized from these data. First, multiple sources of income are a very important component of the household economy in REMACAM communities; therefore, changes in one economic activity will not only affect the total income in the household but will also affect the relative importance of the remaining income streams. Second, although communities appear homogeneous, this is not the case, and awareness of this heterogeneity will help to target the type and content of projects implemented in REMACAM.

Use and perceptions of the mangrove ecosystem

Traditionally, mangrove wood has been a direct source for building, firewood and

Table 11.1. Occupational structure in REMACAM. 'Commerce' refers to small corner shops and 'external help' refers to households that receive money from relatives in other cities or countries.

Main economic activity/income source	Percentage of households
Fishing and cockle gathering	67.7
Cockle gathering (only)	10.0
Fishing (only)	8.3
Agriculture (only)	6.5
Commerce	2.4
External help	1.2
Shrimp farming (only)	0.6
Other occupations (not related to mangroves)	2.9
Other occupations (related to mangroves)	0.6

Table 11.2. Economic profile of each of the communities studied.

Community	No. of households	Fishing (%)	Cockle gathering (%)	Agriculture (%)	Shrimp farm (%)
Gauchal	9	77.8	66.7	11.0	0
Campanita	11	72.7	45.5	63.6	0
Pichangal	11	81.9	81.9	0	0
Bajito	12	30.0	30.0	84.6	0
Viento	14	56.0	87.0	6.0	0
Barca	18	93.7	50.0	12.0	0
San Antonio	26	40.0	70.0	60.0	0
Canchimalero	28	85.7	64.0	21.0	0
Santa Rosa	65	60.0	73.0	30.0	0
Tambillo	130	26.0	63.0	7.0	3.9
Palma Real	184	36.0	40.0	4.0	0
Pampanal	225	76.9	40.0	3.0	0

charcoal production. During the 1950s, the area now covered by REMACAM was the centre of a large-scale exploitation of mangrove wood, which was used by the construction industry, and the bark was used for tannin production. Several logging companies were dedicated exclusively to this exploitation (Labastida, 1995). After 1968, the use of bark stopped because of the collapse of world tannin markets (Snedaker, 1986). The centre of wood exploitation also shifted during the 1960s with the introduction of chainsaws and opening of new roads into the rain and cloud forests.

Charcoal extraction continued until recently, but has almost disappeared in the last 5 years because of the introduction of strict regulations when the area was declared an ecological reserve. Furthermore, the introduction of bottled gas has reduced reliance upon mangrove wood as the primary fuel source to a minimum. Mangrove wood is still used at a subsistence level, especially for building houses. The subsistence use of fuelwood and some medicinal uses were also identified, but one of the most interesting findings is the fact that local communities have a more holistic perception of the mangrove ecosystem. They identified the importance of mangrove more for its life-supporting functions than for its direct uses (Table 11.3).

Local people recognize fully the important role played by mangroves in their local economies. As they see it, mangroves are the source of all life and the most important source of work and protein. When questioned about the importance of mangroves

Table 11.3. Ecological functions of mangroves as identified by respondents in the semi-structured interviews undertaken in REMACAM.

Function	Example given
Nursery	Juvenile fish and shrimp in inlets and mangrove roots
Feeding grounds	Fish and shrimp feed during high waters
Habitat	Cockles, crabs and other molluscs live in the roots
Beach protection	Mangrove protects from erosion
Oxygen producer	Mangrove trees produce and recycle oxygen

for their well-being, community members stressed the importance of the mangrove ecosystem for their livelihood. Common statements during interviews and informal questioning were as follows: 'mangrove is our life, it gives us cockles, crabs, fish'; 'mangrove is the most important thing for us'; 'if we don't have mangrove we will not eat, we will not live'; 'mangrove cockles are the life of the poor'; and 'if mangroves disappear, we will all be finished, mangrove is our life, our source of work'. They believe that the disappearance of the mangroves will inevitably lead to the disappearance of their communities. It is likely that some of the identified environmental functions are borrowed from the environmentalist and development narratives that they have acquired during their struggle against the shrimp-farming industry.

In contrast, the shrimp-farming industry is perceived as producing very few jobs in the area. There is only one household in which shrimp farming is the sole source of income, and only two households obtain work on a temporary basis from the shrimp industry (see Tables 11.1 and 11.2). One of the arguments for the development of the shrimp-farming industry is the creation of local employment, but the perception of the local communities is that the employment provided by the industry is minimal and does not compensate for the loss of livelihoods when mangroves are replaced by ponds. Tobey *et al.* (1998) reported that the range of employment provided by the shrimp industry in Latin America is generally from 0.1 to 1.0 person per ha. Activists contesting the shrimp-farming industry argue that 1 ha of shrimp farming provides only 0.1 job, whereas 1 ha of mangrove produces enough resources for at least ten families (FUNDECOL, 2000b; Greenpeace and Trópico-Verde, 2002). Isherwood (2000) estimated that goods amounting to a total of over US$14,000,000 are extracted annually from the reserve. Non-marketed goods accounted for 14% of this value.

Goods and services from the reserve benefit not only the inhabitants of the reserve but also the fishers, gatherers and businesses based in San Lorenzo and Limones. Most of the marketing of fish and cockles is done through San Lorenzo, from where the products from the mangroves are sent to the rest of the country. San Lorenzo and Limones are also where the mangrove communities obtain external products such as ice, fishing nets, petrol and fresh and preserved foods. Thus, even though San Lorenzo and Limones do not belong administratively to the reserve, they benefit directly from it.

Natural resource use and allocation

In REMACAM, the use and allocation of natural resources are divided along gender and age lines; men fish and women and children gather cockles. When male children reach age 14 to 15, they switch to fishing. When health deteriorates, fishermen (around 60 years of age) take to cockle and mollusc gathering as their main activity.

Small agricultural plots (farms) and coconut plantations are the realm of men, with women helping in their maintenance. This research found only one woman with agriculture as her main economic activity. This division of labour is not so evident in the marketing of the products: some middlewomen buy cockles and sometimes fish.

They normally live in the communities and sell directly to the bigger towns or to external middle-people that visit the communities every week. Because of the lack of infrastructure and electricity in the mangrove communities, fish are normally sold every day to external middle-people from other communities. External middle-people are normally men and, even though some of them live in San Lorenzo or Limones, they are normally *mestizos* (mixed-race people), originally from inner cities.

Artisanal fishery

In economic terms, artisanal fishery is the most important source of income in the Reserve, employing about 2500 fishers (FEPP-Manglares and INP, 2002) and accounting for 77% of the income generated by natural resource extraction (Isherwood, 2000). The fishery is also a vital source of protein for the communities in REMACAM, with smaller fish from the catch generally kept for home consumption.

Fishing is exclusively a male activity. Of the 75 households that reported fishing as their main activity, all were men. Only a few women were found that help their husbands put the nets together when the nets are new. On average, fishermen work 6.3 h per day in a single daily trip, 6 days a week excepting fiestas and bad weather.

A wide variety of fishing equipment is employed in REMACAM. Fishers in the smaller and more remote communities tend to use low-investment equipment and non-motorized canoes, whereas a number of fishers in larger communities own big canoes with outboard motors (Table 11.4). Motorized canoes have appeared in the reserve only in recent years, when petrol for outboard motors was heavily subsidized by the government. Today, twice as many fishers are working from motorized boats as there were in 1995 (Rosales, 1995). Until the arrival of outboard motors, fishing activity was mainly restricted to the estuaries and creeks. Today, fishers can work the coastal waters up to several kilometres out to sea (Table 11.4).

The more economically valuable species are shrimp and prawns. In 2002, shrimp averaged $2 per kg and prawns $9 per kg (FEPP/INP, 2002). The small trawl fishery is the most lucrative, but also the one that needs the most capital investment. The most common fishing gear is the seine and beach gill nets, which can be used with either motorized or paddle canoes, and 73% of the fishermen own their own fishing gear. Small trawls, however, are owned by more wealthy individuals and often a single person owns several that are rented to other fishermen. Men using big and medium-size canoes fish in improvised groups that are formed only for a specific trip. The income from the fishing trip is divided in two (after the expenses for petrol, oil and others are deducted): 50% is for the owner of the boat and the fishing

Table 11.4. Principal fishing equipment used in REMACAM, showing the local name and target species (from FEPP-Manglares and INP, 2002 and fieldwork observations, 2002).

Type of gear	Local name	Type of boat/propulsion	Target species	Fishing area
Small trawl	*Changa*	Big/big motor	Shrimp	Coastal
Seine	*Red de enmalle de fondo*	Big/small motor	Prawn, finfish	Coastal and estuarine
Beach gill net	*Chinchorro de playa*	Small/paddle	Prawn, finfish	Beach and creek mouth
Cast net	*Atarraya*	Small/paddle	Prawn, finfish	Estuarine
Tangle net	*Red de estacada*	Small/paddle	Finfish	Creek mouth
Long line	*Espinel de fondo*	Small/paddle	Finfish	Creek mouth
Shrimp larvae net	*Red larvera*	None	Shrimp postlarvae	Creek and beach
Traps	*Trampas*	Small/paddle	Crabs	Estuarine

gear, the other half for the crew. Fishermen in the small canoes fish on their own in the small estuaries and mangrove creeks.

Fishermen from communities in the south of the reserve (where the number of shrimp farms is higher) have changed their traditional fishing methods and now fish for shrimp postlarvae and pregnant female shrimp to sell to hatcheries and shrimp farms. They have developed several types of improvised gear using monofilament nets with very small mesh size. These are used in the intertidal areas around the beaches and the mangrove creeks. Because there is no need for a boat and the equipment is made at home, the investment for this type of activity is minimal and attracts people from other communities.

The perception among the communities and researchers in the area is that this new activity is causing serious damage to the wild shrimp population and other commercial fishing species. The nets are non-discriminatory and as a result there is a very high by-catch of commercial and non-commercial shrimp and fish larvae, which are normally left to die on the beach. The impact of this new activity has not been assessed, so there is no regulation of the size of the nets used, the number of fishermen or the area they can fish. In communities such as Olmedo, more than 150 of these new types of gear can be found.

Cockle gathering

Mangrove cockles (also called arc-cockles or arc-shells) are harvested commercially and as subsistence food by a large number of people along the Pacific coast from Mexico to Peru. Three cockle species (*Anadara tuberculosa, A. similis* and *A. grandis*) are gathered in REMACAM. *A. tuberculosa* is the most abundant species in the muddy areas around the mangrove roots. Because of its high abundance and its fortitude (it can stay alive up to 8 days after harvest), it is the most commercialized species in the area.

According to MacKenzie (2001), there are at least 15,000 cockle gatherers on the Mexico–Peru seaboard, and Ecuador supports at least 31% of them. According to data from secondary sources and data collected in REMACAM, it is estimated that 79% of Ecuadorian cockle gatherers can be found in REMACAM, even though it protects only 11% of the mangroves in the country.

Customarily, each community in REMACAM has specific gathering grounds. These have been respected by the other communities. Traditionally, cockle gatherers used small wooden canoes with paddles, which limited foraging distances. Furthermore, women prefer gathering grounds close to their communities. Local gatherers used to rotate the grounds, leaving some areas alone for a couple of weeks so that the cockles could recover. Gatherers traditionally left the 'mother' (brooding stock) alone so it could reproduce and also left small shells (less than 5 cm) to grow.

Cockle gathering has traditionally been a female activity. One reason for this is because cockles are picked singly by hand so there is no need for any capital investment. Also, the gathering areas are close to the communities, allowing the women to take their children, making cockle gathering easily combined with housework and other chores. According to the SES, cockle gathering is the most popular female activity in the area. In the 99 households that reported cockle gathering as a main economic activity, 82.4% of the gatherers are female. In addition to the female gatherers, another 2.4% of the women buy cockles to sell to external middle-people. One of the most important findings is that 20% of the female-headed households depend almost exclusively on cockle gathering and, for 10% of the mixed-head households, both the husband and wife gather cockles as the main and, in some cases, the only economic activity. Cockles are gathered for subsistence by 16% of the households in REMACAM. During cockle gathering, other mollusc species are harvested, but, these do not have a market value and are used to provide protein for the household.

Cockles are sold by 'cientos' (units of 100 shells) and, with *A. tuberculosa* fetching up to $3.50 per ciento, cockle gathering can provide substantial household income (Table 11.5). Because of its high value, at least 98% of *A. tuberculosa* is sold, whereas at least 55% of *A. similis* is kept for home consumption.

Changes in resource allocation and practices

Interestingly, 17.6% of the cockle gatherers are now male. According to the interviews, men have taken to cockle gathering because they do not have any other source of income. Men have been displaced from agriculture after selling their land to the shrimp-farming industry. Some of the farms have also been lost because of flooding caused by changes in the hydrodynamics when the shrimp ponds were built. In other cases, it is impossible for local people to travel to their agricultural plots as some of the creeks and estuaries are now blocked by armed guards. Some men who used to work for logging companies that have closed have switched to cockle gathering. Other male gatherers are ex-fishermen who do not fish any longer because of the decrease in fish stocks and the increasing cost of fishing gear.

In some cockle grounds, local gatherers are being displaced by big groups of gatherers from San Lorenzo. These are usually young men, travelling in large fibreglass boats with powerful outboard motors. These are improvised groups: the owner of the boat takes them to the cockle grounds and then buys the cockles they harvest at a much lower price than the market value. The gatherers in these groups do not need to make any investment and after the gathering have no obligation to the boat owner. These young men have been displaced from their traditional logging jobs, as the logging companies are going out of business due to the rapid development of African palm plantations. These groups are attracted to the cockle fisheries because there is no need to invest in any equipment, and because the mangroves are open to all. The cockle-gathering grounds are perceived by these itinerants as a free common resource.

With the use of outboard motors, it is easy for the new itinerant gatherers to go anywhere in the reserve and, because they do not belong to any of the traditional communities, there is no community pressure upon them not to use any gathering grounds left fallow. They argue that mangrove areas belong to all Ecuadorians and, because of that, they are allowed to gather anywhere they wish. These itinerant gatherers use machetes to cut the mangrove roots, making it easier to gather cockles, but thereby destroying the gathering area. According to local people, and direct observations, it takes more than 2 months for the grounds to recover after itinerant gatherers have damaged an area.

These new gatherers are going to have a large long-term impact on the cockle fishery as their practices do not leave brood stock or juveniles. They take everything they find and they often tell the traditional gatherers that they do not know how to gather as they are always leaving cockles behind. Thus, natural replenishment of stocks will take longer. The other problem is that areas that were inaccessible before are now being accessed and gathered. As these areas act as reserve grounds for the cockle populations, their disturbance may lead to a complete failure of the fishery.

The perception that cockle resources are decreasing is corroborated by a monitoring programme started in 2001 by the Mangrove Project and the local communities. A preliminary analysis shows that both the size and abundance of the cockles are diminishing. As

Table 11.5. Income generated from mangrove cockles in REMACAM in 2002 (from preliminary FEPP/INP report, 2002 and fieldwork data).

Species	Average collected per week	Average sold per week	Average consumption per week	Price per ciento (US$)	Weekly income (US$)
Anadara tuberculosa	928	909 (98%)	19 (2%)	3.00–3.50	27.30–32.00
A. similis	307	137 (45%)	170 (55%)	2.00–2.50	2.80–3.40

a consequence, traditional gatherers are having to travel farther to reach the gathering areas and some of the traditional grounds are no longer productive (FEPP/INP, 2002).

These problems are exacerbated by the destruction of gathering grounds by the shrimp-farming industry. The impact of the shrimp farms can be easily observed when travelling in some areas of the reserve, as there are 'no-entry' signs in several of the estuaries and some smaller creeks have been sealed off. Limiting the access to gatherers results in a concentration of gatherers in certain areas; this increases the pressure on the cockle resources and generates conflicts between gatherers from different communities.

Another impact of the shrimp farms is on the use of water. Pond effluent is discharged directly into adjacent estuaries. Local dwellers have reported several cases of massive fish and crab mortality – all, according to them, related to these discharges. No research has been undertaken to verify these claims but research in other countries suggests that pond effluent can produce serious degradation of water quality (Paez-Osuna, 2001).

Community responses

The impact of shrimp farming in Ecuador has led to the creation of several political, ecological and social organizations. Among these is a grass-roots resistance movement in Esmeraldas Province: the Fundación para la Defensa Ecológica (FUNDECOL), which was founded in 1989. At the same time, in the north of the province, the Black Communities Process (BCP) has been fighting to make the north of Esmeraldas an independent territory in the same manner as indigenous territories and districts have become autonomous under national decentralization programmes. This is relevant to REMACAM as the communities are primarily composed of Afro-Ecuadorians who hope to gain collective management rights over the mangroves and their natural resources.

These two movements came together in the north of Esmeraldas to defend the mangrove areas from the shrimp-farming industry. After 5 years of struggle, local and national protest mobilizations, and radio and television campaigns, the Ecuadorian government granted the status of an ecological reserve to REMACAM. As part of REMACAM's creation, the Mangrove Project was approved. The project was conceived by community members, fishermen and women cockle gatherers' associations and is carried out by FEPP, a national NGO with financial support from the Dutch government. The main objective of the project is to preserve the mangroves as a source for people's livelihoods and to look for conflict resolution.

Understanding that one of the major problems was the ill-defined property rights and open access to all of the mangroves, the national coordination for the mangrove defence (C-CONDEM), together with FUNDECOL, the Mangrove Project, and supported by the local communities, devised a novel stewardship practice called '*custodias*'. Under this practice, mangrove areas are allocated to each community for their traditional use and management. Economic practices such as charcoal production or logging are forbidden. Gathering practices are permitted but they have to be carried out by only local gatherers. The custodial permit is given by the Minister of Environment and has a duration of 10 years. After this period, the *custodia* will be inspected by the Ministry, and an extension of 90 years will be granted if the community has appropriately looked after it.

The requisites for getting the *custodias* are very strict. The community needs to have a legally recognized association, which has to present a management plan and a geo-reference map of the *custodia*. Areas eligible for custody are those that have customarily belonged to the community and have been traditionally used for wood or cockle collection. Importantly, traditional fishing grounds are not part of the *custodias*, as the sea and estuarine waters are not part of REMACAM. The Mangrove Project has provided the expertise and money to make the maps and the management plan required by the government. Overall, there has been great success in obtaining custodial areas (Table 11.6).

The success of this stewardship programme can be seen in the slowdown of

Table 11.6. Custodial areas obtained by REMACAM communities up to 2002.

Community name	Size (ha)	Date
Campanita	522.0	18 Dec. 2002
Canchimalero	362.0	18 Dec. 2002
El Bajito	877.0	18 Dec. 2002
El Viento	1207.0	14 Apr. 2000
Guachal	1022.9	18 Dec. 2002
La Barca	785.0	14 Apr. 2000
Olmedo	385.2	7 Nov. 2001
Palma Real	1057.0	8 Aug. 2000
Pampanal	2953.0	14 Dec. 2002
San Antonio	195.7	8 Aug. 2000
Santa Rosa	1114.4	14 Apr. 2000
Tambillo	2576.6	14 Apr. 2000
Total	13,057.8	–

mangrove destruction. It has been calculated that 98% of illegal mangrove removal has been stopped (E. Lemos, Limones, personal communication, 2002). Another important aspect is the strengthening of the local mangrove defence groups and the creation of organized groups in other communities, some of which are now formally requesting their own custodial areas. It is important to point out that this programme is part of a national process to defend the mangrove ecosystem and the traditional livelihoods associated with it, so its success will have national repercussions. For example, in the south around Muisne, FUNDECOL and the National Mangrove Network (*Red Manglar*) have obtained government agreement to administer and rehabilitate 3200 ha of mangrove. This area has also been declared a Protected Area (FUNDECOL, 2003).

The deterioration of the cockle fishery in REMACAM has led to talk of a self-imposed restriction by the local communities. Under this scheme, cockles smaller than 45 mm will be left in the gathering grounds. There are now talks with the Ministry of Environment to implement this scheme, which will be coordinated through the Local Mangrove Committee. This committee was created in 2001 and has representatives from each community group, the Ministry of Environment, the Mangrove Project and the Navy. Under Ecuadorian decentralization laws, this committee acts as the administration authority and is able to engage with municipalities and county authorities in all aspects related to the mangrove ecosystem. One of the most important functions of the committee is the prompt identification and halting of illegal activities such as mangrove clearing and the solution of conflict generated between traditional gatherers and gatherers from San Lorenzo.

The *custodias* used in REMACAM resemble the extractive reserves used since 1990 to co-manage natural resources in Brazil (Glaser and Oliveira, 2004). Both processes demonstrate how integrating local users into the management of their own resources enables the implementation of a better and more viable mechanism for mangrove protection and their sustainable use. Success in REMACAM can be considered part of the growing empirical evidence that local communities are more likely than the state to manage natural resources in a responsible way because their livelihoods depend on this (Hesse and Trench, 2000), and that common property systems can actually work (Berkes, 1989; Ostrom, 1990; Ostrom *et al.*, 2002; Dolšak and Ostrom, 2003).

Conclusions

This study has shown the great importance of the mangrove ecosystem to people living in mangrove areas in Esmeraldas. It has shown that the mangroves are particularly important to the women living there. It has also exposed clearly how shrimp farming is having a negative impact on the livelihoods of the people living there. This impact is exacerbated by displaced people becoming itinerant cockle gatherers. These impacts have led to the creation of several political, ecological and social organizations. Among them are grass-roots resistance movements which are pressing the Ecuadorian government to adopt strategies to defend the mangrove ecosystem and the livelihoods associated with it. The creation of REMACAM has been one of the most decisive steps in this struggle. The stewardship practice based on *custodias* that has been implemented in REMACAM is now being replicated in other regions.

Acknowledgements

This research project was financed by a grant from the Economic and Social Research Council (ESRC), UK. I would like to thank FEPP, FUNDECOL, C-CONDEM and the Mangrove Network; local federations FEDARPOM and FEDARPROBIM; the REMACAM communities; and Dr Jeremy C. Thomason for their help.

References

Bagarinao, T.U. and Flores, E.E.C. (1994) Towards sustainable aquaculture in South-east Asia and Japan. In: *Seminar-Workshop on Aquaculture Development in South-east Asia,* 26–28 July 1994, Iloilo, Philippines, p. 254.

Barbier, E.B. and Cox, M. (2004) An economic analysis of shrimp farm expansion and mangrove conversion in Thailand. *Land Economics* 80, 389–407.

Berkes, F. (1989) *Common Property Resources: Ecology and Community-Based Sustainable Development.* Belhaven Press, London.

Bodero, A. and Robadue, D. (1998) Ecuador working toward a national strategy for mangrove management. *Intercoastal Network,* Mangrove Special edition, pp. 27–30.

Chamberlain, G. and Rosenthal, H. (1995) Aquaculture in the next century. Opportunities for growth: challenges of sustainability. *World Aquaculture* 26, 21–25.

Dolšak, N. and Ostrom, E. (2003) The challenges of the commons. In: Dolšak, N. and Ostrom, E. (eds) *The Commons in the New Millennium: Challenges and Adaptation.* The MIT Press, London, pp. 3–34.

EJF (2003) *Smash & Grab: Conflict, Corruption and Human Rights Abuses in the Shrimp Farming Industry.* Environmental Justice Foundation, London, 77 pp.

EJF (2004) *Farming the Sea, Costing the Earth: Why We Must Green the Blue Revolution.* Environmental Justice Foundation, London, 80 pp.

FAO/NACA (1995) *Regional Study and Workshop on the Environmental Assessment and Management of Aquaculture Development.* FAO and Network of Aquaculture Centres in Asia-Pacific, NACA, Bangkok, 492 pp.

FEPP/INP (2002) *Estudio de la Pesquería Artesanal del Recurso* Andara tuberculosa *y* A. similis *en Tres Comunidades de la Reserva Ecológica Manglares Cayapas-Mataje (REMACAM).* FEPP and INP, Esmeraldas, Ecuador, 54 pp.

FEPP-Manglares (2002) *Censo Camaronero de la Reserva de Manglares Cayapas-Mataje.* FEPP, Esmeraldas, Ecuador, 60 pp.

FEPP-Manglares and INP (2002) *Diagnóstico de la Actividad Pesquera Artesanal de las Comunidades: Limones, Pampanal de Bolívar, Olmedo.* ImpreFepp, Quito, Ecuador, 116 pp.

FUNDECOL (2000a) El camarón se comió al manglar. In: *La Bocina,* Vol. 110, 3 pp.

FUNDECOL (2000b) Usuarios vs. camaroneras. In: *FUNDECOL Diez Años por el Manglar,* Ecuador, pp. 3–4.

FUNDECOL (2003) Los manglares de Muisne, parte del sistema nacional de áreas protegidas. *Martín Pescador* 2, 4–6.

Glaser, M. and Oliveira, R. (2004) Prospects for the co-management of mangrove ecosystems on the North Brazilian coast: whose rights, whose duties, and whose priorities? *Natural Resources Forum* 28, 224–233.

Greenpeace and Trópico-Verde (2002) *El Manglar, un Ecosistema Único.* Trópico Verde, Quito, Ecuador, 8 pp.

Hein, L. (2000) Impact of shrimp farming on mangroves along India's east coast. *Unasylva* 203(4), na.

Hesse, C. and Trench, P. (2000) *Who's Managing the Commons? Inclusive Management for a Sustainable Future.* Russell Press, Nottingham, UK.

Isherwood, I. (2000) Changing rural livelihoods and mangrove management in northwest Ecuador. In: *Center for Tropical Coastal Management Studies,* University of Newcastle upon Tyne, Newcastle, UK, 32 pp.

Labastida, E. (1995) Diagnóstico económico de las actividades relacionadas con la zona de manglar con énfasis en las unidades de producción camaronera. In: Inefan, E. (ed.) *Estudio de las Alternativas de Manejo del Área Comprendida entre los Ríos Cayapas y Mataje, Provincia de Esmeraldas.* EcoCiencia &

INEFAN, Quito, Ecuador. Appendix. 9 pp.
MacKenzie, C.L., Jr. (2001) The fisheries for mangrove cockles, *Anadara* spp., from Mexico to Peru, with descriptions of their habitat and biology, the fishermen's lives, and the effects of shrimp farming. *Marine Fisheries Review* 63, 1–39.
MAG (1987) Acuerdo Ministerial 238. Ministry of Agriculture and Livestock, Quito, Ecuador, p. 5.
Menasveta, P. (1997) Mangrove destruction and shrimp culture systems. *World Aquaculture* 28, 36–42.
Nambiar, K.P.P. and Singh, T. (1997) Sustainable aquaculture. In: INFOFISH (ed.) *INFOFISH-AQUATECH '96 International Conference on Aquaculture.* INFOFISH, Kuala Lumpur, Malaysia, 248 pp.
Ostrom, E. (1990) *Governing the Commons: the Evolution of Institutions for Collective Action.* Press Syndicate of the University of Cambridge, Cambridge, UK, 280 pp.
Ostrom, E., Dietz, T., Dolšak, N., Stern, P.C., Stovich, S. and Weber, E.U.E. (2002) *The Drama of the Commons. Committee on the Human Dimensions of Global Change.* National Research Council, Division of Behavioral and Social Sciences and Education, National Academy Press, Washington, DC, 521 pp.
Paez-Osuna, F. (2001) The environmental impact of shrimp aquaculture: causes, effects, and mitigating alternatives. *Environmental Management* 28, 131–140.
Parks, P.J. and Bonifaz, M. (1994) Nonsustainable use of renewable resources: mangrove deforestation and mariculture in Ecuador. *Marine Resource Economics* 9, 1–18.
Pullin, R.S.V. (1993) An overview of environmental issues in developing country aquaculture. In: Pullin, R.S.V., Rosenthal, H. and Maclean, J.L. (eds) *Environment and Aquaculture in Developing Countries.* ICLARM, Manila, Philippines, pp. 1–19.
Reinertsen, H. and Haaland, H. (1995) Sustainable fish farming. In: Reinertsen, H. and Haaland, H. (eds) *First International Symposium on Sustainable Fish Farming,* 28–31 August 1994, Oslo, 307 pp.
Rosales, M. (1995) La economía ecológica y las formas de propiedad del manglar de la zona norte de Esmeraldas. MSc thesis. Latin American Faculty for Social Sciences, FLASCO, Quito, Ecuador, 120 pp.
Rosero, J.A. (1999) *Valoración de la Reserva Ecológica Manglares Cayapas-Mataje.* INEFAN/GEF, Quito, Ecuador, 105 pp.
Snedaker, S.C. (1986) Traditional uses of South American mangrove resources and the socio-economic effect of ecosystem changes. In: Kunstadter, P., Bird, E.C.F. and Sabhasri, S. (eds) *Workshop on Man in the Mangroves.* United Nations University, Tokyo, pp. 104–112.
Tobey, J., Clay, J. and Vergne, P. (1998) *Maintaining a Balance: the Economic, Environmental, and Social Imapcts of Shrimp Farming in Latin America.* Coastal Resource Center, University of Rhode Island, Kingston, Rhode Island, 60 pp.

12 Interrelations among Mangroves, the Local Economy and Social Sustainability: a Review from a Case Study in North Brazil

U. Saint-Paul
Centre for Tropical Marine Ecology (ZMT), Bremen, Germany,
e-mail: ulrich.saint-paul@zmt.uni-bremen.de

Abstract

The littoral region of coastal Pará in northeastern Brazil is part of the world's second-largest continuous mangrove region. The Bragança peninsula is the specific study area of the currently ongoing joint German/Brazilian interdisciplinary project 'Mangrove Dynamics and Management', which began in 1995. Human use in this mangrove ecosystem is characterized by about 15 important natural resources, which have either subsistence value or otherwise generate monetary income for the local rural population. The significance of these functions for the rural households increases with distance from the urban centre. In the primary production sector, agriculture and artisanal fisheries are the main source of income in the wider Bragantinian region. Both sectors are characterized by many small operators. The industrial sector is very under-represented throughout the region. The control of the allocation of resources within this region currently rests predominantly in the hands of local individuals. This chapter examines the conditions for the successful co-management of diverse species, resource-use patterns and household income portfolios in a mangrove environment. Therefore, stakeholders have been incorporated directly, for example, by participation in workshops. This is a feature introduced by RESEX (*reservas extrativistas*), a Brazilian model of co-management for natural resources.

Introduction

Only minimal research had been carried out in north Brazil until the joint German/Brazilian project on 'Mangrove Dynamics and Management' (MADAM) began in 1995 (Berger *et al.*, 1999). Moreover, although – judging by the sheer number of botanical, zoological and ecological studies – mangroves must be among the most intensely studied tropical ecosystems (Twilley, 1996), progress has been insufficient in the integration of the various studies that would allow for a better understanding of any mangrove system as a whole and of its key processes. As a result, application of the research results obtained by other projects to a new study area is problematic. However, such an integration is clearly imperative for an ecosystem research approach, which intends on the one hand to create the capacity to assess the implications of human resource-use dynamics, or of changes in hydrographic, geomorphological or climatic conditions for the ecosystem, and on the other hand to explore the possible effects of natural or regulative changes on the relevant socio-economic structures. It is also assumed that, if recommendations for ecosystem management are to be elaborated, the socio-economic value of the ecosystem needs to be determined as a guide to decision-making.

The possible scenarios for future mangrove use, as derived from various management approaches and variables such as population growth and employment trends, must be linked in such a way that the scope for activities and decision-making pertaining to the ecosystem becomes evident. To do this, it is necessary in the first place to achieve a minimum interdisciplinary consensus on the precise local meaning of key management goals, such as sustainability. In the identification of management problems and solutions, the involvement of system users and other key stakeholders is considered essential (see Özhan, 1998).

About 15–20 different natural resources that have either subsistence value or generate monetary income for the rural population are derived from mangroves (Fig. 12.1). The first socio-economic surveys showed that approximately 80% of the households live from the diverse products of the mangrove estuary, whereas approximately 68% derive income from the mangrove ecosystem (Glaser, 2003). The economically most important mangrove productive resource is the large, semiterrestrial ocypodid crab (*Ucides cordatus*). Fish, shrimp and other invertebrates, as well as mangrove timber, are also used, the latter predominantly to fire brickwork kilns. Although the ecology of the Caeté ecosystem is considered to be relatively undisturbed by human activities, there are visible trends of expanding tourism, intensification of the fishery industry and urban growth in this area, which further point to the necessity of studying the interrelationship between mangroves and the local populace.

This chapter provides an overview of the resource-related research of MADAM and how management implications are defined under stakeholder participation.

Research Area

The mangrove estuary of the Caeté River is located 150 km southeast of the Amazon delta in northern Brazil. It forms part of the world's second-largest continuous mangrove region, estimated to cover a total area of 1.38 million ha along a coastline of about 6800 km (Kjerfve and Lacerda, 1993). The study area includes the mangrove-covered peninsula on the northwest side of the estuary (Fig. 12.2). It consists of 180 km^2 of mangrove forests and the adjacent 130 km^2 of rural area. The human population consists of the urban area of the city of Bragança, with about 48,000 people, and a rural area of 21 villages with a combined population of about 13,000 people who live and derive their livelihood mainly from this peninsula. Krause *et al.* (2001) give a general and detailed geographic characterization of the region.

Mangrove Products

Multiple income sources at the household level are common in rural areas of Amazonia, especially where occupational specialization at the household level is less viable because of low population densities and limited market size. Dependence on the mangrove by the rural population is also very diverse as specialization in one single target commonly does not meet the subsistence demands of a family (Tables 12.1 and 12.2). Therefore, in the following, the different mangrove products are listed separately.

Crab

The most heavily exploited resource in Brazilian mangroves is the leaf litter-consuming semiterrestrial crab *U. cordatus*. More than 60% of rural subsistence fisher households and over half of the rural commercial fisher households collect crabs for sale (Diele, 2000; Glaser and Diele, 2004).

Ucides cordatus is a relatively large and slow-growing crab living in burrows. In the Caeté estuary, it reaches a size of up to 9 cm

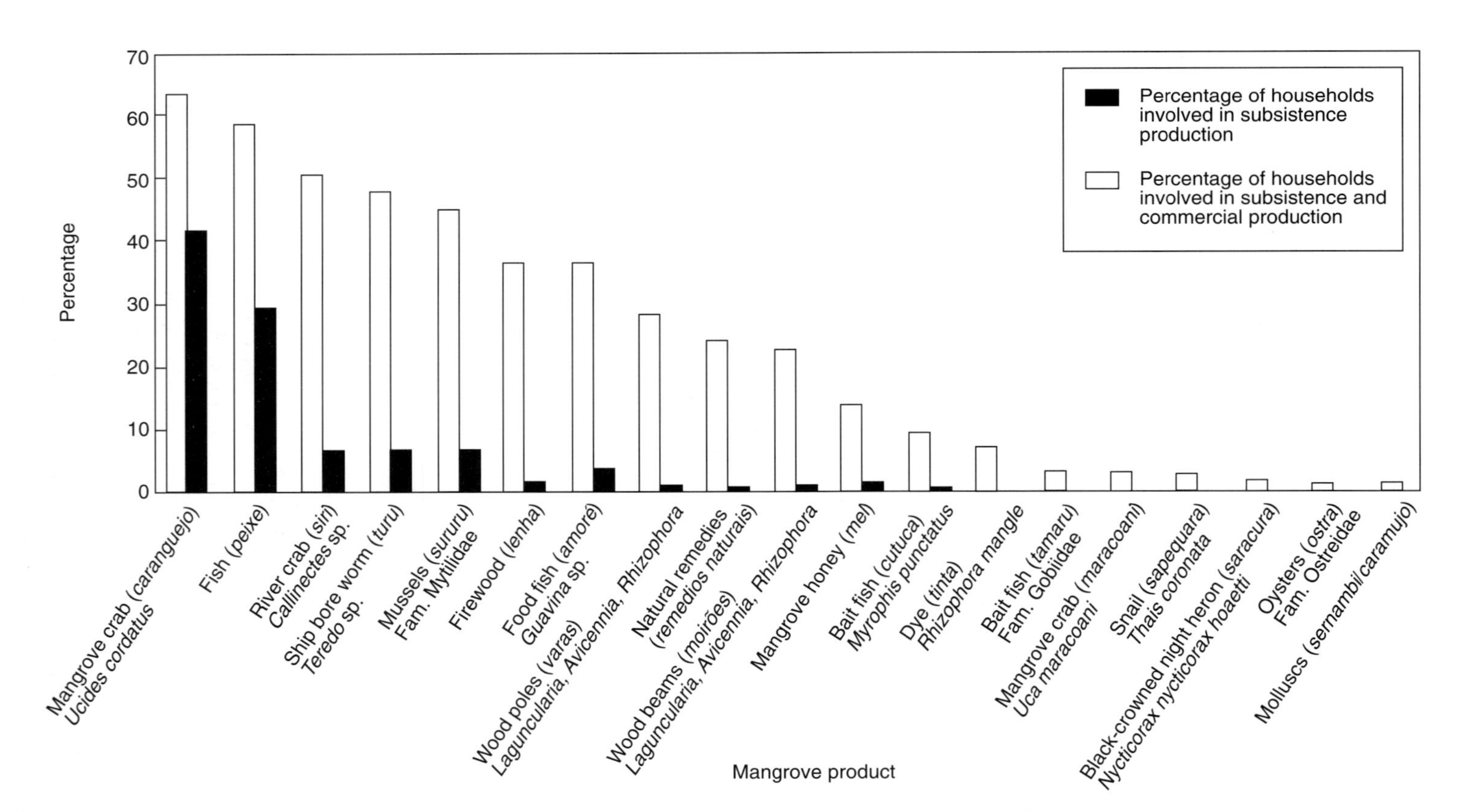

Fig. 12.1. The use of mangrove products in Caeté estuary villages. Local mangrove product categories not corresponding to one specific family, genus or species in biological taxonomy were described by only the local Portuguese expression and an English translation (after Glaser, 2003).

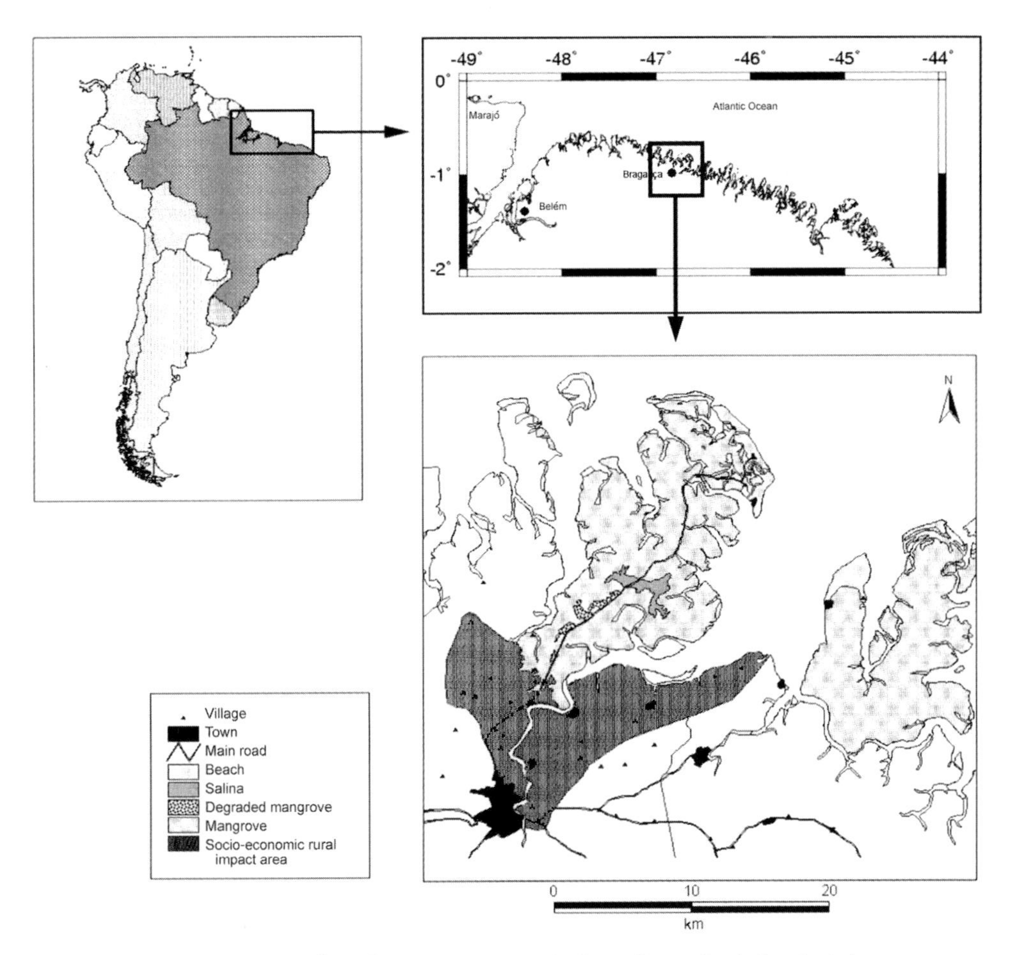

Fig. 12.2. The Bragança peninsula under investigation. In the lower figure, the darker shaded area represents coastal villages where residents derive income directly from the mangroves (after Krause *et al.*, 2001).

in carapace width. Males are approximately 7.1 to 8.7 years old when they reach the commercial size of 6.5 cm carapace width. As *U. cordatus* feed primarily on leaf litter, they play an important role in the nutrient dynamics of the mangrove ecosystem (Koch and Wolff, 2002) and could be a keystone species for the ecosystem.

The spawning season starts at the beginning of the rainy season, with considerable differences in larval output between years. Reproduction follows a strict lunar rhythm. Females spawn in the flooded mangroves around slack spring high tide, from where they are washed out into the tidal channels. From there, they are exported to the estuary and offshore waters. During their 3–4 weeks of development, they remain in coastal waters and it is the postlarva that returns to the estuarine environment (Diele, 2000).

Crab collectors capture mainly the largest males, whereas smaller specimens, including females, are generally rejected because of a lack of market demand. This sex and size selectivity is typical for crab capture aiming for livestock market sales and carries a high potential for sustainability. Collectors catch the crabs by pulling them out of their burrows by hand or with a hooked stick (*gancho*). Crabs are tied to so-called *cambadas*, which are strings of 14 living animals (Diele, 2000).

Table 12.1. Percentage of rural households according to their occupation and income source.

Occupation/income source	Percentage of rural households
Farming	42
Crab collection (*Ucides cordatus*)	42
Fishing (commercial)	31
Retirement pension	19
Fish and/or crab trade	9
Crab processing	9
Other commerce	11
Other occupations (not related to mangroves)	16
Other occupations (related to mangroves)	1

Table 12.2. Percentage of rural households and their linkage to the mangrove ecosystem (after Glaser, 2003).

Linkage to mangrove ecosystem	Percentage of rural households
Overall mangrove dependence (subsistence and/or commercial mangrove production and/or other mangrove-dependent profession)	83
Commercial mangrove dependence (commercial crab collection and fishing, sale of charcoal, wood, fish, crabs and other mangrove products (Fig. 12.1) and crab processing)	68
Fishing (subsistence and commercial)	54
Commercial fishers	32
Subsistence fishers	31
Crab collection (subsistence and commercial)	64
Crab collection (commercial only)	43

The fishery shows seasonal and annual differences in terms of labour input, capture volume and productivity. During the rainy season, more crab collectors work than during the dry season. The number of crabs captured from 1997 to 2001 varied from 500,000 to 789,000 per year. The CPUE (number of crabs collected per person-day) is from 130 to 176. Income declined from 1998 to 2001 by 20%, leading to an average income of 176 R$ (*reais*)/month in 2001, corresponding to 98% of the officially stipulated existence minimum in Brazil at that time (Glaser and Diele, 2004).

The crab population does not seem to be endangered yet. As crab collectors apply only very simple capture techniques and fail in areas with patches of dense *Rhizophora mangle* roots, these patches thus act as a refuge that prevents a total exploitation and consequently rapid depletion of the targeted large males (Diele, 2000).

The regulation of crab collection is the responsibility of the Brazilian Federal Environmental Agency (IBAMA). Crab collection is illegal during the crabs' annual mating days, when crabs walk on the surface rather than hide in their burrows. However, the crab fishery currently operates under a de facto open-access system, under which entry to the mangroves and use of their resources within the valid management laws cannot legally be denied to the many aspirants. Crab collectors are self-employed, many operating within a classic patron–client dependence relation to a crab trader, depending on the latter for production and emergency loans in exchange for the sale of their crab production at lower than market prices (Glaser and Diele, 2004).

Fish

The fishery structure in Bragança is divided into two branches, industrial and small-scale. The highly commercial, long-distance fishing industry contributes significantly to the regional economy; however, it is not linked directly to mangroves. Steel boats usually longer than 12 m are used for fishing trips lasting 20 to 45 days, targeting the more remote areas. Species caught include Pargo (*Lutjanus purpureus*), Pescada amarela (*Cynoscion acoupa*) and pink shrimp (*Penaeus subtilis*), which are sold on both the national and international market.

The small-scale fishery employs simple and traditional technologies. About half of the rural population is engaged in local fisheries. These are subsistence fishermen, who only occasionally sell part of their catch to the local markets. They account for about half of the small-scale fishers (Glaser and Grasso, 1998). They use different types of fishing gear, which can be distinguished by size, type of equipment used and fishing areas. Traditional fishing devices are traps (*corralais*), cast nets (*tarrafas*), a long line (*espinhel*) and harpoons (*arpão*). Captured fish are normally salted in so-called *ranchos*, pile-work barracks located along the bay where fishermen can also stay overnight (Barletta *et al.*, 1998).

The other small-scale fishermen work on commercial fishing boats, equipped with low-power engines (42%), sails (26%) or oars (32%). The catch is normally sold through a network of marketing intermediaries to the local markets. As the Caeté estuary and adjacent coastal waters are principally influenced by marine and fresh waters without any notable impact from the Amazon River, Bragança fishing stocks are dominated by marine or brackish-water species throughout the year: gó (*Macrodon ancylodon*), bagre (*Arius hertzbergii*), uricica (*Cathrops* sp.), amoré (*Eleotiedae guavina*) and pratiqueira (*Mugil* spp.) (Barletta *et al.*, 1998).

Regarding fish, the mangroves play an important role as nursery habitats and tidal-induced fish migration channels between the estuary and tidal creeks. Detailed descriptions and findings regarding the above processes have been discussed in Barletta-Bergan *et al.* (2002), Barletta *et al.* (2003), Krumme and Saint-Paul (2003) and Krumme *et al.* (2004). These studies demonstrate the ecological importance of mangroves for fish populations, and show that a sustainable exploitation depends on an undisturbed and entire ecosystem.

Molluscs

The small bivalves '*sururu*' (*Mytella falcata* and *M. guyanensis*) occur on the large banks that appear during low tide in the rivers and tidal channels. The mussel banks themselves vary from 30 to 3000 m^2. The density can be from 18 to >100 individuals/m^2 (Colin, Bragança, 1999, personal communication). They measure around 35 mm in length as adults.

It is from these banks that the fishermen exhume the *sururu* from the mud using their hands, placing the mussels in vegetable fibre baskets known as *caçuá*. Grasso (2000) assigns market prices to mangrove subsistence values for products such as mussels at US$35/month. However, this activity is more important for household consumption than for commercialization. *Sururu* is not always exploited. Glaser and Grasso (1998) reported that once a *mina de sururu* (which is the local expression for a mussel accumulation) is found, rural producers abandon crab collection to exploit them.

The mangrove oyster (*Crassostrea rhizophorae*) is not commonly found in this research area as generally the mangrove is inundated only during spring tides on average about 9 days/month. However, in other more coastal areas, oysters are found frequently and are enjoyed by the tourists. Attempts to cultivate oysters have been reported to be successful. For cultivation, larvae are captured from mangrove roots and are placed in special devices (rafts) for 3–4 months with periodical immersion during low tides. Commercially viable size (6–8 cm) of the oysters is reached in 16–18 months.

The ship bore worm *turu* (*Teredo* sp.) is also collected by families dependent on mangroves. The worm is collected from the dead wood of fallen mangrove trees for domestic use and not for commercial purposes. *Teredo* is usually consumed alive. Occurrence numbers are not yet available.

Wood

Although mangroves are protected by law, extensive deforestation in the research area, an increase of leather tannin manufacturing that uses mangrove tree bark (mainly *R. mangle*) and the burning of mangrove wood in brick kilns and bakery ovens are widespread and well known (Glaser *et al.*, 2003). Commercial brick kilns burn the largest quantity of mangrove wood in the research area. Wood is also used for house and boat construction and for cooking purposes. *Avicennia* is preferred for burning purposes as it is able to keep the fire alight for longer (Grasso, 2000).

It is estimated that more than 90% of the rural households collect mangrove wood as domestic cooking fuel (Glaser *et al.*, 2003). Villages engaged in industrial crab production also use timber as fuel for cooking. Another common use of mangrove wood is to build fish traps and wires. When used for this purpose, the wood is called poles and posts. Poles used for this purpose must be at least 10 cm in diameter.

Discussion

Federal Brazilian legislation defines forests and other areas that support dune and mangrove growth as areas of permanent protection. This implies the prohibition of all human interference except scientific, educational and ecosystem recovery work. The national coastal management plan makes all activities that lead to the degradation of ecological stations and reserves punishable by law. Glaser *et al.* (2003) provide a detailed overview on the legislation relevant to mangroves. This makes the discrepancy between legislation objectives and reality very evident.

However, an important outcome from our investigations is that the dependence of the primary producers on the natural resources of the mangrove area is clear. Even if the ecology of the system can still be considered tolerably undisturbed by human activities, there is a considerable increase in remote and local anthropogenic interference (Krause *et al.*, 2001). Mangrove resources constitute subsistence food security for the poorest coastal households and the sale of fish and crabs is central to rural livelihood strategies in mangrove-adjacent areas. More than 83% of rural households harvest natural resources associated with mangroves (Glaser *et al.*, 2003). Use pressure on these products is increasing because of coastward migration, a lack of alternative income options for coastal dwellers and a rising urban demand for mangrove products.

Glaser *et al.* (2003) pointed out that users are typically aware of their own long-term need to maintain mangroves and are interested in sustainable, long-term management of mangroves within their area.

Because previous attempts of centralized, state-managed conservation to achieve effective mangrove management have mostly been unsuccessful, co-management-oriented concepts are being considered as an alternative.

Extractive reserves (RESEX) have been considered as a legal option for the co-management of natural resources in Brazil since 1988 and have come into focus as a solution to coastal mangrove management problems. Brazil Federal Decree No. 98.897 of January 1990 established that RESEX are 'territorial spaces destined for the self-sustained exploration and conservation of renewable natural resources by user populations' (Glaser and Oliveira, 2004). Under the RESEX concept, rule-making authorities, resource-use rights and management duties are assigned to local associations of interested residents who, with official capacitation and training and with intermittent implementation assistance from conservation authorities, are expected to generate the control and management quality that top-down, coercive state management has been unable to provide (Glaser *et al.*, 2003). RESEX were discussed for the first time in 1985 under the leadership of Chico Mendes to meet the interests of *seringueiros* in Amazonia. For aquatic resources of the Amazon flood plains (*várzea*), this concept had been applied successfully in the early 1990s. This concept proposed the transfer of fishing rights and their management to riverine communities with the aim of guaranteeing fish supply for subsistence (McGrath *et al.*, 1994). Rather than fence people away from the mangrove, extractive reserves are supposed to permit people to manage the forest without destroying it.

According to Glaser and Oliveira (2004), the aims of RESEX can be summarized as follows:

- protection of nature through use,
- improvement in living conditions for the traditional users of natural resources,
- integration of traditional users into national development processes, and
- promotion of user cooperation, that is, collective decision-making and action.

Effective co-management depends on the active participation of local ecosystem users and must not be contrary to the current environmental legislation. If successful, RESEX may turn the current, non-viable, open-access regime into a viable form of user-regulated and use-monitored common-pool management (Glaser and Oliveira, 2004). However, no data are available so far on how many people can be supported on a sustainable basis by 1 ha of mangrove area. And, experiences from the rainforest extractive reserves show clearly that waves of migrants entering the coastal zone cannot be absorbed by such a system. Therefore, RESEX are best viewed as part of the mosaic of land-use systems in the region, rather than as a social and ecological panacea.

Conclusions

These studies have revealed the relationship between the rural population in Caeté Bay and the mangrove ecosystem. The results make it evident that the preservation of the mangrove system in the region is fundamental to maintenance of the households' quality of life. The mangroves provide subsistence products for nutrition, housing and fuel, as well as commercial products from which income is generated. Further mangrove degradation will result in significant income loss for the locals and increase the potential for social conflicts.

It is still unclear whether the actual use of mangrove resources is sustainable. The RESEX concept is an attempt to turn the Caeté area into user-regulated and user-monitored common-pool management. A final evaluation will show the success or failure of such co-management.

Acknowledgement

This research was conducted under the Brazilian–German Scientific Cooperation Agreement and financed by the German Federal Ministry for Education and Research (BMBF) as part of the programme 'Research focus on the ecology of tropical coastal areas: mangrove dynamics and management' (No. 03F0383A5). This is MADAM contribution no. 87.

References

Barletta, M., Barletta-Bergan, A. and Saint-Paul, U. (1998) Description of the fishery structure in the mangrove dominated region of Bragança (State of Pará, North Brazil). *Ecotropica* 4, 41–53.

Barletta, M., Barletta-Bergan, A., Saint-Paul, U. and Huboldt, G. (2003) Seasonal changes in density, biomass, and diversity of estuarine fishes in tidal mangrove creeks, of the lower Caeté Estuary (northern Brazilian coast, east Amazon). *Marine Ecology Progress Series* 256, 217–228.

Barletta-Bergan, A., Barletta, M. and Saint-Paul, U. (2002) Structure and seasonal dynamics of larval fish in the Caeté River Estuary in North Brazil. *Estuarine, Coastal and Shelf Science* 54, 193–206.

Berger, U., Glaser, M., Koch, B., Krause, G., Lara, R.J., Saint-Paul, U., Schories, D. and Wolff, M. (1999) An integrated approach to mangrove dynamics and management. *Journal of Coastal Conservation* 5, 125–134.

Diele, K. (2000) Life history and population structure of the exploited *Ucides cordatus* cordatus (Linnaeus, 1763) (Decapoda: Brachyura) in the Caeté estuary, North Brazil. PhD thesis, University of Bremen (ZMT Contribution 9), Germany.

Glaser, M. (2003) Interrelations between mangrove ecosystems, local economy and social sustainability in Caeté Estuary, North Brazil. *Wetland Ecology and Management* 11, 265–272.

Glaser, M. and Diele, K. (2004) Asymmetric outcomes: assessing the biological, economic and social sustainability of a mangrove crab fishery, *Ucides cordatus* (Ocypodidae), in North Brazil. *Ecological Economics* 49, 361–373.

Glaser, M. and Grasso, M. (1998) Fisheries of a mangrove estuary: dynamics and inter-relationships between economy and ecosystem in Caeté Bay, northeastern Pará, Brazil. *Boletim do Museu Paraense Emílio Goeldi, Série Zoologia* 14, 95–125.

Glaser, M. and Oliveira, R. da S. (2004) Prospects for the co-management of mangrove ecosystems on the North Brazilian coast: Whose rights, whose duties and whose priorities? *Natural Resources Forum* 28, 224–233.

Glaser, M., Berger, U. and Macedo, R. (2003) Local vulnerability as an advantage: mangrove forest management in Pará state, north Brazil under conditions of illegality. *Regional Environmental Change* 3, 162–172.

Grasso, M. (2000) Understanding, modeling and valuing the linkages between local communities and the mangroves of the Caeté River bay (Pa-Brazil). PhD thesis, University of Maryland, USA.

Kjerfve, B. and Lacerda, L.D. (1993) Mangroves of Brazil. In: Lacerda, L.D. (ed.) *Mangrove Ecosystems Technical Reports*. ITTO TS-13(2), 245–272.

Koch, V. and Wolff, M. (2002) Energy budget and ecological role of mangrove epibenthos in the Caeté estuary, North Brazil. *Marine Ecology Progress Series* 228, 119–130.

Krause, G., Schories, D., Glaser, M. and Diele, K. (2001) Spatial patterns of mangrove ecosystems: the Bragantinian mangroves of North Brazil (Bragança, Pará). *Ecotropica* 7, 93–107.

Krumme, U. and Saint-Paul, U. (2003) Observation of fish migration in a macrotidal mangrove channel in Northern Brazil using 200 kHz split-beam sonar. *Aquatic Living Resources* 16, 175–184.

Krumme, U., Saint-Paul, U. and Rosenthal, H. (2004) Tidal and diel changes in the structure of a nekton assemblage in small intertidal mangrove creeks in northern Brazil. *Aquatic Living Resources* 17, 215–229.

McGrath, D., Castro, F. and Futema, C. (1994) Reservas de lago e o manejo comunitário da pesca no Baixo Amazonas: uma avaliação preliminar. In: D'Inaco, M.A. and Silveira, I.M. (eds) *A Amazônia e a Crise de Moderização*. MPEG, Belém, Brazil, pp. 389–402.

Özhan, E. (1998) Estuaries and coastal waters: research and management – introduction. *Journal of Coastal Conservation* 4, 2–6.

Twilley, R.R. (1996) *The Significance of Nutrient Distribution and Regeneration to the Recovery of Mangrove Ecosystems of South Florida in Response to Hurricane Andrew*. Report, University of Southwestern Louisiana, Lafayette, Louisiana.

13 Mangrove: Changes and Conflicts in Claimed Ownership, Uses and Purposes

M.-C. Cormier-Salem
UR169-IRD/MNHN, Département Hommes, Natures, Sociétés, Paris, France, e-mail: cormier@mnhn.fr

Abstract

From its early discovery, mangrove has inspired ambivalent feelings among Westerners, ranging from delight to repulsion. It has been considered in turn as an unhealthy and hostile milieu, as a source of multiple resources or as a fragile, diversified and rich ecosystem. Management policies have also varied between extremes: from periods of degradation and conversion to periods of rehabilitation, restoration and protection. This contribution is centred on claims and conflicts over mangrove wetlands.

First, through the case study of West Africa, we will show the very early (at least, the 15th century) multiple-use system controlled by peasant-fishermen communities, then the disruption of this remarkable and sustainable system with the 'white' penetration and colonization, ending with the recent process of rehabilitation, restoration and protection or 'heritage construct', that leads to conflicts.

Second, this contribution will question the relevance of the concept of natural heritage, and set out the different interpretations and constructions that can be made from it: Which are the living objects designated as heritage? Who are the decision-makers, the stakeholders, the managers and/or the caretakers? Is mangrove a world heritage, a communal territory, a private capital or a public good?

Third, to answer these questions, multidisciplinary approaches and new tools of management, not only integrated but also concerted, have to be developed. The diverse relationships in mangrove societies call for a variety of legal and management policies. In addition, a diachronic approach is necessary, investigating several time intervals (from a few years to eras).

This contribution also discusses the implications of innovative tools such as the institutionalization of heritage and new ways of valuing nature, such as product labelling, eco-certification or ecotourism.

Introduction

When mangroves were designated as part of humankind's common heritage in Agenda 21 at the Rio Earth Summit (1992), the reversal of attitudes and policies regarding these sites was institutionalized. Long considered, at least in Western eyes, to be impenetrable swamps, mangroves are now considered as rich ecosystems, fragile and threatened by human activity, to the extent that they must urgently be protected. Making mangroves part of our heritage, or patrimony, is justified by the desire to preserve this ecosystem, which is indispensable for maintaining both terrestrial and marine biodiversity.

Yet, the idea that mangroves are a unique heritage, collectively shared by all of

 Environment and Livelihoods in Tropical Coastal Zones (eds C.T. Hoanh, T.P. Tuong, J.W. Gowing and B. Hardy)

humankind, is, from our point of view, not only scientifically false but also socially unacceptable. This notion of a common heritage engenders a threefold question: Can mangrove marshes in their entirety constitute a heritage, and, if not, then which elements are worth conserving? This issue is connected to the question of stakeholders involved in these processes: from local users to international bodies, from managers to decision-makers. It is clear that mangroves lend themselves to a range of projects, expressing the multiple values and functions that are associated with them. What mangroves represent for some is not what they are for others; and this is why mangroves are a disputed heritage, the subject of many conflicts related to representations, uses and access. The third question is that of mangroves' legal status: To whom does this patrimony belong? Is it a public, private or community good? Who are the beneficiaries, the stakeholders and the guardians?

This chapter aims to show that treating mangroves as patrimony often does not meet ecological imperatives, nor does it satisfy the needs of the communities involved. Most protection policies, or 'heritage-building' processes, are failures, and exacerbate conflict between stakeholders, at the local as well as at the regional, national and international levels. We try to illustrate the major causes of these failures, show the multiple issues – ecological, economic, political and social – that are at stake in mangrove marshes, and analyse the dynamics and strategies at work.

In the first section, we analyse the change in attitude towards mangroves, and the prevailing conditions that led to them being considered as a heritage. In the second section, using three case studies, we analyse local strategies that have been elaborated to meet the challenges of biodiversity conservation and sustainable development. In the third section, we demonstrate the effects induced by heritage-designation processes, their limitations and contradictions, as well as the innovative changes they have brought about. In our conclusions, we emphasize the main questions and perspectives in terms of research and action to achieve integrated and concerted management of mangrove areas.

Mangroves: from Depreciation to Heritage Building

Changing attitudes towards mangroves testify to the progression of knowledge, but above all they reveal the ideological and economic ramifications of policies aimed at managing nature in tropical areas. In this section, we show that ignorance of local know-how and the destruction of genuine mangrove civilizations are the corollaries of the expansionism of colonial powers, and that our way of perceiving and managing mangroves is deeply rooted in this colonial mindset (Haraprasad, 1999). The present-day tendency to preserve, even to monumentalize, them as sanctuaries is only one of the more recent examples of these politics of domination (Neumann, 1998; Zimmerer and Bassett, 2003).

Dense and long-standing occupation until the whites arrived

Shell debris and middens in Asia, Latin America and Africa attest to the long-standing occupation of coastal mangroves and their early exploitation for multiple purposes. Research by Higham (1988) in South-east Asia highlights the long tradition of coastal life in which rice growing is associated with the gathering of seafood (fish, crab, molluscs, turtles), as far back as 5000 years before our era. Several authors underscore the nutritional importance of food derived from mangroves and adjacent coastal zones in the ancient Americas (Bearez, 1996; Craighead, 1997; Musset, 1998; Kneip, 2001). In West Africa, excavation of kitchen middens in the region between the Saloum and present-day Guinea Bissau attests to the presence of human groups that lived in the mangroves and exploited their resources (oysters, shellfish, fish, salt and rice), and traded them (Cormier-Salem, 1999a).

From the 15th century onward, with the major discoveries and exploration of the

New World, written sources are more and more numerous and specific (Saenger and Bellan, 1995; Cormier-Salem, 1999a). The first descriptions of the West African coast testify to its settlement in ancient times, as well as to the density of the coastal population, to the importance of products drawn from the mangroves and to the highly elaborate implantation of rice-growing areas in the mangroves.

Mangrove, the 'white man's grave'

The arrival of the white man – navigators and explorers from the Old World – constituted a major break in the history of mangroves. With the great seafaring voyages, followed by colonial conquest, indigenous coastal populations were unable to resist foreign aggression, while at the same time the development of maritime trade profoundly and durably altered ancient relationships of exchange and multiple-use systems (Rodney, 1970; Brooks, 1993; Cormier-Salem, 1999a).

The predominant image of mangrove swamps, the one that emerges from the accounts of French voyagers and missionaries in the 17th century, echoed throughout the 18th and 19th centuries, is that of a repugnant, hostile, unhealthy and impenetrable environment. Two conceptions about mangroves are particularly tenacious. The first notion is of their wild and inhospitable nature: living conditions are so difficult that only primitive peoples who are forced to take refuge there can live in mangroves. In the writings of explorers, and of later colonial agents, it is easy to move from the idea of marginal status in space to marginal status in society, from the view of a malignant environment to the perception of a malignant population.

The second notion is that mangroves are unhealthy. The hot and humid coasts of 'Northern Rivers', in particular Sierra Leone, were commonly called the 'white man's grave' because of the various fevers that were rampant in these regions (Carlson, 1984). Mangroves were preferred terrain for malaria, a disease perceived as linked to humours emanating from swamps. The term *paludism* is derived from *palud*, meaning swamp, and in Italian malaria means 'bad air'. It is true that Africa is a vast foyer for malaria. But, mangrove forests are not particularly pestilential, and might even be healthier than places inland. Entomologists have shown that the dominant *Anopheles* species (*Anopheles melas*) in the mud flats of African brackish waters is not a very efficient vector of malaria (Mouchet *et al.*, 1994). The most effective vectors are freshwater species (*A. gambiae*, *A. arabiensis*). The numbers of these species rise with increasing rainfall and the construction of freshwater ponds and lakes, and thus, paradoxically, with colonial hydraulic and agricultural works.

In practice, repugnance towards mangroves led to vast colonial operations intended to drain, cleanse and improve mangrove marshes for productive purposes. For a long time, mangrove soils, which are heavy, fluid and subject to acidification and salinization, were deemed unsuitable for agriculture, except at the cost of massive investment. Demographic pressure and technological progress in the first half of the 20th century lifted some of these constraints, leading to a change in thinking, and maritime mud flats were seen as immense reserves of arable land.

In Latin America, mangroves were converted to sugarcane and palm plantations, while West African mangroves were meant to become the rice bowl of French West Africa. Most of these conversions were spectacular failures, economically as well as ecologically and socially (Rue, 1998). Yields were low, and the civil works involved were not reappropriated by farmers because they were not adapted to agricultural and pedological conditions. The modification of natural outflows has caused lasting disturbances in ecosystem functioning. Vast expanses of mangrove forest, cleared and transformed into solid land, are now abandoned. Ultimately, the multiplication of freshwater collection points favours new outbreaks of malaria in these areas (Mouchet and Brengue, 1990).

The hygienist and productivist aims of the colonial period were followed in the 1950s by the newly independent countries' desire to acquire substantial industrial and port infrastructure, and the need to expand the limits of urban areas. Drying and filling mangrove mud flats was a response to the demand for land that accompanied maritime development and urbanization of coastal regions.

With the boom in shrimp farming in the 1980s and 1990s, the productive function of mangroves was once again to the fore. Mangrove forests, but above all the mangrove backcountry, are now increasingly becoming preferred sites for shrimp farming. The commercial value of shrimp is on another scale altogether compared to other mangrove products. This activity constitutes a growing, if not a predominant, foreign-currency contribution to the GNP of many Southern countries. Today it is one of the major factors causing the disappearance of mangrove systems (Barbier and Satirathai, 2004).

Mangrove rehabilitation

While repugnance for mangroves is deeply rooted in our imagination and their utilitarian functions are given precedence in many Southern countries, new and radically different conceptions appeared in the course of the 20th century. These ideas converged towards rehabilitation of mangroves and have given rise to protection and restoration programmes (Williams, 1990; Dugan, 1992; Mitsch, 1995).

Starting in the late 1960s, international organizations such as FAO and UNESCO began to pay close attention to the situation of mangroves, aware of their accelerating disappearance and concerned about the risks incurred for coastal environments, both locally and globally. This change in attitude can be attributed to the rising influence of environmental concerns and movements, as well as to greater knowledge of the multiple functions and values of humid coastal wetlands and the importance of their preservation in order to conserve land and marine biodiversity. As shown in these studies (Saenger *et al.*, 1983; UNESCO-UNDP, 1986; Hook *et al.*, 1988), mangroves are stabilizing factors for the coastline, protecting beaches and river banks, retaining shorelines, acting as breaks against storms, hurricanes and waves,[1] etc. In addition to this physical role, they have a biological role in enriching coastal waters and soils. They provide ecological niches for micro flora and fauna; spawning and nursery grounds for fish, shrimp and shellfish; refuges for migratory birds; and natural habitats for various forms of animal life. Mangroves are henceforth presented as rich, complex and fragile ecosystems that must be protected from human impact.

In this rehabilitation of mangroves, although the scientific community played an essential part in raising consciousness, and international organizations such as UNESCO and FAO were quick to launch research and reforestation programmes, the pressure exerted by certain NGOs must be underscored. Groups such as the World Wide Fund for Nature (WWF) and the International Waterfowl and Wetlands Research Bureau (now Wetlands International), along with the World Conservation Union (IUCN), were the principal driving forces behind the International Convention on Wetlands signed in Ramsar (Iran) in 1971. The Ramsar Convention was a major milestone in the implementation of new policies for mangrove management, or heritage building.

Mangroves: from resource to heritage

Once merely zones for exploitation of resources, mangrove systems have become assets, project spaces and even a projection of our dreams. The notion of 'natural heritage', although scientifically ambiguous, is increasingly popular, corresponding to the desire to enact effective environmental protection policies (Cormier-Salem and Roussel, 2000). Originating in the Western world, this notion is now spreading around the world.[2] Red lists of endangered species

[1] The tsunami that occurred on 26 December 2004, in South-east Asia, dramatically testifies to the mangroves' buffer role.

[2] The term 'patrimony' is part of the vocabulary of Western civil and religious rights. From the Latin *patrimonium*, it designates collectively the goods and rights inherited from the father (Rey, 1992).

keep getting longer, and protected areas (national parks, Biosphere Reserves, etc.) are expanding. This process, which is relatively recent and remained limited in spatial terms and in its objectives up to the 1990s, is now experiencing an explosion concerning mangroves. According to Spalding *et al.* (1997), only 685 protected areas contain mangrove forest sections. The first patrimonial reserves to be constructed had purely ecological aims, sometimes related only to fauna. Mangroves were at first viewed as refuge habitats for animal species deemed to be of patrimonial value, either emblematic ones such as the Bengal royal tiger (Haraprasad, 1999) or those of international significance such as migratory species. Accordingly, the first West African coastal areas to be registered, starting in 1976, were migratory bird sites, initially recognized as Ramsar sites, then as UNESCO Biosphere Reserves (Saloum Delta with the Île aux Oiseaux in Senegal) or national parks (the Arguin Bank and the Barbary Point in Mauritania, Djoudj in the delta of the Senegal River and Île de la Madeleine off Dakar) (Fig. 13.1).

Alongside these official asset-building processes, for the most part exogenous in origin, our research highlights the existence of endogenous nature conservation processes (Cormier-Salem *et al.*, 2002). We observe characteristics habitually attributed to patrimonial objects, as described elsewhere (Babelon and Chastel, 1994; Cormier-Salem and Roussel, 2000). These characteristics are three in number. For a natural object to be given the status of a heritage, it must first of all be inherited from ancestors; second, it must be destined to be bequeathed to following generations (which assumes sustainable management); and third, it must embody a collective identity.

The objects officially designated as constituting a heritage and the dynamics that underpin these choices are rarely those of the local human community. Three case studies illustrate the divergences among ways of viewing and conceiving mangroves as a heritage.

Diversity of Local Strategies for Confronting Conservation Issues

The Kaw estuary: a sanctuary

The Kaw estuary in French Guiana is the largest French wetlands zone and constitutes a remarkable natural sanctuary in terms of the diversity of its vegetation, richness of fauna and minimal human impact. The dense forest is made up of mangrove trees, and also pinot palm, epiphytes and vines. It is home to rare animals such as the black cayman (*Melanosuchus niger*) and the matamata turtle (*Chelus jimbratus*); endemic species, including a legless aquatic frog (*Typhlonectes compressicaudus*); and remarkable birds such as the crested hoatzin (*Opisthocomus hoazin*) and the red ibis (*Eudocimus ruber*). Kaw shows very little trace of human presence. Excepting the village of Kaw itself, built at the end of the 19th century by Noir-Marron, descendants of slaves, and numbering about ten houses, there are no permanent dwellings. Woody mangrove stands are not particularly exploited; grassland is somewhat more exploited, as seasonal pasture land for grazing herds. Mangroves are also a strategic location for the reproduction of fish species such as atipa (*Hoplosternum littorale*) that build their nests there. Exploitation of coastal and marine resources is very limited, except for shrimp fishing. Fishing in the estuary and the river is more common: certain native American communities (Wayapi) specialize in fishing, and small groups of itinerant fishers travel along the rivers. Fishing, hunting and grazing animals are the main uses of the mangrove swamps; these are all itinerant and extractive activities.

In spite of recent and limited human settlement, the creation of the Kaw-Roura Swamplands Natural Reserve (by government decree in March 1998 and registered as a Ramsar site in November 1993) on some 98,500 ha and its ecotourism operations have generated much tension. The inhabitants of Kaw are afraid that their hunting and fishing practices may be threatened by restrictive measures for the protection of wild fauna and by competition from 'metro' (mainland

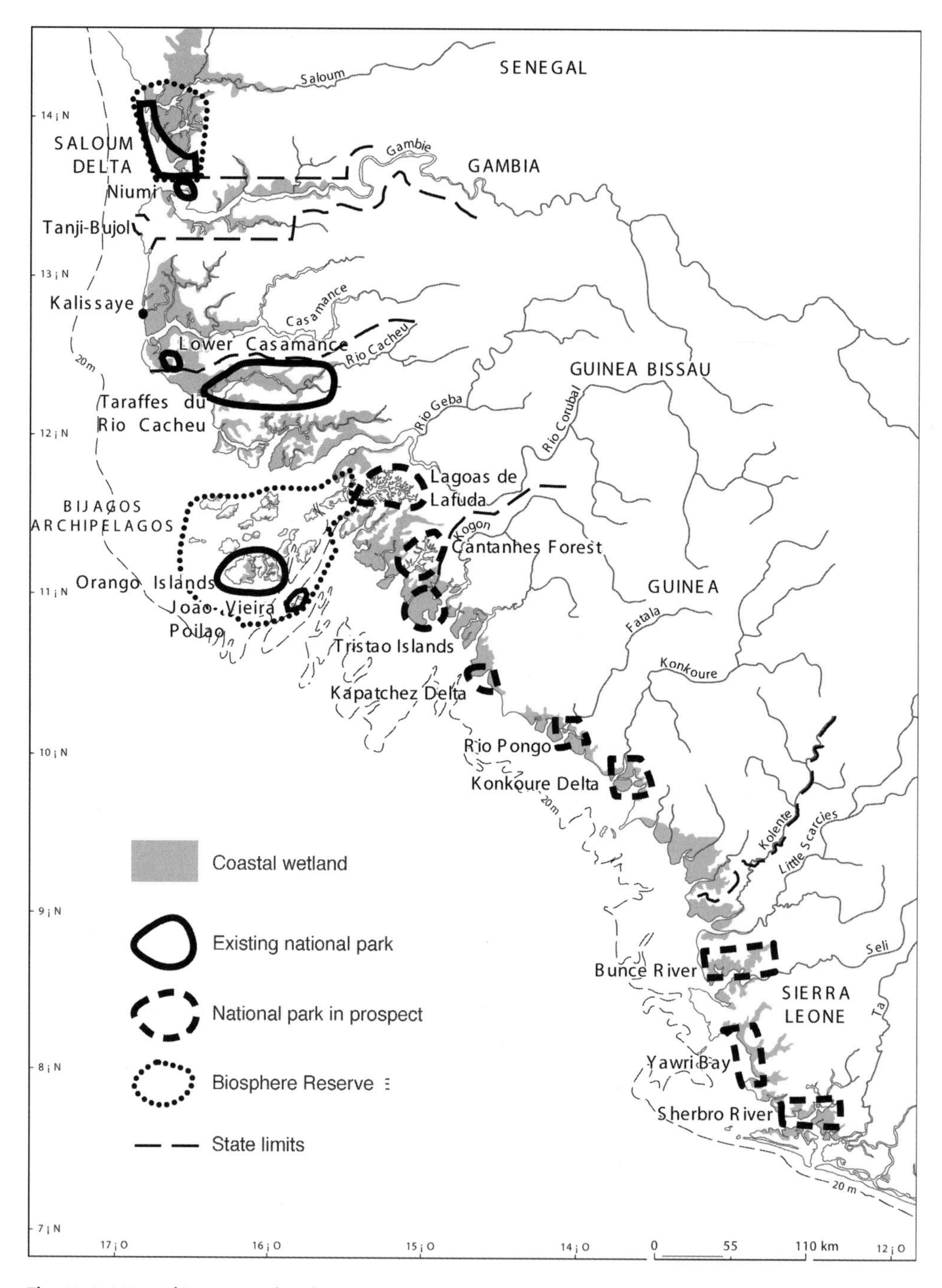

Fig. 13.1. West African coastal and marine protected areas.

France) hunters out of Cayenne. Hunting, a sporting activity for foreigners, is of prime importance to local nutrition and social structure. Competition between fishing and animal husbandry is also a source of conflict. Kaw is attached to Regina Township, where the mayor possesses large herds of cattle that graze throughout the swamplands in the dry season. These herds are entrusted to the care of Brazilian herders, who are underpaid immigrants. They are seen as responsible for the trampling of atipa nests, necessary for the reproduction of this fish. Sale of atipa in Cayenne is the main source of income for the people of Kaw. These competing uses are mirrored by power struggles within the village: some support and some oppose the mayor of Regina. Lastly, the Noir-Marron population is not indigenous. The people have neither property rights nor recognized rights of usage for mangrove resources. Unlike Native Americans, they can make no claim to traditional naturalist knowledge or mythical ancestors. And yet they are just as fully entitled as French citizens.

The Menabe mangrove wetland in Madagascar: a pioneer frontier

Madagascar is one of the places where the contradictions between discourse and action plans are the most glaring, where the conflicts of interest among economic development, preservation of traditions and protection of biodiversity are the most blatant. A 'natural paradise' endowed with exceptional biodiversity, ranking fifth among the 17 Global Diversity Hotspots according to the United States Agency for International Development, the Grande Île (Madagascar) is one of the poorest and most heavily indebted states in the world. International pressure (World Bank, UNDP, etc., and major NGOs) and new policies of openness to the outside world, economic renewal and social development, followed since 1992, have led to the implementation of the National Environmental Action Plans. These plans are intended to preserve the remarkable Malagasy biodiversity, while at the same time fighting poverty. Among all the measures for protection and coordinated management of biodiversity, one cannot help but underscore the meagre attention paid to mangroves. There are only two protected coastal or marine areas, both located on the eastern coast, where there are no mangroves. Only one protected zone on the west coast, Kirindy near Belo-sur-Mer, includes a small section of mangrove.

This relative lack of interest contrasts with the high stakes attached to the degradation of continental biodiversity: forest loss (68–85%), migratory dynamics and territorial pressure are all related to the dominant mode of agricultural production: itinerant slash-and-burn cultivation (Razanaka *et al.*, 2001). Deforestation is also due to the creation of economic value from the heritage of the remarkable landscapes and unique sites of the Merina kingdom in the High Plateau region, and from emblematic species such as orchids and lemurs, which embody ecological and aesthetic value far greater than that of mangrove trees or crabs. The lack of interest in mangrove conservation is also explained by the economic importance of coastal resources and the prime goal of turning them into sources of foreign currency. When shrimp are declared to be strategic national resources (Goedefroit *et al.*, 2002), mangroves become frontiers to be conquered and converted into shrimp farms and saltworks.

The coastal wetlands of this region were long used essentially as grazing lands for herds of zebu raised by seasonal Sakalava herders. Human settlements were limited to small fishing or wood-gathering camps. From the mid-19th century to the 1960s, several waves of migration contributed to the settlement and improvement of the marshes. Starting in the 1970s, shrimp fishing, then farming, engendered an influx of population.

The increasingly sedentary population and the development of agriculture since the end of the colonial period, the search for land and water with the arrival of migrants, and the increasingly rapid clearing of fields and diminution of fallow lands have profoundly disturbed the organization of the Sakalava territories. Since the 1970s, a worsening of the climate and modification of river drainage hastened the decline of animal raising and the abandonment of rice growing. This crisis in ancient agrarian systems has elicited strategies for the diversification of activities, reconversion and migration, leading to doubts as to the survival of the Sakalava identity.

In seashore villages such as Bosy, fishing is now the sole activity, along with maritime trade. Fishing techniques have become more diversified. In particular, we must take note of the use of large-mesh dragnets for catching swordfish, tuna and, above all, shark. Another alarming trend is the conversion of Vezo fishers into mere intermediaries who recover the by-catch from trawling vessels and sell it on Malagasy markets. While sea fishing from Bosy is diminishing, migration to fishing camps farther north has increased, and the extraction of mangrove resources during the rainy season is increasing (particularly for shrimp, fishing and woodcutting for the making of poles).

In mangrove backwater villages such as Ampataka, commercial exploitation of mangrove crab is becoming a major activity (Montibert, 2001). This specialization is linked to the collection system established by the Société de Pêche de Morondava (SOPEMO) that transports the crab to its factory in Morondava, the Menabe regional capital. The crab is then exported to Asian markets.

Today, the Menabe mangrove marshes are subject to multiple pressure from local users, migrants and foreigners (*vahiny*). The GSM saltworks, located some 15 km from Ampataka, and the Aquamen shrimp farm farther north at Tsangajoly, are steadily expanding their occupation of land and intensifying their production, leading to sharp tension in the local communities. The force of local claims based on heritage and territory can be read in the phenomena of trances, the resurgence of ancient taboos in reaction to the Ampataka saltworks and sabotage of shrimp ponds.

Northern Rivers: endangered amphibious territories

In the Northern Rivers region, the communal territories carved out in the mangrove marshes are today threatened because ancient social and technical systems of managing these spaces are in jeopardy, and because of environmental changes. Hard labour in the rice fields, the burden of familial constraints, and isolation and difficult living conditions in mangrove villages without fresh water or electricity are all factors that drive people away. Migrations that were initially seasonal now tend to become permanent. In the zones affected by rural population flight, the hydraulic and agricultural infrastructure is no longer maintained. Because of the lack of labour, many rice fields have been abandoned. This retreat, visible as early as the 1950s, has accelerated because of the worsening climate and salinization of soil and water. In Ziguinchor (Casamance, Senegal), average rainfall that was over 1500 mm per year before the 1960s fell below 1000 mm in the 1970s and dropped as low as 690 mm in 1983. The consequent rapid increase in salinity level has led to a simplification of ecosystems and depletion of the surrounding environment. In Guinea, Sierra Leone and the south of Guinea Bissau, the impact of drought is less severe, but, in contrast, political instability has caused massive emigration and the collapse of local and national economies (Cormier-Salem, 1999a).

To confront this many-faceted crisis and compensate for the deficit in rice production, the people in Northern Rivers have developed strategies to diversify crops and broaden the range of rural activities. Growing rice in the highlands, tree cultivation and vegetable farming have made considerable progress. The old activities of gathering, often done by women, have been rehabilitated and integrated into merchant networks. Mangrove resources (mangle wood, salt, oysters) as well as the products of highland forests (cashew, palm) are exploited for small-scale commercial production and sent to rural and urban, even international, markets. In just a few years, Guinea Bissau has become one of the top-ranking producers of cashew nuts.

The development of estuary and sea fishing within local communities of rice growers, herders and fisher-farmers is without a doubt one of the major phenomena of the last 20 years. For some socio-cultural groups, rice growing, once their main identity, has been superseded by fishing.

Overall in Northern Rivers, family strategies have become diversified, leading to a redistribution of tasks among sexes and age groups. The centre of gravity of rice growing has, by and large, shifted from the flood plains to the highlands. But rice, whether imported or produced locally, remains the basis of the local diet, and, all in all, the people are still culturally rice farmers.

Against this background, several local heritages can be identified: clearly, rice fields can be considered as a heritage by lineage, and a first definition of this patrimony can be proposed: 'the set of rice cultivation techniques, practices and institutions transmitted by ancestors that shape the community space and from which the community draws its identity, and which are to be kept fruitful and enriched'. Other components of biodiversity are valued as a heritage: animals (monkeys, manatees, turtles, flamingoes, etc.), plants (baobab, fromager – *Ceiba pentadra* – stands of mangrove and palm), products (palm wine, honey and mead) and landscapes (dyked sea marshes). The amphibious mangrove lands and the multiple-use and traditional management systems that maintain them are identified as both a bio-ecological and cultural heritage, in which the people of Northern Rivers find their identity (Cormier-Salem, 2002).

With the creation of protected areas, such as the Saloum National Park, most uses of mangrove resources have become illegal, with hardly any compensation or employment in return. Many animal species (e.g. turtles) are henceforth fully protected. Taking mangrove resources is allowed only for household use and a permit is required. Conflict with national park guards was frequent in the early years (1976–1979) but has now fallen off: schools and dispensaries have been built, wells dug and water towers and electric generators installed. The islands that once were isolated, proud of their unique features and keenly opposed to outside influence, are opening up, at the price of a profound questioning of ancient ways, notably the social control exercised by the elders.

Learning from the case studies

These case studies highlight that mangroves are far from being homogeneous and stable entities. First of all, they are a patchwork of ecosystems, of widely varying size, composition and structure, depending on their own geographical and historical context. Second, mangroves lend themselves to differing socio-spatial units. These differences stem from the history of these regions, and from the conditions of settlement to their integration into colonial or postcolonial economic systems. They are also linked to the techniques, practices, knowledge and rules elaborated by the surrounding communities that leave their mark on the landscape. Lastly, mangroves depend on the capacity of these same communities to adapt to change in their environment (climate, policy, economy, etc.).

Moreover, the aspects of mangrove areas that are considered to be worth preserving, in other words the heritage, are very different depending on the stakeholders, from local to international levels, and the criteria change over time. Most local heritages (by lineage, community, etc.) are in disagreement with those defined by outside bodies and enshrined in international conventions. Furthermore, these heritages appear to be increasingly threatened by two major trends. One is the privatization of resources and spaces, the expropriation of the property of traditional holders in favour of public or private enterprises, and the search for value in a monoculture perspective. The second is the tendency towards individualism and the abandoning of 'traditional' systems.

Ambivalent Policies for Mangrove Heritage

In the last 20 years, multiple initiatives have been launched to fight the accelerating disappearance of mangrove forests and overexploitation of coastal resources. An initial phase, marked by the desire to protect the habitats of migrating waterfowl, led to the creation of integral reserves and mangroves being designated as Ramsar sites. This was followed by a mangrove rehabilitation phase with replanting of mangle trees, and then a landscape preservation phase characterized by the creation of Biosphere Reserves. These reserves (under the UNESCO *Man and Biosphere* programme) do not preclude a human presence and uses, at least in their peripheral and buffer zones. This policy evolution, from the creation of sanctuaries against human presence to conservation with and for humankind, and from protection of a biodiversity compartment to integrated sustainable management of coastal areas, echoes the larger international debate on environmental issues and new approaches that seek to be more ecosystemic (including humans and their knowledge) and ecoregional (Cormier-Salem *et al.*, 2002).

Another notable change on the international scene is the legal status granted to heritage. The status of 'the common good' of humanity was officially adopted in 1982 at the Montego Bay Convention on the rights of the sea (Humbert and Lefeuvre, 1992). This status was granted to mangrove forests in Agenda 21, the set of resolutions submitted for approval by the countries attending the Rio Earth Summit in 1992. But, the Convention on Biological Diversity, drawn up at this same summit meeting, retreated from this stance on international environmental law. While the convention affirms in its preamble that 'conservation of biological diversity is a common concern of mankind', it none the less reiterates the sovereign control of states over their biological resources. At the same time, on the question of *in situ* conservation, this same convention recommends taking into account 'knowledge, innovations and practices of indigenous and local communities' (article 8, §J). Since 1996, reference to Traditional Ecological Knowledge (TEK) has been a constant feature of all biodiversity talks (Posey, 1999; Chouvin *et al.*, 2004). Henceforth, 'indigenous and local communities' are considered to be the prime beneficiaries of benefit sharing (Cormier-Salem and Roussel, 2002). The way is now open for the recognition of local heritages. The proliferation of patrimonial constructions, varying widely in nature and status, leads to questions about their effects.

Instruments that remain defective

Research is increasingly questioning the ecological legitimacy and the economic and social acceptability of regulatory instruments, such as the red-listing of endangered species or the designation of protected areas. In addition to questionable selection criteria for species and spaces to be designated as heritage, such designations can lead to imbalances and ultimately to a loss of biodiversity. Coastal biodiversity in its entirety cannot realistically be conserved as heritage. The choices made translate into a compartmentalization or parcelling out, disturbing fluxes and exchange between ecosystems. Many ecologists have denounced this conception of a two-tiered protection system, with exceptional natural objects on the one hand and 'ordinary' nature on the other, the latter with a precarious status or even seen as a 'waste dump' that is strongly at risk of being neglected, if not destroyed (Génot, 1998). Another risk is that these processes may stall at a certain point, obstructing the remarkable dynamics of mangrove marshes, and perpetuating inadequate regulatory measures that fail to take environmental changes into account. Lastly, field studies reveal the incoherence of these policies and the difficulty encountered in respecting principles of equity – inter- and trans-generational – due in particular to the multiplicity of partners and lack of coordination.

Mangroves, an ecotone between sea and land, are particularly vulnerable to the harmful effects of the compartmentalization of government agencies and administrations, dispersed among different ministries. The administrative jigsaw puzzle observed at the national level is even more evident on a regional scale. Taking the example of the Saloum delta on the Senegalese seashore, the various territorial authorities do not have the same boundaries, and stakeholders are legion. In addition to Water and Forest Department employees and those from the

Fishery Department, there are personnel from the National Parks Division (for the Saloum Delta National Park), the World Conservation Union and UNESCO (for the Saloum Delta Biosphere Reserve), representatives of village associations and federations, producer groups, and international cooperation missions (Japanese, German, Belgian, Swiss, French, etc.), not to mention private investors increasingly attracted by the potential for tourism in the Saloum delta (Cormier-Salem, 1999a).

In addition to this jumble of conservation territories, management plans keep piling up: the same section of mangrove is alternatively designated for reforestation, conversion to shrimp farm ponds, rehabilitation as a recreation site, etc.

To comply with the Convention on Biological Diversity, management projects must reconcile environmental protection with the well-being of the local population. While these projects are in practice implemented at the local level, quite often they are still drawn up by centralized institutions that are too far removed from reality in the field. Local actors find themselves invested with a new 'manager' status that they are not always ready to assume, especially since the transfer of responsibility rarely comes with the corresponding financial means. Furthermore, recognition of indigenous people's priority in terms of access and usage, if not of their exclusive rights, quite often leads to the exclusion of other users, notably migrants, for whom mangroves and their resources have a fundamental role.

Seeking sustainable alternatives

In the wake of the frequent failures of public policy on mangrove conservation, new alternative and sustainable pathways are being explored. The development of non-merchant economic value in mangroves, such as nature-based tourism, or commercial value via labelling of local products, is emerging as a way of reconciling environment and development, maintaining biodiversity and sharing benefits equitably (Dugan 1992; Perrings, 2002).

The aesthetic and visual features of mangroves make them attractive sites. This natural spectacle draws an ever-greater audience, often from Northern countries. Tourism at developed sites is a much-appreciated source of foreign currency for developing countries, and can be seen as an alternative to the depletion of resources, overexploitation of ecosystems and rural population flight. Ecotourism has many harmful side effects, however, in terms of local development and the conservational management of biodiversity. Experience in various mangrove marshes gives rise to questions about the carrying capacity at these sites, the most appropriate types of infrastructure and the outcomes that can be expected, particularly for local users (Blangy, 1999; Young, 2003).

The reputation of mangrove products (honey, rice, salt, crab, oysters, wood, etc.) is well established, and long-distance trade in these products is attested to as early as the first centuries of our era. These products relied on ancient know-how, rooted in a *terroir* (communal territory). Creating economic value and legal protection for these products is at the core of Geographical Indications. This approach, inspired most notably by French *Appellations d'Origine Contrôlées* (AOCs), is meeting with growing success. Creating a heritage is now a goal, in addition to the initial objectives of protection for product names and commercial promotion. Labelling is seen as an instrument for conserving the complexity of the living world, at all levels (biological and cultural), and for upgrading and enriching the status of the lands involved. But, for geographical indicators to perform well, there must be quality and hygiene (Bérard and Marchenay, 2004). In practice, some operators will not be able to afford the cost (economic, but also social and cultural) of complying with standards. Some consumers may no longer have access to these products, at a revalued price. In parallel, standardization of production may lead to a crystallization of practices, and ultimately to a loss of capacity for adaptation and innovation in local knowledge (Chouvin *et al.*, 2004).

Although they have not met with the

anticipated success, the alternatives that are proposed – ecotourism, labelling, eco-certification – are innovative and dynamic, opening up a vast vista for investigation.

By Way of Conclusion: for Integrated and Concerted Mangrove Management

With increasing pressure on coastal areas, mangrove ecosystems have become the stage for a number of claims and confrontations. The new institutional status of natural heritage and new forms of value derived from coastal biodiversity have exacerbated tension among stakeholders. The divergence of views, knowledge, practices and issues at stake is particularly evident between local communities and those who come in from the outside. Yet, even at the local level, strong conflicts of interest have arisen between indigenous users (fisher-farmers) and migrants (woodcutters, nomadic herders, sea fishers), as well as between the older and younger generations, and between men and women. Regionally, local social groups that want to see mangroves recognized as their territory are opposing public and private enterprises. Nationally, public institutions share management responsibilities in these areas: among others, water and forestry, fisheries, tourism, environment ministries, etc., and they often hesitate between exploitation and sanctuary establishment. Finally, at the international level, while environmental activists lobby for protection of species, groups in favour of indigenous peoples' rights are fighting for recognition of their unique cultural traits and for preservation of their knowledge.

In mangrove marshes as in other coastal wetlands, cohabitation is difficult between certain specific, and sometimes antagonistic, economic practices of farmers, herders, hunters, oyster farmers, etc. Coupled with the incoherence of public environmental policies, the complexity of accumulated and overlapping legal and regulatory provisions, and local users' identity crises, the future of coastal wetlands is in question.

The dynamics and complexity of interaction among all the stakeholders in mangrove ecosystems are an incentive to develop integrated and concerted management of coastal zones. The global or integrated approach aims to take into account the full range of stakeholders and factors on diverse spatial and temporal scales. Comparative study of mangroves must also be encouraged in order to highlight the specific features of local contexts, draw up appropriate management scenarios and better guide public policy, both national and international (Cormier-Salem, 1999b).

At the same time, to better understand the motives underlying claims (patrimonial, territorial, identity-driven) to mangroves, more in-depth research should be devoted to the history of these coastal regions. Shell middens have not yet been sufficiently excavated and inventoried. Toponymy, founding myths and other oral traditions also deserve closer linguistic analysis to establish the unique nature of mangrove peoples and possibly to unearth forgotten civilizations.

This three-pronged approach – integrated, comparative and historical – is intended to foster mangrove projects (planning, zoning schemes, etc.) that are adapted to local socio-spatial contexts and negotiated among all stakeholders. Elaboration of these projects is thus a 'theatre' of concertation, where researchers in social sciences can act as mediators, or relays, to establish standards.

References

Babelon, J.P. and Chastel, A. (1994) *La Notion de Patrimoine.* Liana Levi, Paris.

Barbier, E.W. and Sathirathai, S. (2004) *Shrimp Farming and Their Mangrove Loss in Thailand.* MPG Book Ltd., Bodmin, UK.

Bearez, P. (1996) Ictyofaunes marines actuelles et holocènes et reconstitution de l'activité halieutique dans les civilisations précolombiennes de la côte du Manadi Sud (Equateur). Thèse Ichtyologie Générale et Appliquée, Muséum National d'Histoire Naturelle, Paris.

Bérard, L. and Marchenay, P. (2004) *Les Produits de Terroir entre Cultures et Règlements.* CNRS editions, Paris.

Blangy, S. (1999) Tourisme autochtone et communautaire. *Courrier de l'UNESCO,* juillet-août 1999, Paris, pp. 32–33.

Brooks, G.E. (1993) *Landlord & Strangers: Ecology, Society and Trade in Western Africa, 1000–1630.* Westview Press, San Francisco, California.

Carlson, D.G. (1984) *African Fever: A Study of British Science, Technology, and Politics in West Africa, 1787–1864.* Science History Publications. Watson Publishing, Massachusetts.

Chouvin, E., Louafi, S. and Roussel, B. (2004) *Prendre en Compte les Savoirs et Savoir-faire Locaux sur la Nature: les Expériences Françaises.* IDDRI, Les documents de travail de l'IDDRI, Paris.

Cormier-Salem, M.-C. (ed.) (1999a) *Rivières du Sud: Sociétés et Mangroves Ouest-africaines.* 2 volumes. IRD, Paris.

Cormier-Salem, M.-C. (1999b) The mangrove: an area to be cleared ... for social scientists. *Hydrobiologia* 413, 135–142.

Cormier-Salem, M.-C. (2002) Mouvantes mangroves. In: Baron-Yelles, N., Goeldner-Gianella, L. and Velut, S. (eds) Le *Littoral: Regards, Pratiques et Savoirs.* Éditions rue d'Ulm, Conservatoire du Littoral, Paris, pp. 269–284.

Cormier-Salem, M.-C. and Roussel, B. (2000) Patrimoines naturels: la surenchère. *La Recherche* 333, 106–110.

Cormier-Salem, M.-C. and Roussel, B. (2002) Patrimoines et savoirs naturalistes locaux. In: Martin, J.Y. (ed.) *Développement Durable? Doctrines, Pratiques, Évaluations.* IRD, Paris, pp. 125–142.

Cormier-Salem, M.-C., Juhé-Beaulaton, D., Boutrais, J. and Roussel, B. (eds) (2002) *Patrimonialiser la Nature Tropicale: Dynamiques Locales, Enjeux Internationaux.* IRD, Paris.

Craighead, J.G. (1997) *Everglades Wildguide: The Natural History of Everglades National Park, Florida.* US Department of the Interior, National Park Service, Handbook 143. Washington, DC.

Dugan, P.J. (ed.) (1992) *La Conservation des Zones Humides: Problèmes Actuels et Mesures à Prendre.* UICN, Gland, Switzerland.

Génot, J.C. (1998) *Ecologiquement Correct ou Protection Contre Nature?* Edisud, Paris.

Goedefroit, S., Chaboud, C. and Breton, Y. (eds) (2002) *La Ruée vers l'Or Rose: Regards Croisés sur la Pêche Crevettière Traditionnelle à Madagascar.* Co-édition PNRC/DID/IRD, Paris.

Haraprasad, C. (1999) *The Mystery of the Sundarbans.* A Mukherjee Co. Pvt. Ltd., Calcutta.

Higham, C.F.W. (1988) *The Prehistory of Mainland South-east Asia: From 10,000 BC to the Fall of Angkor.* Cambridge University Press, London.

Hook, D.D., Me Kee, W.H. and Smith, H.K. (eds) (1988) *The Ecology and Management of Wetlands. II. Management, Use and Value of Westerns.* Timber Press, Portland, Oregon.

Humbert, G. and Lefeuvre, J.-C. (1992) A chacun son patrimoine ou patrimoine commun? In: Jollivet, M. (ed.) *Sciences de la Nature, Sciences de la Société: les Passeurs de Frontières.* CNRS, Paris, pp. 287–294.

Kneip, L.M. (2001) *O Sambaqui de Manitiba I e Outros Sambaquis de Saquarino.* Dpto. de Antropologia, Museu Nacional, Universidade Federal do Rio de Janeiro, coll. Documento de Trabalho Serie Arqueologia. Rio de Janeiro.

Mitsch, W.J. (ed.) (1995) Restoration and creation of wetlands: scientific basis and measuring success. *Ecological Engineering* 4(2), 61–162.

Montibert, N. (2001) *Enjeux, Usages et Appropriation de la Mangrove en Menabe Central.* Mémoire de DEA EMTS, Muséum National d'Histoire Naturelle, Paris.

Mouchet, J. and Brengue, J. (1990) Interfaces agriculture-santé dans les domaines de l'épidémiologie des maladies à vecteurs et de la lutte antivectorielle. *Bulletin de Sociopathie et Pathologie Exotique* 83, 376–393.

Mouchet, J., Faye, O. and Handschumacher, P. (1994) Les vecteurs de maladie dans les mangroves des Rivières du Sud. In: Cormier-Salem, M.C. (ed.) *Dynamique et Usages de la Mangrove dans les Pays des Rivières du Sud (du Sénégal à la Sierra Leone).* Orstom, Paris, pp. 117–123.

Musset, A. (ed.) (1998) *Les Littoraux Latino-américains: Terres à Découvrir.* IHEAL-CREDAL, Paris.

Neumann, R.P. (ed.) (1998) *Imposing Wilderness: Struggles over Livehood and Nature Preservation in Africa.* University of California Press, Berkeley, California.

Perrings, C. (2002) Sustainable and equitable use of biodiversity. In: ISEE, *Environment and Development,* Keynote for the 7th Biennial Conference of the International Society for Ecological Economics, 6–9 March 2002, Sousse, Tunisia.

Posey, D.A. (ed.) (1999) *Cultural and Spiritual Values of Biodiversity: A Complementary Contribution to the Global Biodiversity Assessment.* UNEP, Nairobi, Kenya.

Razanaka, S., Grouzis, M., Milleville, P., Moizo, B. and Aubry, C. (eds) (2001) *Sociétés Paysannes, Transitions Agraires et Dynamiques Écologiques dans le Sud-ouest de Madagascar.* CNRE/IRD, Antananarivo, Madagascar.

Rey, A. (ed.) (1992) *Dictionnaire Historique de la Langue Française.* Le Robert, Paris.

Rodney, W. (1970) *A History of the Upper Guinea Coast, 1545 to 1800.* Clarendon Press, Oxford, UK.

Rollet, B. (1981) *Bibliography on Mangrove Research (1600–1975).* UNESCO, Paris.

Rue, O. (1998) *L'Aménagement du Littoral de Guinée (1945–1995): Mémoires de Mangroves.* L'Harmattan, Paris.

Saenger, P. and Bellan, M.F. (1995) *The Mangrove Vegetation of the Atlantic Coast of Africa: A Review.* Laboratoire d'Ecologie Terrestre, Université de Toulouse III, Toulouse, France.

Saenger, P., Hegerl, E.J. and Davie, J.D.S. (eds) (1983) Global status of mangrove systems. *Environmentalist* 3, suppl. 3, 88 pp.

Spalding, M., Blasco, F. and Fields, C. (1997) *World Mangrove Atlas.* The International Society for Mangrove Ecosystems, Okinawa, Japan.

UNESCO-UNDP (1986) *Workshop on Human-Induced Stresses on Mangrove Ecosystems,* Bogor, Indonesia, 2–7 October 1984, New Delhi.

Williams, M. (ed.) (1990) *Wetlands: A Threatened Landscape.* Basil Blackwell, Oxford, UK.

Young, E. (2003) Balancing conservation with development in marine-dependent communities: is ecotourism an empty promise? In: Zimmerer, K.S. and Bassett, T.J. (eds) *Political Ecology: An Integrative Approach to Geography and Environment-Development Studies.* The Guilford Press, New York, pp. 29–49.

Zimmerer, K.S. and Bassett, T.J. (eds) (2003) *Political Ecology: An Integrative Approach to Geography and Environment-Development Studies.* The Guilford Press, New York.

14 Comparing Land-use Planning Approaches in the Coastal Mekong Delta of Vietnam

N.H. Trung,[1] L.Q. Tri,[2] M.E.F. van Mensvoort[3] and A.K. Bregt[4]

[1] *College of Technology, Can Tho University, Can Tho City, Vietnam, e-mail: nhtrung@ctu.edu.vn*

[2] *College of Agriculture, Can Tho University, Can Tho City, Vietnam*

[3] *Laboratory of Soil Science and Geology, Wageningen University, Wageningen, The Netherlands*

[4] *Center for Geo-Information, Wageningen University, Wageningen, The Netherlands.*

Abstract

This chapter presents the application and comparison of three land-use planning (LUP) approaches in the coastal area of the Mekong Delta (MD), Vietnam. The land use of the studied area is diverse, quickly shifting and strongly contrasting. The contrast is not only in terms of resources but also in economic profitability and environmental sustainability. We wanted to use LUP approaches representing various levels of complexity and computation intensity, from empirical and qualitative to mechanistic and quantitative. From the variety of methods available, we selected a participatory LUP (PLUP) methodology, the guidelines for LUP by FAO enhanced with multi-criteria evaluation (FAO-MCE) and the land-use planning and analysis system (LUPAS) using interactive, multiple-goal linear programming. We used the same planning goal, worked in the same study area and the same period and produced three land-use plans. We compared the credibility, which is the technical and scientific appropriateness of the approach, and the stakeholder acceptability, which is the perception of the stakeholders of its practical value. The LUPAS map was best appreciated by stakeholders, but it also was the most expensive method. When comparing land-use plans of 2003 with actual land use of 2004, the PLUP map, which is disagreed with most strongly by the scientists, agrees best with the actual land use by the farmers. In the dynamic and contrasting land-use systems of the coastal MD, PLUP seems the most suitable approach for short-term advice, but for longer-term planning a combination of methods will probably work best.

Introduction

In the period 1975–1986, Vietnam had a centrally planned economy decreed by 5-year plans with production targets. In the southern part of the country, every province operated a number of large-scale state communal farms. These farms produced industrial crops such as sugarcane or pineapples but also rice or, in the coastal zone, shrimp. Private farmers had to sell predetermined quantities of rice for fixed prices to the government. Land use was planned by local and provincial authorities, guided by the Ministry of Planning and Investment (MPI) and supported by the National Institute for

 Environment and Livelihoods in Tropical Coastal Zones (eds C.T. Hoanh, T.P. Tuong, J.W. Gowing and B. Hardy)

Agricultural Planning and Projection (NIAPP).

In 1986, economic liberalization was accomplished through the Vietnamese *doi moi* (renovation) policy. State farms were reformed and became cooperatives, where land-use decisions were left to the farmers. Private farmers negotiated long-term lease contracts for land-use rights with the local authorities and were free to decide about land use themselves. These changes had consequences for the role of NIAPP and MPI: from top-down centralized planning agencies they became advisers in land-use planning.

The FAO Framework for Land Evaluation was used most widely as a methodology for land-use advice. Mekong Delta-wide studies and studies at the district level were carried out. Some studies were a purely biophysical assessment of crop growth possibilities, others were enhanced with economic data on the evaluated land-use systems. The FAO method was also the basis for the NIAPP land-use planning study of Vinh Loi District (NIAPP, 1999). The strongly contrasting land-use types (extensive and intensive shrimp, mangrove forestry, double or single rice), with often unknown or hard-to-determine requirements (saltwater *versus* freshwater, tidal movement, growth conditions of mangrove trees), made a reliable assessment difficult. The approach was rather top-down, with limited interdisciplinary interaction and weak communication among stakeholders. This resulted in conflicting interests between stakeholders (Hoanh, 1996). Other problems also surfaced, such as the environmental effects and unreliability of shrimp cultivation and the acceptance of proposed land uses by the local people. Therefore, to ensure a more sustainable development, it is essential to introduce a land-use planning approach that can overcome these problems and better support the land-users and other stakeholders in the coastal area of the Mekong Delta (MD).

A multitude of recent land-use analysis and planning methods is available, for example, the land-use planning (LUP) guidelines by FAO (1993), the participatory land-use planning (PLUP) methodology (Amler *et al.*, 1999), the conversion of land use and its effects (CLUE) of Veldkamp and Fresco (1996), the trade-off model (Stoorvogel, 2001) and the land-use planning and analysis system (LUPAS) described by Hoanh *et al.* (2000), van Ittersum *et al.* (2004) and Roetter *et al.* (2005). They vary in degree of complexity from empirical to mechanistic, and in degree of computation from qualitative to quantitative. We believe that the methods can be grouped into: (i) participatory qualitative empirical methods using farmer and expert knowledge such as PLUP; (ii) semi-quantitative and mechanistic methods such as the FAO guidelines; and (iii) quantitative and computational methods such as CLUE (based on regression) or LUPAS (based on multiple-goal programming).

It is hard to choose a 'most suited' method for the coastal MD. Each method is developed for a certain purpose or covers only a part of the land-use planning sequence set out by FAO (1993), but all methods are developed to advise on land use. Instead of choosing one single method, we decided to select three different approaches, apply them to the same study area and try to recognize the pros and cons of each approach. We used approaches with a different level of complexity and computation intensity, we compared the approaches in credibility (the technical and scientific appropriateness) and acceptability (the perception of the stakeholders on their practical value) and we hoped to come to an objective judgement for application in the coastal MD.

The Study Area

This study was carried out in an area of approximately 10,700 ha in two coastal villages of the MD: Vinh My and Vinh Thinh (Fig. 14.1). The study area has contrasting degrees of saltwater intrusion and land-use systems. Most of the area is now (2005) used for shrimp cultivation. There are two cultivation techniques in these villages: improved extensive and semi-intensive, with a difference in level of applied technology, feeding strategy, recruitment system (natural recruitment for improved extensive or stocking with larvae from hatcheries for semi-intensive), stocking density and use of chemical inputs. The risks involved in shrimp cultivation come from shrimp disease, water pollution, lack of technology and capital, bad farm management and the quality of shrimp larvae (Kempen, 2004).

There is a thin strip of strongly exploited mangrove forests along the coast. Farmers practice the government-controlled forest–shrimp system. Behind the mangrove zone, salt production occurs, sometimes in combination with shrimp. Rainfed rice can still be found in the northern and central part of the area. Most farmers can grow only one high-yielding variety per year in the wet season because salinity intrusion in the dry season inhibits a second crop. Sometimes the single rice crop is combined with vegetables in the dry season, but this is limited to areas where fresh groundwater is available for irrigation. Combined rice–shrimp systems were also expected, but no farmers were practising the rice part of this system during the rainy season of 2003 because there had not been enough rainfall to flush the accumulated salts from the soil at the beginning of the rainy season (Kempen, 2004).

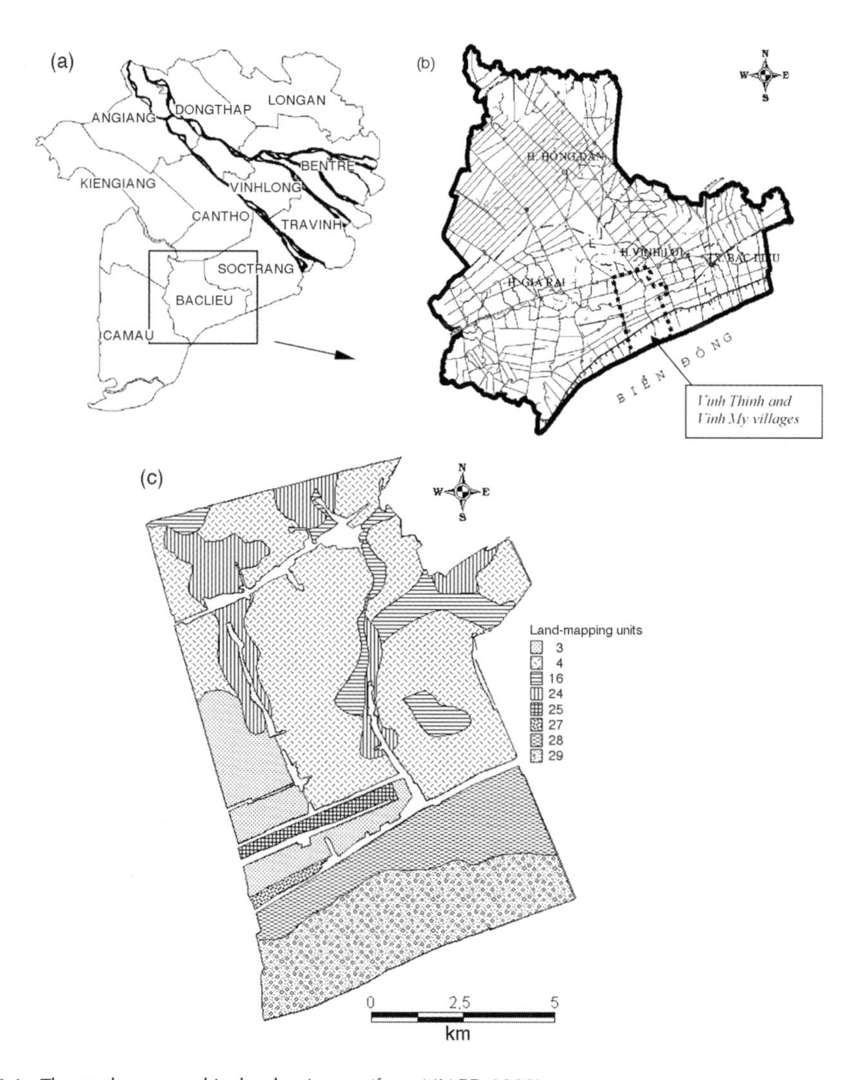

Fig. 14.1. The study area and its land unit map (from NIAPP, 1999).

Four soil types are found (NIAPP, 1999). More than half of the area (5025 ha) has non-acid or weak-acid alluvial soils, which are slightly saline in the dry season. Severe acid sulphate soils (ASS), which are strongly saline in the dry season, can be found along the coast (3134 ha). These soils extend approximately 4 km land-inward. The remaining two soil types, found further inland, are weakly ASS (685 ha) and severe ASS (968 ha), both with moderate salinity in the dry season. NIAPP did not clearly define the degree of salinity (Kempen, 2004).

Methodology

The study has been carried out in four steps: (i) review and select LUP approaches for the study; (ii) apply the selected LUP approaches in the studied area; (iii) compare the credibility and acceptability of the LUP approaches; and (iv) discuss a proper LUP approach for the coastal areas of the MD.

We took the most widely used methodology (FAO-MCE) as a starting point. We wanted methods of different levels of complexity and computation intensity, so we also used a participatory land-use planning (PLUP) method and the land-use planning and analysis system (LUPAS). Why these? The FAO-MCE uses a semi-quantitative and mechanistic approach. In the FAO guidelines (FAO, 1993) as applied in Vietnam (NIAPP, 1999), the biophysical evaluation and socio-economic analysis are carried out in ten steps. The multi-criteria evaluation serves to investigate a number of choice-possibilities in the light of multiple criteria and conflicting objectives or development targets. In doing so, it is possible to generate rankings of alternatives according to their attractiveness (Jansen and Rietveld, 1990).

PLUP has gained increasing recognition as an important tool for reaching sustainable resource management by local communities (Amler *et al.*, 1999). PLUP mobilizes local knowledge and resources for self-reliant development and thus reduces the cost to governments for development assistance. People's participation is also recognized as an essential element in strategies for sustainable agriculture since the rural environment can be protected only with the active collaboration of the local population (FAO, 1991; WB, 1996).

LUPAS is a computerized decision support system for strategic planning based on interactive multiple-goal linear programming (IMGLP) (De Wit *et al.*, 1988). LUPAS uses a set of tools for yield estimation, quantification of input–output relations and optimization of land use at the regional scale under alternative sets of multiple objectives and constraints (van Ittersum *et al.*, 2004; Roetter *et al.*, 2005). LUPAS addresses the questions 'What would be possible?' and 'What has to be changed?' It can be applied for scenario analysis of complex problems such as conflicts in land use (Hoanh *et al.*, 2000). The LUPAS methodology was developed under the Systems Research Network for Eco-regional Land Use Planning in Tropical Asia (SysNet) project (1996–2000). So, LUPAS meets our technological requirements and has also been developed and applied in Asia.

Application of the FAO-MCE Approach

Figure 14.2 presents what steps were performed in this study. There were eight land mapping units (LMU) in the study area (Fig. 14.1) (NIAPP, 1999). For each LMU, eight land-use types (LUTs) were evaluated. The result of this biophysical land evaluation by NIAPP is shown in Table 14.1.

In the socio-economic and environmental assessment (Fig. 14.2, step 3), secondary data from NIAPP (2001) and data gathered from farm households by a questionnaire are assessed in order to quantify socio-economic and environmental indicators of the land-use systems (LUS). Each indicator is transformed by means of a standardization scheme into a criterion score. The following socio-economic indicators were taken into account: gross income, investment costs, variable costs, total costs, benefit–cost ratio, labour days, environmental impact and financial risk. The environmental assessment was done qualitatively because of a lack of data. The impact of an LUT on the surrounding

Table 14.1. Land suitability classification after biophysical land evaluation.

LMU	S1	S1/S2	S2	S2/S3	S3	N
3	SR, DR, RV		RS		i-e S, s-I S	SS, FS
4	SR, DR, RV	RS			i-e S, s-I S	SS, FS
16, 24	i-e S		RS, s-I S		SR, DR, RV	SS, FS
25			RS, i-e S, s-I S			SR, DR, RV, SS, FS
27, 28	i-e S	SS	i-e S, FS			SR, DR, RS
29	FS			SS	i-e S, s-I S	SR, DR, RS

S1, highly suited; S2, moderately suited; S3, marginally suited; N, unsuited; SR, single rice; DR, double rice; RV, rice–vegetable; RS, rice–shrimp; i-e S, improved extensive shrimp; s-I S, semi-intensive shrimp; SS, salt–shrimp; FS, forest–shrimp.

environment is estimated using six indicators, notably sedimentation, salinization, groundwater use, water pollution with organic wastes and nutrients, the use of fertilizer and chemicals, and (irreversible) terrain adjustments. The terrain adjustments are taken into account only for semi-intensive shrimp, improved extensive shrimp and salt. The terrain adjustments needed for the development of these LUTs are the most severe. The degree of environmental impact of each indicator is determined by the results of the farmer interviews, expert knowledge and literature consultation.

The accessibility analysis (Fig. 14.2, step 4) was performed using the 'Accessibility Analyst' extension of the ArcView software package developed by the International Center for Tropical Agriculture (CIAT). It calculates the travel time (road or, in our case, boat) from any given geographical location to the nearest local market where farmers can sell their products. Farmers can reach the markets within 90 min from any location. This is thought to be enough to keep the products fresh and makes accessibility no limiting factor for socio-economic development.

Figure 14.2, step 5 – generation of land-use scenarios by applying MCE – requires the decision-maker to formulate development targets for which the LUTs are to be evaluated. This person decides what socio-economic and environmental indicators affect the development targets and to what degree. Impact weights have to be assigned to the indicators for each development target.

A land-use scenario can be defined on the basis of development targets with their priority weights (e.g. a 25% priority for economic development and a 75% priority for environmental conservation) and the weighted linear combination has to be applied to the alternative LUTs. For each LMU, the LUT that has the highest final evaluation score for a given development target can be regarded as the most suited. The generation of land-use scenarios based on socio-economic and environmental assessments is a trade-off between LUTs: the best is chosen for a given set of priority weights. Table 14.2 presents the priority weighting sets that were applied to the development targets. In this table, both the second and third scenarios put an accent on social security but the second emphasizes job creation (with a high impact weight for the labour day indicator) while the third minimizes financial risk (high impact weight for this indicator).

When a high priority is given to economic development, most of the LMUs are assigned to semi-intensive shrimp (Fig. 14.3, scenario 6). However, the feasibility of this scenario in terms of labour, capital and technology is not well known. When social security has a high priority, most land is assigned to rice–vegetable or single rice. However, according to farmers, the vegetable market is small in the study area. When environmental sustainability gets a high priority, most land is assigned to single rice. Because of the low income from rice, farmers may not accept this scenario. When all targets have the same priority, single rice is the main LUT; forest–shrimp is mainly suitable near the coast, with rice–vegetable and rice–shrimp in the highland area and near canals (Fig. 14.3, scenario 5).

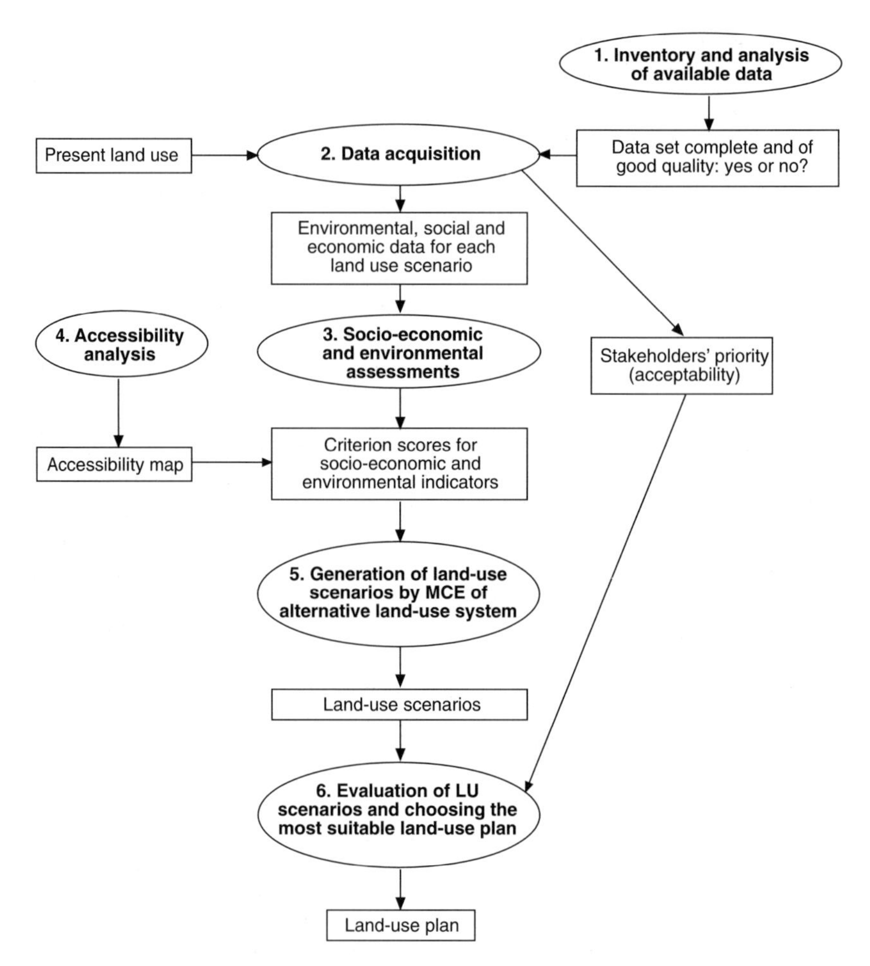

Fig. 14.2. Integrating the FAO approach with MCE (from Kempen, 2004).

The last step is evaluation and selection of the most suitable land-use plan. The evaluation must not only take place on the basis of the development targets but also take into account the wishes and needs of the involved farmers and communities. The final decision must be a compromise between top-down and bottom-up approaches to satisfy the needs of both government and participating stakeholders, communities and municipalities.

Figure 14.3 also presents a proposed LUP based on land-use scenario analysis. The proposed LUP gives priority to economic development while attempting to reduce environmental and financial risk. Shrimp farms should be located close to a canal for proper access to a saltwater source. Rice–shrimp is a safer alternative in the north of the study area, close to the Quan Lo Phung Hiep salinity protection area. The reason is that polluted water from the water control activities may make single-shrimp cultivation risky (Kempen, 2004).

Application of the PLUP Approach

A modified participatory rural appraisal (PRA) was used based on the toolbox designed by Ticheler *et al.* (2000) and on experiences from an earlier study in the same area by Feitsma *et al.* (2002). We took care to avoid their difficulties regarding com-

Table 14.2. Twelve scenarios with weight sets for the development targets (FAO-MCE approach).

Scenario	Economic development	Social security	Environmental sustainability
1	1.0	0	0
2	0	1.0	0
3	0	1.0	0
4	0	0	1.0
5	0.33	0.33	0.33
6	0.75	0.25	0
7	0.25	0.75	0
8	0.25	0	0.75
9	0.75	0	0.25
10	0.60	0.20	0.20
11	0.50	0.25	0.25
12	0.25	0.50	0.25

munication, lack of secondary data, large and scattered hamlets and limited time. Groups of about ten key informants (experienced farmers) were formed in each hamlet. The PLUP was repeated twice, in 2002 and 2003. To have a thorough set of perspectives, agriculture farmers and aquaculture farmers were grouped separately. In each group, farmers participated in reviewing the hamlet's land-use history, described their land conditions and production systems, explained the reasons for land-use change, defined the socio-economic factors that affect the change decisions, drew a sketch map showing the land use and land constraints of their hamlets and proposed the preferred future land use. Transect walks were also conducted to verify the farmers' resource map. During the transect walk, farmers were asked for information on the land and also the land-use types they practised. We analysed land-use change (actual use compared with use in the previous year), the realization of preferences (actual land use compared with what farmers indicated as their preferred land use during the previous year), the preferred change (what they would hope to do next year) and preferences conflicting with those of neighbours.

Results show that land use in the studied area is very dynamic. Within 1 year, more than half of the studied area (58%) changed use, mostly from agriculture to aquaculture (Fig. 14.4a). The land-use change in 2003 was more than could be expected from the preference expressed by both agriculture and aquaculture farmers in 2002. Half of the preferences were realized, mostly in aquaculture (Fig. 14.4b). In the areas where plans could not be realized, aquaculture or mixed agriculture–aquaculture was practised instead of the preferred agriculture. In other locations, aquaculture was also practised instead of the preferred combination of agriculture–aquaculture.

The major change in farmers' preference was the increased preference for aquaculture at the expense of agriculture (Fig. 14.4c). While in 2002 the farmers' preference for agriculture covered 27% of the area, in 2003 this was only 4%. The preference change from agriculture to aquaculture or to mixed agriculture–aquaculture was about 23.6% of the total area. The preference change from mixed agriculture–aquaculture to mono aquaculture covered 17.6% of the area. The main reasons for those changes were that aquaculture has a higher profit than rice, and that increasing saltwater intrusion due to the expansion of aquaculture forces farmers to plan for aquaculture as other agricultural practices become virtually impossible because of the lack of fresh water (Kempen, 2004). Moreover, according to the adjustment plan for the coastal areas of Bac Lieu, the government was advised to invest in dredging existing canals and excavating new canals for aquaculture development (NIAPP, 2001).

Conflicts in preference were analysed in seven hamlets with both agriculture and aquaculture groups. The 2003 preference maps of both groups were overlaid to delineate the areas of preference conflict. The difference in preference was classified into five levels (Fig. 14.4d): (i) same land-use preference; (ii) partly different preference based on natural conditions; (iii) partly different preference based on economic considerations; (iv) completely different preference based on natural conditions; and (v) completely different preference based on economic considerations. In most of the cases, the aquaculture groups wanted to convert part of the agricultural land into shrimp land while the agriculture groups wanted to continue

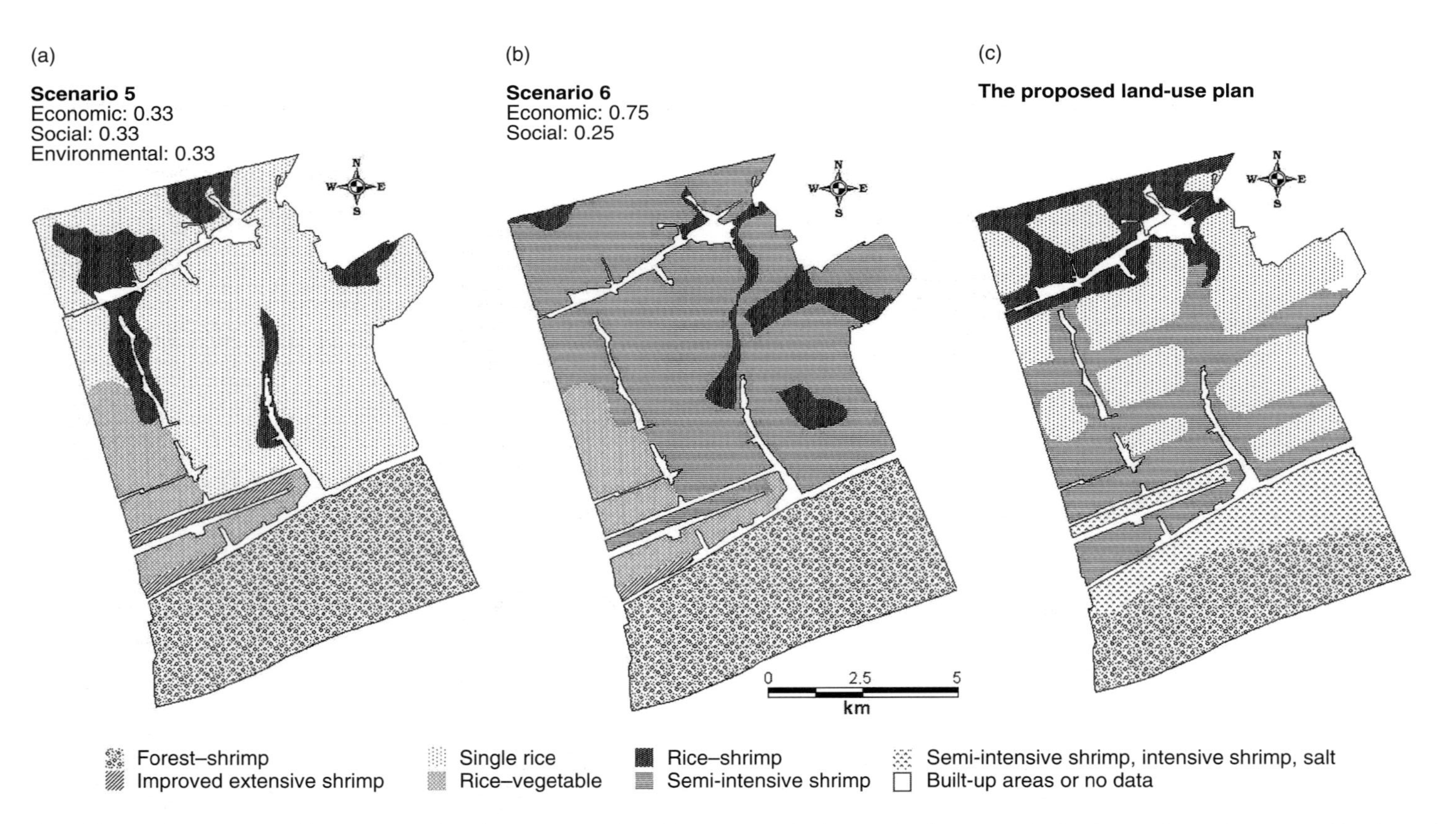

Fig. 14.3. Land-use scenario 5 (same priority for all development targets), scenario 6 (higher priority for economic development) and the proposed land-use plan (from Kempen, 2004).

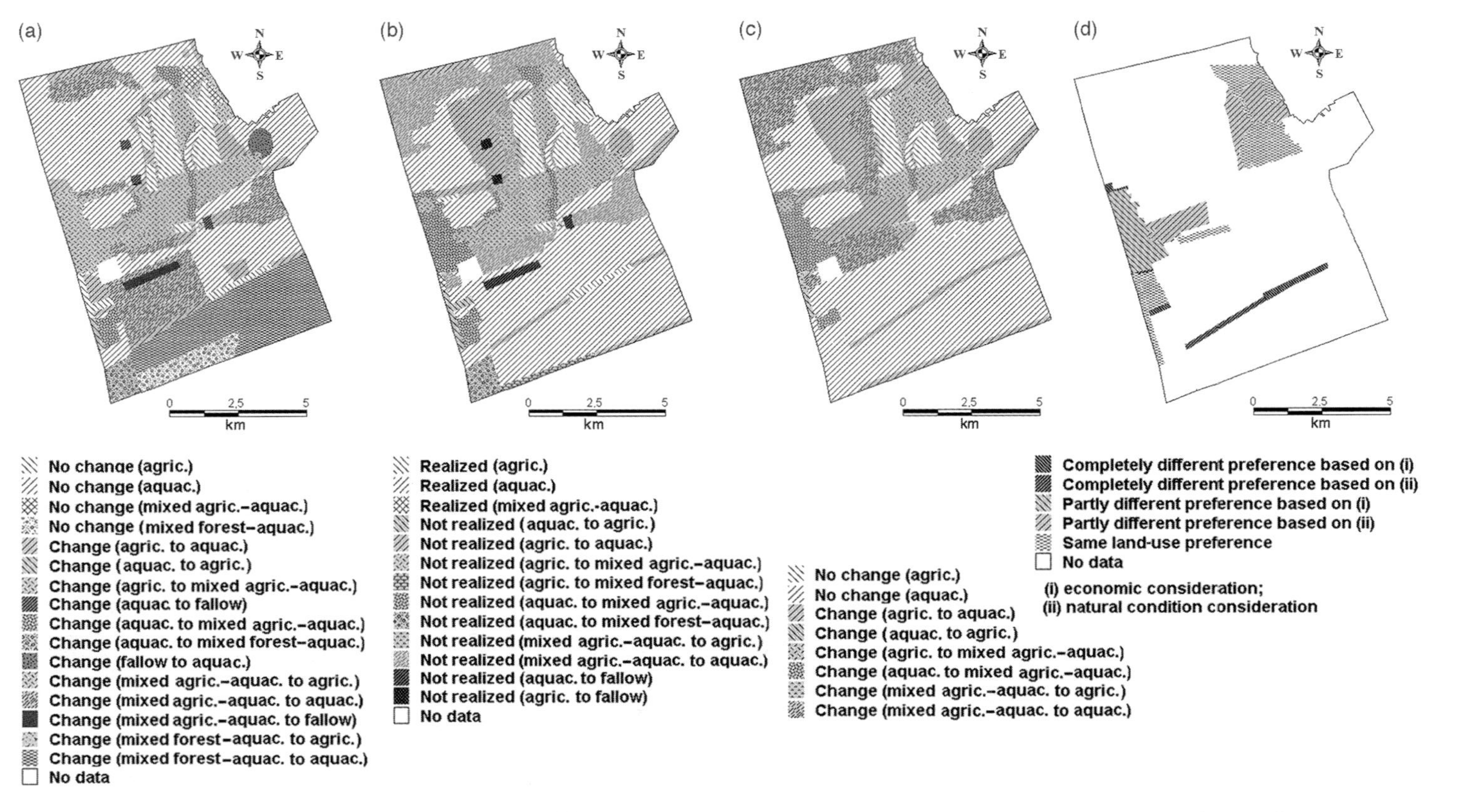

Fig. 14.4. (a) Land-use change 2002–2003; (b) realization of the farmers' preferences in 2002 to the actual land use in 2003, (c) preference change 2002–2003, and (d) preference conflict in 2003.

cultivating their crops. The agriculture groups either lacked capital and knowledge of aquaculture or believed that rice and vegetables were less risky and still profitable.

Application of the LUPAS Approach

The structure of LUPAS is illustrated in Fig. 14.5a and consists of a resource balance and land evaluation, a yield estimation, an input–output estimation and an interactive multiple-goal linear programming (IMGLP) part (van Ittersum *et al.*, 2004).

Resource balance and land evaluation

Similar resource requirements create competition between LUTs. It is critical to determine the resource availability and subsequently the potential or limits of production (Ismail *et al.*, 2000). In LUPAS, the studied area is divided into land units (LU), unique combinations of an agro-ecological and an administrative unit. The LU map is used for spatial display as well as for the IMGLP. In this study, the LMUs of NIAPP (1999, 2001) are overlain with the village boundary map, resulting in 30 LUs.

Eleven promising LUTs considered in this study were single rice, rice–vegetable, vegetable, extensive shrimp, modified extensive shrimp, semi-intensive shrimp, salt, salt–shrimp, forest–shrimp, rice–shrimp and mangrove forest.

Resource needs such as labour, capital, land and water were assessed. Available land for production (10,700 ha) was determined by excluding built-up and protected areas. Available labour was based on the total population, 17,700 in Vinh My and 10,480 in Vinh Thinh (NIAPP, 1999). Tri *et al.* (2002) claim that 60% of the population in the study area is between 18 and 60 years old. We observed many labourers under 17 and from adjacent villages. Because data on capital used are not available, they were estimated from the current input cost for actual land use plus available credit.

Yield and input–output estimation

The main tools and techniques used for input–output and yield estimation are crop yield simulation or statistical models, expert judgement and farm surveys (Hoanh *et al.*, 2000). Input–output is described by total input cost, labour requirement and revenue of each promising LUT per LU. In this study, yield and input–output of LUTs are determined at the current technical level and at an improved technical level. These levels are based on recent average and recent maximum values from a field survey or previous studies.

Interactive, multiple-goal linear programming

Based on existing land-use planning (NIAPP, 1999), annual development strategy documents and actual land use, objectives and goal restrictions are distinguished. They are to maximize the total regional income from agriculture and aquaculture products, to realize the strategic rice, shrimp, salt and vegetable production quotas and to protect the mangrove forest. Since our objective is to compare LUPAS with PLUP and the FAO-MCE, the case study was narrowed to maximizing the total income in two scenarios: (i) all farmers apply the actual technical level (this refers to the production techniques currently practised by the majority of the farmers in the studied area); and (ii) all farmers apply an improved technical level that refers to a higher level of production, the 'attainable yield' (Tawang *et al.*, 2000), by advanced farmers in the study area.

The study area had the following resource constraints and goal restrictions:

- The total area of all LUTs allocated in an LU must be less than or equal to the total area available of that LU;
- The total labour needed for all planned production activities in a village must be less than or equal to the labour available in that village;
- The total capital need for the allocated LUTs in an LU must be less than or equal to the total available capital of that LU;

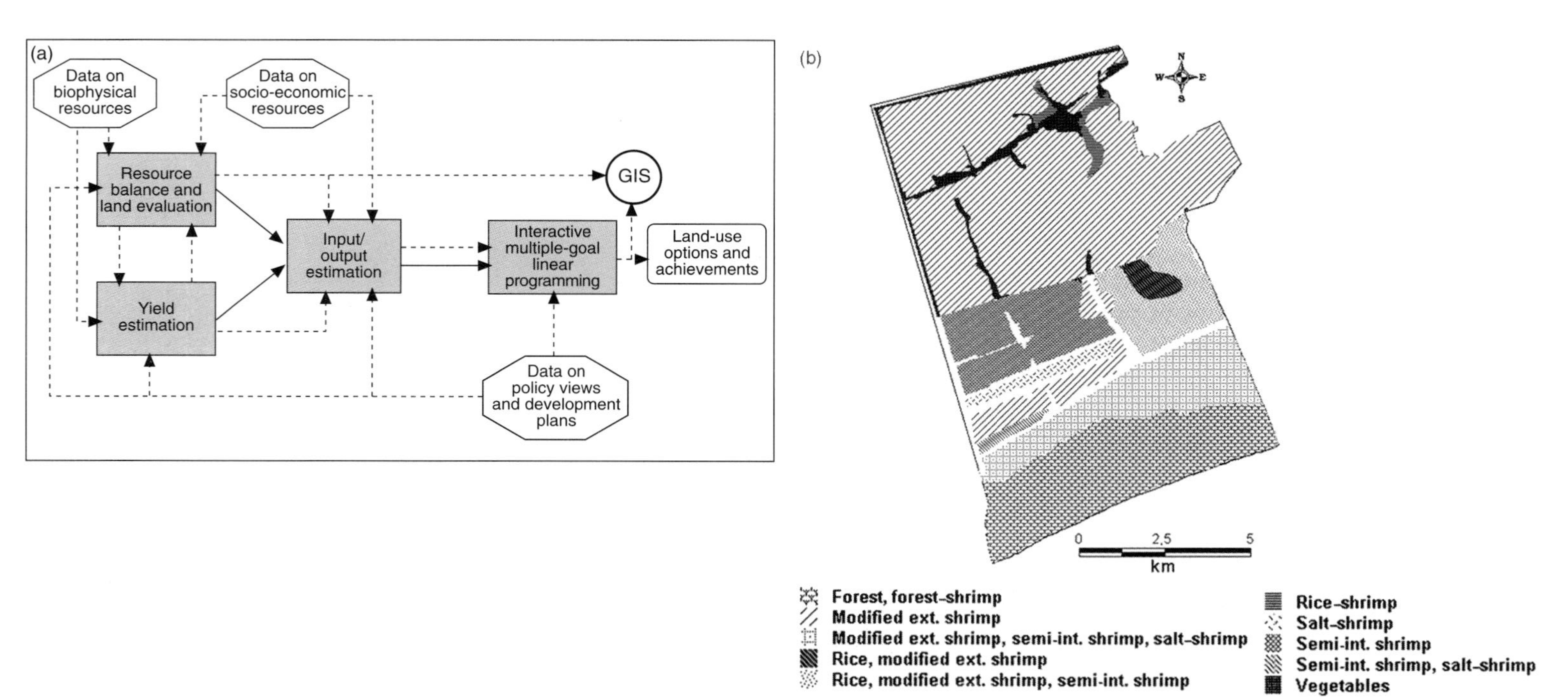

Fig. 14.5. (a) Components of LUPAS (from Roetter *et al.*, 2000) and (b) land-use planning for scenario 9 (with rice, vegetable, salt and forest goals) at current technical level.

- The total rice, shrimp, salt and vegetable production of the study area must be greater than or equal to the rice, shrimp, salt and vegetable production required/ targeted by the local government; and
- The total mangrove forest areas allocated in the study area must be greater than or equal to the mangrove forest area targeted by the local government.

A LUPAS model for the case study was developed using the GAMS software (GAMS Development Corporation, <http://www.gams.com>). Table 14.3 shows the results of the nine feasible scenarios based on combinations of objectives with sets of constraints and/or goal restrictions. The first scenario represents the most favourable conditions, when only biophysical suitability of the land and the land area constraint apply. It generates the potential biophysical production, which is hard to achieve. However, it can be used to evaluate the potential income from agriculture and aquaculture under the actual biophysical conditions. The subsequent scenarios demonstrate what constraints and goal restrictions affect the overall goal most, so that the tradeoffs between goals can be analysed and a more feasible and sustainable land-use plan made.

In general, the results (Table 14.3) imply that:

- For maximizing the total income of the study area, the model assigns a high proportion of land area to shrimp LUTs. This is very risky in case a drop in shrimp yield is experienced;
- Twice the income can be achieved by improving cultivation technology to the existing maximum level;
- Capital and cultivation techniques are the main constraints. Labour problems can be solved by using machines, especially in land preparation and harvesting; and
- Goal restrictions (upper limit of production targets) slightly affect the total income but strongly influence land allocation. Thus, by changing goal restrictions, the risk can be reduced, for example, reducing the shrimp production target.

Figure 14.5b presents the land-use planning resulting from scenario 9 when targets of rice, vegetable, salt and forest area requirement were taken into account at the present technical level.

Comparing the Three LUP Approaches

We used acceptability (agreement of stakeholders with the results and requirements of the three methods) and credibility (prediction quality and uncertainty of the plans) as criteria to compare the LUP approaches. The acceptability and credibility criteria were derived from a framework presented by van der Molen (1999).

Acceptability

The agreement was prepared by asking 25 stakeholders their opinion (agree, partly agree, disagree) on the three LUP maps for 2004 that had resulted from our approaches. They had no prior knowledge of which method produced which map: results are in Table 14.4. The stakeholders are managers (provincial and district politicians), local experts (staff of extension services) and scientists (university staff and researchers in LUP). LUPAS scores best, with 21/25 partly agreeing and 4/25 fully agreeing with the LUPAS plan. Most answered partly because they feared insufficient provision of fresh water for rice and rice–shrimp. Few people fully agree with FAO-MCE and PLUP but far more people disagree with these plans than with the LUPAS plan. The reasons for this are expected insurmountable problems for adequate irrigation (both plans) and far too scattered land allocation (PLUP). The three groups judge LUPAS similarly (and most favourably) but local experts judge PLUP higher than the managers and scientists. In general, the results suggest that the perception of the local experts and local managers is somewhat closer to the farmers' perception than that of the scientists

In terms of resources required for implementation of LUP methods, PLUP is much cheaper (4640 VN dong/ha) than FAO-MCE (20,650) and LUPAS (24,000). PLUP takes the most time and person-months but it requires

Table 14.3. Results of maximizing income at actual technical level (LUPAS approach).

Constraints and targets	Income	Production (t)				Environment		Resource used	
	(10^9 VND)	Rice	Shrimp	Salt	Veg.	Forest (ha)	Land (%)	Capital (10^9 VND)	Labour (10^6 days)
Land	1,944	351	22,611	4,296	15,115	1,096	100	1,241	1.85
Land, labour	1,694	1,474	20,485	11,295	7,208	1,050	84	1,188	1.39
Land, capital	1,111	351	9,123	4,296	15,115	1,096	87	186	1.41
Land, labour, capital	974	738	8,025	30,390	15,115	1,096	83	178	1.29
Land, labour, capital, rice	972	2,940	7,990	27,678	15,115	1,096	86	177	1.29
Land, labour, capital, salt	974	738	8,025	30,390	15,115	1,096	83	178	1.29
Land, labour, capital, vegetable	969	738	8,071	28,821	3,000	1,096	76	174	1.17
Land, labour, capital, rice, vegetable	968	2,940	8,045	28,105	3,000	1,096	80	175	1.18
Land, labour, capital, rice, vegetable, salt, forest	930	2,940	7,845	16,000	3,000	1,360	81	181	1.10

VND, Vietnamese dong.

Table 14.4. Stakeholders' opinions on land-use plans generated by LUPAS, FAO-MCE and PLUP.

Group	Agreement	Number of stakeholders		
		LUPAS	FAO-MCE	PLUP
Managers (9)	Agree	2	3	0
	Partly agree	7	2	5
	Disagree	0	4	4
Local experts (9)	Agree	1	1	3
	Partly agree	8	4	2
	Disagree	0	4	4
Scientists (7)	Agree	1	0	0
	Partly agree	6	2	2
	Disagree	0	5	5
All (25)	Agree	4	4	3
	Partly agree	21	8	9
	Disagree	0	13	13

only simple software (statistics, GIS) for analysis. FAO-MCE requires modelling and statistics skills, whereas, for LUPAS, specialized training and much more expert time for analysis are needed.

Credibility

We compared the land-use plans made in 2003 with the actual land use of 2004. PLUP looks most like the actual use (75% of the surface agrees) and next are LUPAS (62%) and FAO-MCE (33%). The areas planned for aquaculture agreed best (96%, 96% and 76% of the planned area actually had aquaculture the next year). For agriculture (rice, vegetables), FAO-MCE agreed best but this is because in this plan a much larger area was reserved for these activities than in the other two plans. In all three plans, the areas that did not agree were largely mixed systems of

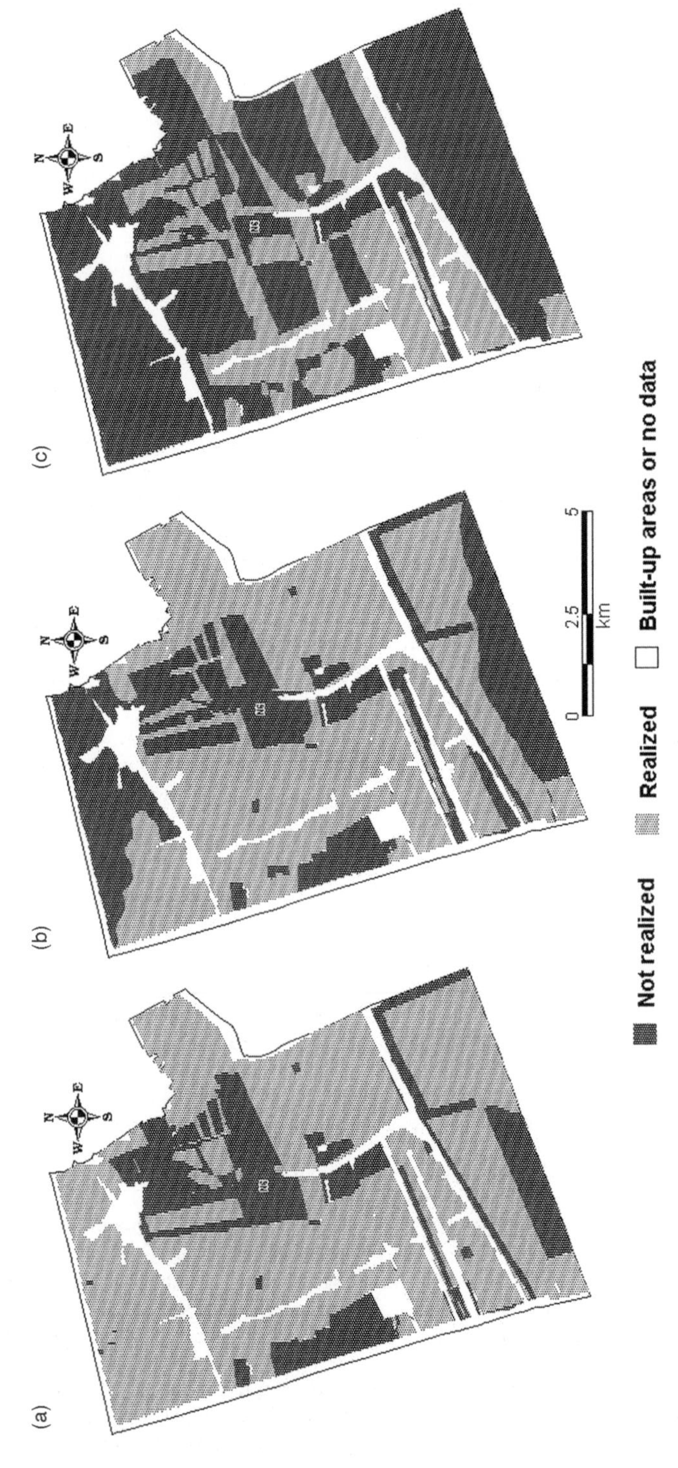

Fig. 14.6. Realization of land-use plans for 2004 of (a) PLUP, (b) LUPAS and (c) FAO-MCE approaches.

rice–shrimp, vegetable–rice or forest–shrimp. In most cases, farmers preferred to go for one crop only, aquaculture, as this one is far more profitable than the other planned activities. So, most mixed systems turned into single aquaculture. Figure 14.6 shows where the plans were realized and where not. Areas where plans were not realized are along the coast (the planned mixed forest–shrimp was turned into shrimp only by the farmers) and in the central-northeastern part, where planned single rice and mixed rice–shrimp turned mostly into single aquaculture.

Discussion and Conclusions

The LUPAS map is the most accepted by managers, scientists and local experts. While scientists do not agree with either the FAO-MCE or the PLUP maps, the proportion of local experts and local managers that fully or partly agree with PLUP is higher than with FAO-MCE.

When comparing land-use plans made in 2003 with the actual land use in 2004, the PLUP map, which is disagreed with most strongly by the scientists, is the most realized. The reason for this may be that the PLUP map, drawn by the farmers, is based on actual land use and for the short term, whereas the other approaches aim at optimized resource use and the long term.

In PLUP, the attitude of the key farmers and the skill of the discussion facilitator during the PRA are the most important factors. In the FAO-MCE approach, it is crucial to detect the right key indicators for each discipline, qualify them and determine their importance. Land evaluation studies supply the input for both LUPAS and FAO-MCE approaches. However, their static description of biophysical conditions seems unsuitable for describing the rapid changes in the coastal area as explained above. Accurate transfer of the socio-economic characteristics of the study area into the model and obtaining precise data on production systems are challenges when applying both LUPAS and FAO-MCE in the coastal area of the MD.

The analysis shows that with dynamic and contrasting land uses, as in the coastal zone of the MD, PLUP seems the most suitable approach since it is capable of acquiring up-to-date information on actual land conditions and of presenting the farmers' land-use preference. In PLUP, the places and causes of land-use conflicts can be defined. This can help land-use planners in finding solutions to achieve an acceptable land-use plan. However, for a sustainable land-use plan over a longer term that can optimize the use of resources and balance different stakeholders' priorities, these LUP approaches should be integrated.

References

Amler, B., Betke, D., Eger, H., Ehrich, C., Hoesle, U., Kohler, A., Kösel, C., Lossau, A.V., Lutz, W., Müller, U., Schwedersky, T., Seidemann, S., Siebert, M., Trux, A. and Zimmermann, W. (1999) *Land Use Planning Methods, Strategies and Tools.* GTZ, Berlin.

De Wit, C.T., Vankeulen, H., Seligman, N.G. and Spharim, I. (1988) Application of interactive multiple-goal programming techniques for analysis and planning of regional agricultural development. *Agricultural Systems* 26, 211–230.

FAO (1991) Plan of action for people's participation in rural development. *FAO Conference Sessions* 26, Rome, p. 19.

FAO (1993) Guidelines for land-use planning. *FAO Development Series* 1, FAO, Rome, p. 96.

Feitsma, M., Vincent, L. and Can, N.D. (2002) Understanding the challenges of PRA for farming systems research: lessons from coastal Vietnam. In: Van Mensvoort, M.E.F. and Tri, L.Q. (eds) *Selected Papers of the Workshop on Integrated Management of Coastal Resources in the Mekong Delta, Vietnam.* C.T. de Wit Graduate School for Production Ecology & Resource Conservation (PE & RC), Wageningen, Netherlands.

Hoanh, C.T. (1996) Development of a computerized aid to integrated land use planning (CAILUP) at regional level in irrigated areas: a case study for the Quan Lo Phung Hiep region in the Mekong Delta, Vietnam. PhD thesis, Wageningen University, Wageningen, Netherlands.

Hoanh, C.T., Roetter, R.P., Aggarwal, P.K., Bakar, I.A., Tawang, A., Lansigan, F.P., Francisco, S., Lai, N.X. and Laborte, A.G. (2000) LUPAS: an operational system for land use scenario analysis. In: Roetter, R.P. *et al.* (eds) *Systems Research for Optimizing Future Land Use in South and South-east Asia.* SysNet Research Paper Series No. 2, International Rice Research Institute, Los Baños, Philippines, pp. 39–53.

Ismail, A.B., Ahmad Shokri, O., Laborte, A.G., Aggarwal, P.K., Lansigan, F.P. and Lai, N.X. (2000) Methodologies for resource balancing and land evaluation as applied in the SysNet case studies. In: Roetter, R.P. *et al.* (eds) *Systems Research for Optimizing Future Land Use in South and South-east Asia.* SysNet Research Paper Series No. 2, International Rice Research Institute, Los Baños, Philippines, pp. 121–131.

Jansen, R. and Rietveld, P. (1990) Multicriteria analysis and GIS: an application to agricultural use in the Netherlands. In: *Geographic Information System for Urban and Regional Planning.* Kluwer Publishers, Dordrecht, Netherlands.

Kempen, B. (2004) Multi-criteria evaluation as a decision-making tool in land use planning on district level: integrating biophysical land evaluation with socio-economic and environmental assessments in a decision support system for land use planning, with applications for the coastal zone of Vinh Loi district, Mekong Delta, Vietnam. MSc thesis, Wageningen University and Research Center, Wageningen, Netherlands.

NIAPP (1999) *Report on Land Evaluation for Agriculture, Forestry and Fishery in Bac Lieu Province* [in Vietnamese]. National Institute of Agriculture Planning and Projection, Vietnam.

NIAPP (2001) *Adjustment Plan for Agriculture, Aquaculture, Salt and Forest Production in South of National Road 1A, Bac Lieu Province* [in Vietnamese]. Ho Chi Minh City, Vietnam.

Roetter, R.P., Keulen, H.V., Laborte, A.G., Hoanh, C.T. and Laar, H.H.V. (eds) (2000) *Systems Research for Optimizing Future Land Use in South and South-east Asia.* SysNet Research Paper Series No. 2, International Rice Research Institute, Los Baños, Philippines, 266 pp.

Roetter, R.P., Hoanh, C.T., Laborte, A.G., Van Keulen, H., Van Ittersum, M.K., Dreiser, C., Van Diepen, C.A., De Ridder, N. and Van Laar, H.H. (2005) Integration of systems network (SysNet) tools for regional land use scenario analysis in Asia. *Environmental Modelling and Software* 20, 291–307.

Stoorvogel, J.J. (2001) *The Tradeoff Analysis Model Version 3.1: A Policy Decision Support System for Agriculture: User Guide.* Wageningen University, Laboratory of Soil Science and Geology, Wageningen, Netherlands.

Tawang, A., Bakar, I.A., Kamaruddin, A.A., Yahya, T.M.T., Abdullah, M.Y., Jaafar, H.B., Othman, A.S., Yusof, A., Ismail, Z., Hamzah, A.R., Jaafar, A.M., Aziz, Z.A., Ghani, M.Z.A., Majid, N.A. and Rashid, A.A. (2000) Developing and appying LUPAS in the Kedah-Perlis Region, Malaysia: methodologies, results and policy implications. In: Roetter, R.P. *et al.* (eds) *Systems Research for Optimizing Future Land Use in South and South-east Asia.* SysNet Research Paper Series No. 2, International Rice Research Institute, Los Baños, Philippines, pp. 71–89.

Ticheler, J., Defoer, T. and Kater, L. (2000) ResourceKit for participatory learning and action research. Detailed field tools for PLARUser's guide for the ResourceKIT, CD-ROM *Managing Soil Fertility in the Tropics.* KIT, Amsterdam, p. 1.

Tri, L.Q., Sanh, N.V., Ha, V.V., Loi, L.T. and Binh, N.S. (2002) Social-economic aspects of farming systems in Vinh Loi, Thanh Phu, and Dam Doi districts, Mekong Delta, Vietnam. In: Van Mensvoort, M.E.F. and Tri, L.Q. (eds) *Selected Papers of the Workshop on Integrated Management of Coastal Resources in the Mekong Delta, Vietnam.* C.T. de Wit Graduate School for Production Ecology & Resource Conservation (PE & RC), Wageningen, Netherlands, pp. 17–27.

van der Molen, D. (1999) The role of eutrophication models in water management. PhD thesis 2626, Wageningen University, Wageningen, Netherlands.

van Ittersum, M.K., Roetter, R.P., van Keulen, H., de Ridder, N., Hoanh, C.T., Laborte, A.G., Aggarwal, P.K., Ismail, A.B., and Tawang, A. (2004) A systems network (SysNet) approach for interactively evaluating strategic land use options at sub-national scale in South and South-east Asia. *Land Use Policy* 21, 101–113.

Veldkamp, A., and Fresco, L.O. (1996) CLUE: a conceptual model to study the conversion of land use and its effects. *Ecological Modelling* 85, 253–270.

WB (1996) *The World Bank Participation Sourcebook.* Environmentally Sustainable Development Publications, World Bank, Washington, DC, 259 pp.

15 Applying the Resource Management Domain (RMD) Concept to Land and Water Use and Management in the Coastal Zone: Case Study of Bac Lieu Province, Vietnam

S.P. Kam,[1] N.V. Nhan,[2] T.P. Tuong,[3] C.T. Hoanh,[4] V.T. Be Nam[2] and A. Maunahan[3]

[1] *WorldFish Center, Penang, Malaysia (formerly with the International Rice Research Institute, Philippines), e-mail: s.kam@cgiar.org*

[2] *Integrated Resources Mapping Center, Sub-National Institute for Agricultural Planning and Projection, Ho Chi Minh City, Vietnam*

[3] *International Rice Research Institute, Metro Manila, Philippines*

[4] *International Water Management Institute, Regional Office for South-east Asia, Penang, Malaysia*

Abstract

Because of changing hydrological conditions due to infrastructure development to prevent salinity intrusion into the coastal zone, local authorities in Bac Lieu Province, Vietnam, faced complex natural resource management issues concerning managing saline and freshwater resources to support diverse production activities in the coastal zone while farmers had to adjust their production strategies. The resource management domain (RMD) concept was applied, using geospatial techniques, to delineate spatial clusters of hamlets that reflected the influence of key environmental factors on land-use changes and the resulting socio-economic conditions of the rural communities. While some socio-economic differentiation was discernible among the hamlet clusters, the clustering was mainly dominated by land-use change and hydrological characteristics. The results, interpreted on a broader scale, supported the identification of land-use and water management zones to accommodate rice-based, shrimp combined with rice, and shrimp-based production systems in the area, thereby reversing an earlier policy of intensifying rice cultivation. The hamlet clusters also provided a sampling frame for selecting pilot sites for evaluating improved rice production techniques with farmers. The RMD approach is meant to provide an analytical platform to support an adaptive land-use planning process to support the use and management of coastal resources regionally and locally.

Introduction

Coastal areas, particularly in the tropics, have historically been the favoured areas for human habitation. The diversity of coastal ecosystems offers a wide range of resources that lend themselves to a multiplicity of uses for people to make a living. Some uses are compatible, whereas others are potentially conflicting. Increasing population pressure

 Environment and Livelihoods in Tropical Coastal Zones (eds C.T. Hoanh, T.P. Tuong, J.W. Gowing and B. Hardy)

invariably brings about more conflicts on the use of natural resources, threatens the sensitive ecological systems in the coastal zone and increases the vulnerability of these systems to natural and man-made disasters.

National and regional authorities take broader viewpoints of national demands and long-term benefits in steering the direction of development. At the same time, local communities, tending to focus on shorter-term benefits, expect to play more proactive roles in determining how they use coastal resources. Both need to adjust and respond to the dynamism of the coastal environment, but often this is not done in consonance. There is therefore a need for some means of bridging between the two and arriving at some common understanding and recognition of the problems encountered in order to identify rational development options whereby actions and activities taken will lead to effective use and wise management of coastal resources.

The delineation of zones, or spatial units of land, is considered fundamental for agricultural planning and natural resource management as it is impractical to deal with individual plots of land or households. Early adoption of this zoning approach for agricultural development, as in the agro-ecological zoning (AEZ) projects of the FAO, entails identifying natural land units that are characterized by biophysical potential and limitations (FAO, 1976). The delineation of AEZs does not take into consideration the human element. The underlying assumption in applying the AEZ approach is that these land units are relatively homogeneous, with inherent suitability (or limitations) for specific uses, and that production systems and technologies proven successful within a particular AEZ can be adopted by farmers in other similar AEZs.

Arguing that natural potential is not a criterion that dominates farmers' choice of activities or enterprises they embrace, Collinson (1996) pointed out that any serious effort at technology transfer requires people-based domains at the local level. This harks back to the concept of the 'farm household systems' hierarchy described by De Kartzow *et al.* (1992), whereby production activities of a farm are grouped into farming household systems (FHHS); several FHHS with similar characteristics make up a farming system (FS), thereby leading to the identification of farming systems zones (FSZ) at the regional level. This concept is based on the principle that farmers with similar problems and development potentials have similar objectives, resource availability and use, strategies and practices. An attempted application of this concept by FAO is in the context of promoting fish farming in Zambia (De Kartzow *et al.*, 1992). However, the feasibility of spatially delineating FSZ, that is, determining homogeneous groups of farm households (based on similarities in their socio-economic circumstances) that occupy distinct and mutually exclusive geographic areas, is questionable (FAO, 1999).

Neither of these approaches is currently considered satisfactory, considering contemporary insight into the interplay of environmental and human factors governing decisions on land use and resource management. In the wake of the 1992 United Nations Conference on Environment and Development (UNCED), popularly known as the Earth Summit, there has been increasingly expressed concern over sustainability issues in development, prompting calls for promotion of more holistic approaches to integrated land-use planning and natural resource management. One key aspect is the integration of environmental, social and economic issues; another is the engagement of stakeholders, particularly local communities, and institutional strengthening for implementing Agenda 21.

In the context of rural development, there remains an underlying need to delineate spatial units, whether for planning or for management of land and other resources. Hence, those involved in researching natural resource management have promoted the concept of resource management domains (RMDs). In general terms, these are domains defined for managing resources at the disposal of stakeholders, embodying the hierarchy of users (farmers, farming communities) to managers (local, regional and national authorities). To serve current orientation towards more holistic approaches, these spatial units should share not only certain common biophysical properties but also socio-cultural-economic characteristics to capture the human dimension in resource use and management.

The RMD concept is a construct of researchers concerned with sustainable natural resource management. From the researchers' viewpoint, it is a means of formalizing and integrating information about the main driving factors of resource use for (rural) development. Modern information management technologies offer opportunities to operationalize the RMD concept. Spatial delineation of RMDs, whose biophysical and socio-economic characteristics can be quantified and mapped, is facilitated by geographic information systems (GIS) technology, although data, scale and other issues related to geospatial modelling need to be considered (Jones, 1996; Syers and Bouma, 1996).

However, resource use and management lie in the realm of national/local authorities and communities that recognize their own problems, have demands for changes and are in a position to bring about these changes and improvements if there is collective will and some mechanism for collective action. In this chapter, we argue the case for using the RMD concept and approach as an objective basis for planning and targeting development strategies to help local/national authorities and farming communities come to common terms on the awareness, sustainable use and management of coastal resources for livelihood. Through a case study of land and water use and management in the coastal province of Bac Lieu in the Mekong Delta in Vietnam, we illustrate how we have attempted to apply the RMD concept in identifying opportunities for management decisions to resolve issues at various scales, with participation and interaction among researchers and local users.

Applying the RMD Concept in the Case of Bac Lieu Province in the Mekong Delta of Vietnam

Located in the eastern coastal zone of the Mekong Delta in Vietnam, about 61% of Bac Lieu Province lies inland from a series of sluices, constructed with the original intention of saline water exclusion for intensifying rice cultivation (see inset of Fig. 15.1). Over a 7-year period (1994–2000) of the phased construction of sluices, various developments occurred that diminished the prospect and promise of development of the rural economy entirely through rice intensification (Tuong *et al.*, 2003). First, the expansion of brackish-water pond culture in the western part of the target area made shrimp production much more lucrative, albeit more risky, than rice production in the eastern part where acid sulphate soils pose constraints to rice cultivation. Second, the decline in the world rice price became a disincentive for Vietnam to intensify rice production. The threat of conflict in the use of the land and water resources loomed as rice farmers in the eastern part and shrimp pond operators in the western part made demands on the freshwater and brackish-water supply, respectively. A workable compromise solution had to be sought rather quickly to diffuse the growing tension that became evident in early 2001.

Resolving such a conflict required a rethinking of the land-/water-use policy and the management implications arising. Rationalizing the spatial land use needs to be based on a systematic evaluation of the prevailing circumstances and the main drivers of land-use change in the area of interest. Investigations by a multidisciplinary team of researchers identified water and soil characteristics as the two most important biophysical factors influencing land use in the area (Tuong *et al.*, 2003). The changes in hydrological dynamics brought about by the westward phased operation of sluices over the 1997–2000 period were captured by shifts in the isolines (contours) representing the threshold salt concentration for rice (7 dS/m) during the dry-season months of January to May each year (Hoanh *et al.*, 2001). A map of soil types provided information on areas with acidity and salinity constraints to cropping. The rapid land-use changes resulting from the phased protection of areas from salinity intrusion were captured by land-use maps created over this period from the inter-

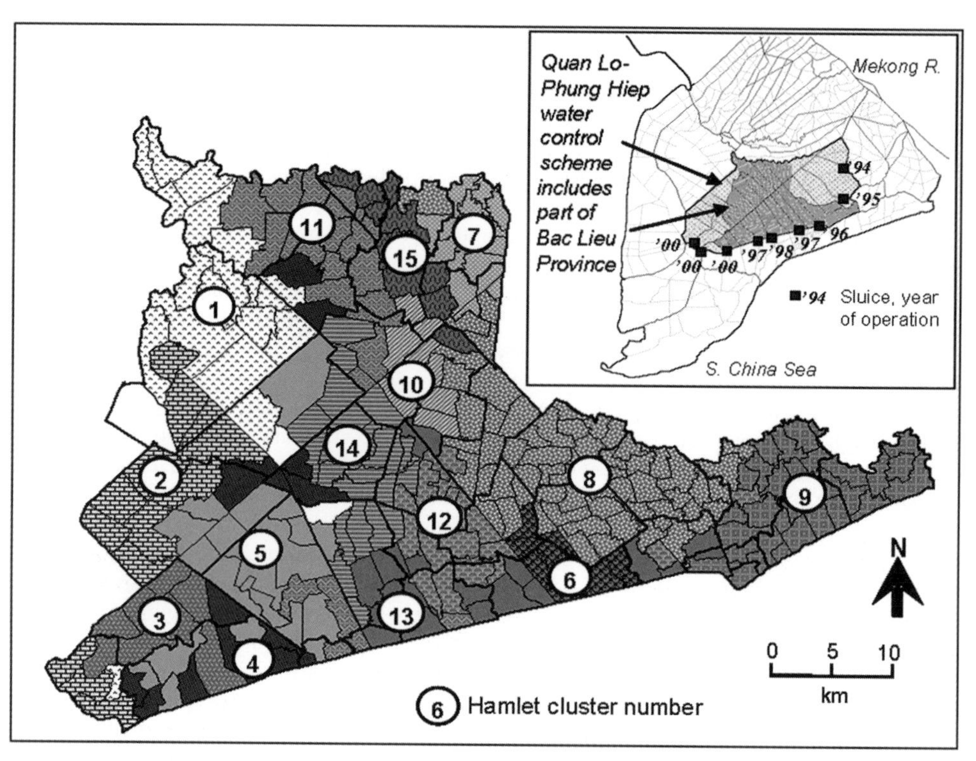

Fig. 15.1. Hamlet clusters within Bac Lieu Province, based on hydrological, soil and land-use characteristics.

pretation of aerial photographs and satellite imagery, supplemented by ground information (Kam *et al.*, 2000). To determine the impact of the hydrological and accompanying land-use changes on the livelihoods of the rural communities, baseline surveys and participatory rural appraisals (PRA) were conducted in 2000 in 14 sample villages representing different hydrological, soil and land-use regimes (Tuong *et al.*, 2003). The surveys indicated considerable variation in socio-economic conditions even at the level of the commune (a collection of hamlets, or villages), suggesting that it would be necessary to characterize socio-economic conditions at the hamlet level.

Phase 1: application of the RMD concept at a broad level

To respond to the urgent needs of the local authorities for a broad characterization of the salinity-protected area that would help formulate a new land-use zoning plan, we performed a preliminary delineation of RMDs based on available hydrological, soil and land-use information using GIS techniques. The collection of socio-economic data to cover all affected hamlets required a comprehensively broad-scale survey. Using the hamlet as the basic spatial unit of mapping, the soil and the temporal hydrological and land-use attributes were read off geo-registered map layers for each of 305 hamlets within the area of interest. The attributes used included percentage of hamlet area that was: (i) under various rice-, shrimp- and rice–shrimp-based production systems over the years 1997 to 2000; (ii) under potential acid sulphate, deep acid sulphate, shallow acid sulphate, saline and alluvial soils; and (iii) affected by saline water, that is, February water salinity exceeding 7 dS/m from 1997 to 2000.

Multivariate clustering was carried out on the hamlets based on these attributes, resulting in the 15 clusters as shown in Fig. 15.1. The clusters in the eastern part (15, 7, 10, 12, 8, 6 and 9) are associated with rice-based land-use systems, whereas the clusters in the western part (1, 2, 3 and 4) are associated with predominantly shrimp-based land-use systems. Except for cluster 4, these clusters are also spatially contiguous. The remaining clusters (5, 11, 13 and 14) show more spatial dispersion, indicating the heterogeneity of conditions in the transitional zone between the eastern 'rice belt' and the western 'shrimp belt'.

The results of the preliminary RMD analysis lent support to the zoning exercise conducted jointly by researchers and provincial authorities as a first step in addressing the emerging land- and water-use conflict issues. Rather than planning the entire area targeted for salinity protection for rice intensification, due recognition was given to the diversity of economic activities as a basis for delineating land-use zones. The zoning exercise, regarded as a scaling-up of the RMDs, was performed jointly by researchers and provincial authorities, who took into consideration the main features of RMD clusters.

Three main classes of economic activities were identified:

- rice-based production systems (two or three rainfed and/or irrigated rice crops) occurring in the eastern 'rice belt' described above,
- shrimp-based production systems (one or two crops of shrimp culture) occurring in the western 'shrimp belt' described above, and
- rice–shrimp rotational production systems (one rice + one shrimp, one rice + two shrimp) occurring mainly in the transitional zone described above.

Further consideration of the hydrological and soil conditions, and the kinds of production systems that can potentially be supported, led to a delineation of six broad land-use and water management zones depicted in Fig. 15.2. These zones could be regarded as a generalization of the hamlet clusters, with some boundary realignment to conform to the canal network configuration.

This zoning provided the basis for determining the biophysical, mainly water, requirements of each production system. For example, the intensified rice areas in the eastern part require a freshwater supply throughout the year, whereas the shrimp-producing areas in the western part require the maximum possible length of saltwater supply. The intermediate zone for rice–shrimp production systems requires a careful balance of saltwater supply to support shrimp culture in the dry season and fresh water for rice cultivation during the rainy season.

The management implications for such a land-use zoning hinge upon the ability to control the operation of the series of sluices to enable a dual regime of saline and fresh water in different parts at different times of the year. Various scenarios of sluice operation were simulated using a hydraulic model to identify the configuration that best satisfies the water requirements of the proposed production systems (Hoanh *et al.*, 2003). The hydraulic model became an operational decision support system (DSS) tool used by the Bac Lieu provincial authorities and the water company controlling the sluice operations in order to implement the land-use proposal.

This policy and accompanying water management intervention at the provincial level opened up opportunities for diverse economic activities, which constitutes a departure from the original policy focusing on rice intensification. These opportunities also meant that the local communities had to respond to the rapidly changing situation and alter their economic activities accordingly. Researchers and local agricultural extension authorities face the challenge of introducing relevant production systems and supporting technologies to help farmers cope with these changes.

Phase 2: application of the RMD concept at a detailed level

We explored the use of the RMD concept at a more detailed scale to help differentiate the conditions for which different production systems and management practices might be suited. We attempted a more detailed charac-

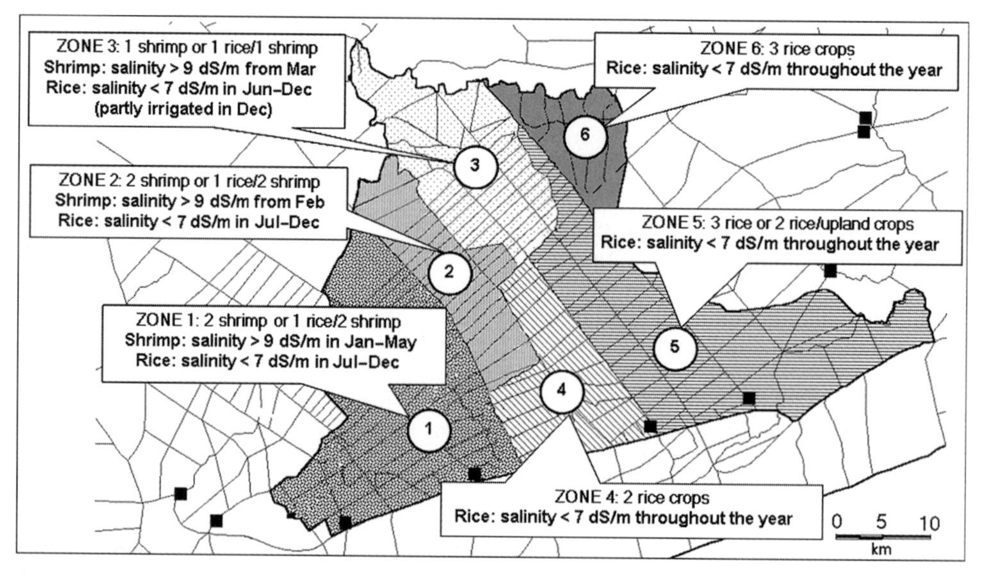

Fig. 15.2. Land-use zones for water management.

terization of the hamlets by incorporating water acidification due to the disturbance of acid subsoil layers by land preparation, as well as socio-economic factors. A broad-scale survey of key socio-economic variables was carried out through a questionnaire survey covering all hamlets within the salinity-protected area in 2001, whereby hamlet heads and key informants were chosen as respondents. The variables selected had been identified as important determinants of household economic activities, based on the findings of earlier PRA (Gowing *et al.*, this volume) and baseline, detailed socio-economic surveys done in sample villages (Hossain *et al.*, this volume).

Because of the diversity of socio-economic conditions among hamlets, only a few socio-economic factors contributed to the distinct spatial pattern of hamlet clustering shown in Fig. 15.3. These included population density, ethnic composition, household landholding size and proportions of agricultural and landless households. The main environmental, land-use and socio-economic characteristics of the 14 hamlet clusters are given in Table 15.1. The spatial clusters obtained do not differ substantially from those obtained earlier, based only on biophysical and land-use characteristics (Fig. 15.1). It is to be noted that the hamlet clusters do not coincide with the communes to which they belong administratively, suggesting that hamlets within a commune can have rather different conditions and needs.

The spatial differentiation of hamlets remains driven mainly by land use, which is in turn influenced by hydrological changes (i.e. shifting of the salinity front as more sluices became operational from east to west) and soil conditions (especially acid sulphate conditions). None the less, addition of the socio-economic variables did alter the membership of hamlets in the clusters. Examples are the separation of cluster 6, consisting of two urban centres, as well as the tendency of hamlets along the southern border to fall into clusters 12 and 13 because of the higher population densities of these hamlets. The southern border of the salinity-protected area is marked not only by the sluices but also by the highway running across Bac Lieu Province, and hence the concentration of settlements along this highway. The additional socio-economic dimension is also reflected in the differentiation of the hamlet clusters of the western 'shrimp belt'. For example, hamlets in clusters 1 and 2 tend to be more agriculturally based, with lower population densities and more agricultural land per

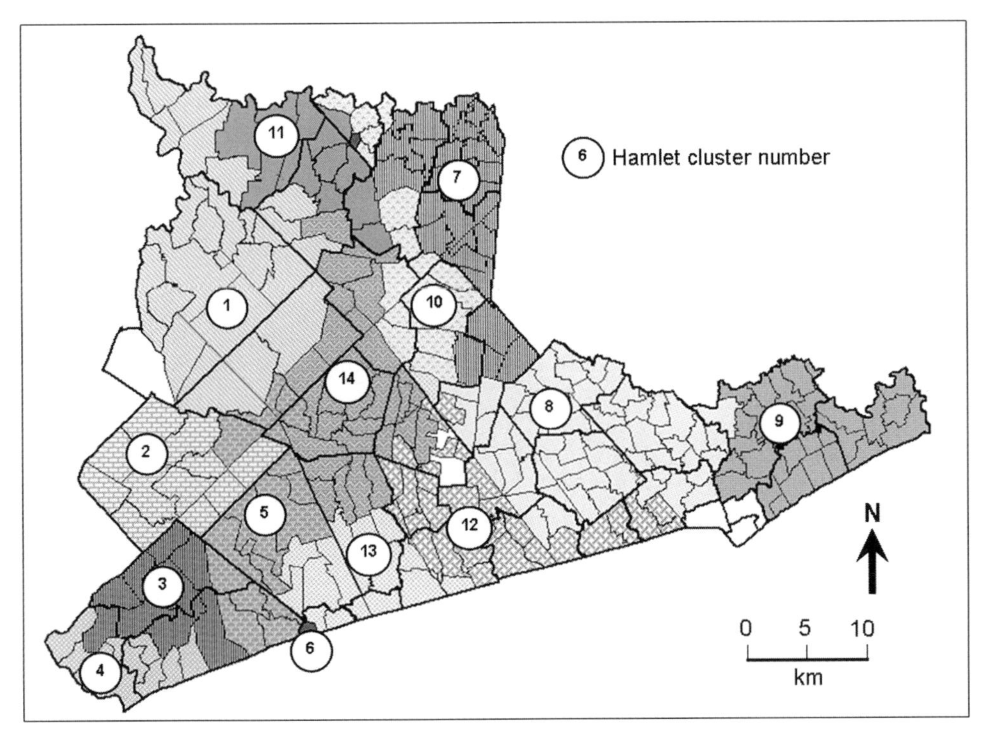

Fig. 15.3. Hamlet clusters within Bac Lieu Province, based on hydrological, water acidification, soil, land-use and selected socio-economic characteristics.

household, whereas in clusters 3 and 4 hamlets are not only more densely populated but also tend to be less rural in character, with higher proportions of households not involved in agriculture. These characteristics are likely to influence household strategies and deployment of resources for agricultural/aquacultural development of these hamlets.

The hamlet clusters provided a sampling frame for selecting hamlets for developing and evaluating new production systems and farming practices. In 2004, 11 hamlets in Bac Lieu Province were selected for conducting on-farm pilot testing of five different production systems. These production systems involve the cultivation of agricultural crops (e.g. system of multiple cropping of rice only, or rice in rotation with upland crops), aquaculture production (e.g. polyculture of shrimp, crab and fish) and combined agriculture and aquaculture production (e.g. system of rice-cum-fish production in rotation with upland crops, rotational shrimp and rice system, and polyculture of shrimp, crab and fish in rotation with upland crops). The results obtained from on-farm tests in the selected hamlets will be evaluated for feasibility and introduced more widely.

Production systems and technologies for particular rice- and shrimp-based production systems successfully tested at selected hamlets would be considered most likely suited for other hamlets within the same cluster, based on the common characteristics that these hamlets share. In other words, the hamlet clusters could constitute the spatial extrapolation domains for promising technologies and management practices. However, variations among farms and households that are not mapable at the hamlet level would still influence the degree of adoption and success of new production systems, practices and technologies. More detailed socio-economic differentiation within hamlets would be more appropriately

Table 15.1. Selected biophysical and socio-economic characteristics of hamlet clusters in Bac Lieu Province, Vietnam.

Cluster	Soil	Water quality	Land use	Socio-economic
1	Acid shallow and deep, with some saline	Saline-free in Jan–Feb, 1999–2000; saline in Apr–May; some hamlets acid-affected in 2001–2002	Single rice area largely maintained; double rice appeared in 2000 but declined in 2001; shrimp farming expanded since 1997 and steadily increased	Lower population density
2	Acid shallow and deep	Saline-free since 2000; acid-affected 2001–2002	Switch from single to double rice in 2000 but decline of double rice in 2001; shrimp farming expanded since 1997 and steadily increased	No non-agricultural households
3	Acid shallow and deep	Saline-free in Jan–Feb, 2000; saline in Apr–May; acid-affected in 2001–2002	Switch from single rice to shrimp–rice and shrimp since 1997	Lower population density; very few non-agricultural households; more agricultural land per household
4	Saline and acid deep	Saline-free in Jan–Feb 2000; acid-affected in 2001–2002	Switch from single to double rice in 2000, with slight decline in 2001; extensive shrimp cultivation in 2001	Higher population density among rural hamlets; higher proportion of non-agricultural households; significant proportion of landless households
5	Acid deep and shallow, with some saline	Saline-free in Jan–Feb, 1999–2000; saline in Apr–May; some hamlets acid-affected in 2001–2002	Switch from single to double rice in 2000 but decline of double rice in 2001; shrimp farming started in 1998, with sharp increase in 2001	Significant proportion of landless households
7	Alluvial and acid deep	Saline-free before 1997; not acid-affected	Switch from single to double rice in 1999; almost entirely triple rice by 2000	Significant presence of Khmer population
8	Saline and acid deep	Saline-free before 1997; not acid-affected	Switch from single to double rice in 1998; with intensification to triple rice since 2000	
9	Saline and potential acid	Saline-free since 1998; not acid-affected	Switch from single to double rice since 1998; with some intensification to triple rice since 2000	
10	Saline, acid deep and shallow	Partly saline-affected until 1998; not acid-affected	Partial shift from double to triple rice in 2000	
11	Mainly acid deep, with some saline	Saline-free since 1999; acid–affected in 2001–2002	Switch from single to double rice in 1998	High concentration of Khmer population

Table 15.1 *Continued.* Selected biophysical and socio-economic characteristics of hamlet clusters in Bac Lieu Province, Vietnam.

Cluster	Soil	Water quality	Land use	Socio-economic
12	Saline, with some acid deep	Saline-free since 1998; not acid-affected	Switch from single to double rice in 1998	Higher population density
13	Saline and acid deep	Saline-free since 1998–1999; not acid-affected	Switch from single to double rice in 1998	Higher population density
14	Saline and acid deep	Saline-free since 1998–1999; partly acid-affected in 2001–2002	Switch from single to double rice in 2000; shrimp started in a few hamlets in 2001	

determined, spatially or otherwise, at the local community level. This work is being continued and expanded to the entire Bac Lieu Province.

RMDs and the Natural Resource Planning and Management Process

Fundamentally, the RMD approach provides a means of organizing relevant information to meet some natural resource use and management purposes. In practice, the RMD concept became subjected to nuances in interpretation and usage by different groups for different purposes.

In the realm of agricultural planning and development, the RMD is considered as an expansion of the FAO's AEZ concept, by bringing in the socio-economic context (Antoine, 1996). RMDs are used to mean 'areas within a broad physico-biotic zone that have similar socio-economic conditions' (FAO, 1995). An example of how this definition is applied can be seen in the delineation of 35 RMDs for southern Africa (covering ten countries) by GIS map overlay of farming system zones, AEZ, national boundaries and urbanization level (as an indicator of access to markets) (FAO, 2004). These RMDs are then used for identifying and analysing crop and livestock practices, as well as constraints to and opportunities for improving water use.

Benites *et al.* (1997) consider that RMDs provide the basis for spatial zoning for determining land (quality) change indicators to monitor processes affecting the natural resource base resulting from agricultural development, and to evaluate the sustainability of these changes. Further along in the context of integrated natural resource management (INRM), Campbell *et al.* (2001) interpret RMDs as a typology of land-use systems, whereby land use is an expression of human response to the interactions among biophysical, economic, social and technological components operating in an environment at a particular time. They therefore consider that the RMD provides the spatial basis for assessing system performance and the impact of INRM research.

Dumanski and Craswell (1996) offer a definition of RMD that recognizes both the environmental and socio-economic characteristics of a definitive unit of land, as well as emphasizing natural variability as an inherent characteristic rather than the notion of homogeneity of the area. They further argue that RMDs should be amenable to address NRM issues that are multiscale in nature – for farm-level management by households to regional-level management by policymakers.

We have attempted to illustrate the multiscale use of RMDs with the Bac Lieu case. RMDs are applied at a broader scale to address NRM implications of policies and programmes, and at a more detailed scale to differentiate management practices employed by communities and farmers. As a further step, if RMDs delineated at various spatial scales are linked, they permit analysis of cross-scale interaction loops between broader-level policy interventions and the actual local-level implementation and community response, as depicted in Fig. 15.4.

Using the Bac Lieu example, RMD analysis at a more detailed scale helps in formally recognizing the biophysical and socio-economic differentiation of the hamlets, and through that in identifying their specific requirements for introducing production systems and technologies, as well as other interventions and necessary support from local authorities. Scaling up the results from the analysis to the level of policymakers helps in influencing policy decisions that are more amenable to the objective conditions and aspirations of the target communities. In the Bac Lieu case, changes in agricultural production policy and land-use planning based on an improved understanding of NRM issues and the implementation of these changes led to a major change in water management and control. The opportunities emerging for diverse production systems required a variety of production practices and technologies to be introduced to farmers, based on their natural resource endowment and circumstances. RMDs delineated at the most detailed spatial level possibly help to identify target domains for these practices and technologies.

As objective conditions change, resulting from the adoption of innovations or for other

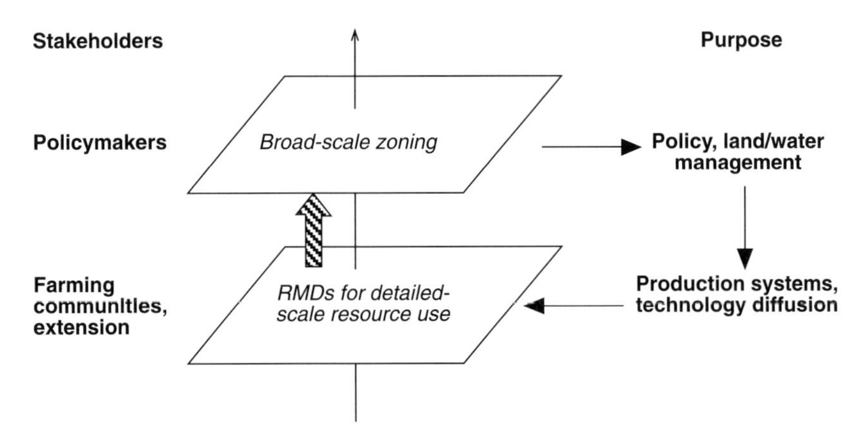

Fig. 15.4. Cross-scale roles of resource management domains (RMDs).

reasons, the hamlet groupings and their cluster characteristics will alter accordingly. Cluster membership of hamlets would change as: (i) conditions change with time, hence the values of selected variables change; or (ii) the variables selected, that is, those considered to be important, change. For example, land use and production systems are likely to be fluid in the transitional zone that is subjected to the more dynamic interface between saltwater and freshwater movements. At sites where the saltwater supply becomes better assured, farmers may prefer shrimp-only cultivation over the rice–shrimp rotation if labour costs and low rice prices are disincentives for the latter. Such changes, if substantial, will be reflected in how the hamlets cluster. Therefore, the clustering of hamlets is not meant to be permanent, but will be reassessed and reworked. Major changes emerging, as reflected in spatial clustering, would indicate that conditions are appropriate for a review of existing land-use plans. This approach acknowledges the dynamism of livelihood systems and community change in the complex coastal environment. Accordingly, the planning and management of coastal resources for sustainable development should be a continuously evolving process involving local authorities and communities, with technical and research support.

The RMD process, applied in the dynamic and adaptive manner described above, would be most effectively implemented if it is integral to and supports the formal land-use planning and management process instituted locally and nationally, as depicted in Fig. 15.5. The RMD process supports analysis at, and bridges across, two major levels with respect to natural resource planning and management – the level of policymakers and resource managers, and the level of the community and resource-users (farmers). Interpretation of the RMDs helps in developing land-use scenarios at a broad level, and proposals for specific economic activities and production systems, with accompanying knowledge and technologies, at the local level. These proposals would be refined, rechecked and selected based on economic/financial analysis of production scenarios. Land-use proposals legalized through the formal planning process would receive the necessary government support for implementation, for example, public works, credit schemes, processing and marketing facilities. The RMD process, besides functioning as the data and knowledge gathering, analysis and interpretation hub in the planning process, provides yet another conduit of information flow across the social hierarchy of stakeholders concerned with NRM for sustainable development.

Conclusions

The RMD concept lent itself well to delineating domains that reflected the influence of

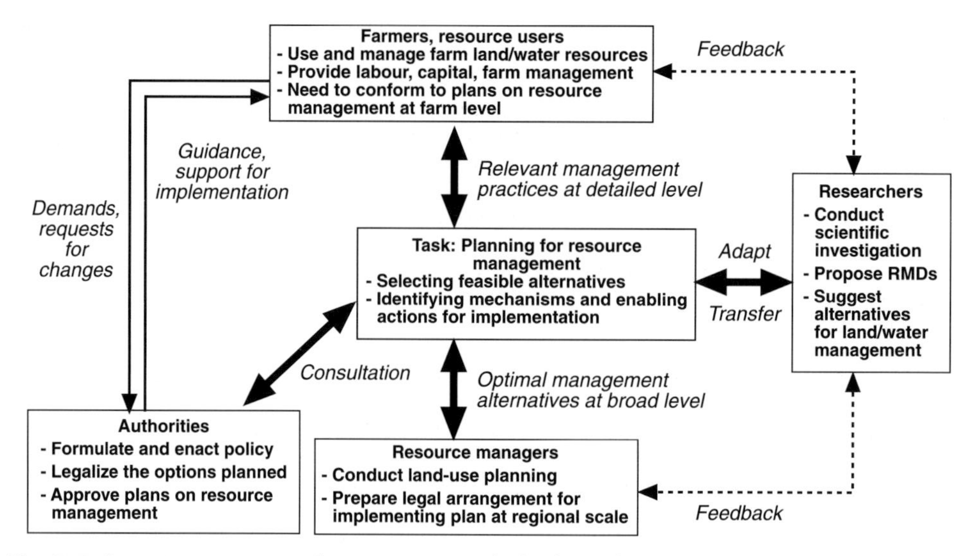

Fig. 15.5. Resource management domains (RMDs) in the land-use planning process.

key environmental factors on land-use changes in the part of Bac Lieu Province that was targeted for salinity protection. The results, interpreted at a broader scale, supported the identification of land-use and water management zones to accommodate rice-based, combined shrimp- and rice-based, and shrimp-based production systems in the area, thereby reversing an earlier policy of intensifying rice cultivation through full salinity exclusion. The utility of the RMDs at a more detailed scale for supporting farm management decisions is still being tested. The experience of the Bac Lieu case suggests that the RMD approach could be developed as a multiscale analytical platform to support an adaptive land-use planning process, and could also be regarded as an effective support tool for using and managing coastal resources.

Acknowledgements

The authors are grateful to the Department for International Development (DfID) of the United Kingdom for financial support for the study, and to the Bac Lieu Provincial Government and Bac Lieu Department of Agricultural Development for field support and consultations in carrying out the study.

References

Antoine, J. (1996) Resource management domains in relation to land-use planning. In: Syers, J.K. and Bouma, J. (eds) *Proceedings of the Conference on Resource Management Domains*, Kuala Lumpur, Malaysia, 26–29 August 1996. IBSRAM Proceedings No. 16.

Benites, J.R., Shaxson, F. and Vieira, M. (1997) Land condition change indicators for sustainable land resource management. In: *Land Quality Indicators and Their Use in Sustainable Agriculture and Rural Development*. FAO Land and Water Bulletin 5. <http://www.fao.org/docrep/W4745E/w4745e09.htm>

Campbell, B., Sayer, J.A., Frost, P., Vermeulen, S., Ruiz Pérez, M., Cunningham, A. and Prabhu, R. (2001) Assessing the performance of natural resource systems. *Conservation Ecology* 5(2), 22. <http://www.consecol.org/vol5/iss2/art22/>

Collinson, M. (1996) Social and economic considerations in resource management domains. In: Syers, J.K. and Bouma, J. (1996) *Proceedings of the Conference on Resource Management Domains*, Kuala Lumpur, Malaysia, 26–29 August 1996. IBSRAM Proceedings No. 16.

De Kartzow, A., Van der Heijden, P. and Van der Schoot, J. (1992) *Integration of Fish Farming into the Farm-Household System in Luapula Province, Zambia.* Fisheries Department FAO Non-series publications No. 16. <http://www.fao.org/docrep/005/ad011e/ado11e03.htm>

Dumanski, J. and Craswell, E. (1996) Resource management domains for evaluation and management of agro-ecological systems. In: Syers, J.K. and Bouma, J. (1996) *Proceedings of the Conference on Resource Management Domains*, Kuala Lumpur, Malaysia, 26–29 August 1996. IBSRAM Proceedings No. 16.

FAO (1976) *A Framework for Land Evaluation.* FAO Soils Bulletin No. 32, FAO, Rome.

FAO (1995) Planning for sustainable use of land resources: toward a new approach. Background paper to FAO's Task Managership for Chapter 10 of Agenda 21 of the United Nations Conference on Environment and Development (UNCED). FAO Land and Water Bulletin No. 2, FAO, Rome.

FAO (1999) *Poverty Alleviation and Food Security in Asia: Land Resources.* Regional Office for Asia and the Pacific (RAP) Publication 1999/2.

FAO (2004) A perspective on water control in southern Africa. Support to regional investment initiatives. Land and Water Discussion Paper No. 1. FAO, Rome. <http://www.fao.org/documents/show_cdr.asp?url_file=/docrep/006/y5096e/y5096e1k.htm>

Hoanh, C.T., Tuong, T.P., Kam, S.P., Phong, N.D., Ngoc, N.V. and Lehmann, E. (2001) Using GIS-linked hydraulic model to manage conflicting demands on water quality for shrimp and rice production in the Mekong River Delta, Vietnam. *Proceedings MODSIM 2001 International Congress on Modelling and Simulation*, Vol. 2, 221–226.

Hoanh, C.T., Tuong, T.P., Gallop, K.M., Gowing, J.W., Kam, S.P., Khiem, N.T. and Phong, N.D. (2003) Livelihood impacts of water policy changes: evidence from a coastal area of the Mekong River delta. *Water Policy* 5, 475–488.

Jones, P. (1996) Geographic information systems in natural resource management domains. In: Syers, J.K. and Bouma, J. (1996) *Proceedings of the Conference on Resource Management Domains*, Kuala Lumpur, Malaysia, 26–29 August 1996. IBSRAM Proceedings No. 16.

Kam, S.P., Tuong, T.P., Hoanh, C.T., Ngoc, N.V. and Minh, V.Q. (2000) Integrated analysis of changes in rice cropping systems in the Mekong River Delta, Vietnam, by using remote sensing, GIS and hydraulic modeling. *CD-ROM Proceedings of the XIX International Congress for Photogrammetry and Remote Sensing (ISPRS),* July 2000, Amsterdam, TP VII-07-18.

Syers, J.K. and Bouma, J. (eds) (1996) International Workshop on Resource Management Domains. *Proceedings of the Conference on Resource Management Domains,* Kuala Lumpur, Malaysia, 26–29 August 1996. IBSRAM Proceedings No. 16.

Tuong, T.P., Kam, S.P., Hoanh, C.T., Dung, L.C., Khiem, N.T., Barr, J. and Ben, D.C. (2003) Impact of sea-water intrusion control on environment, land use strategies and household incomes in a coastal area. *Paddy Water Environment* 1, 65–73.

16 Developing a Consultative Bayesian Model for Integrated Management of Aquatic Resources: an Inland Coastal Zone Case Study

E. Baran,[1] T. Jantunen[2] and P. Chheng[3]
[1]*WorldFish Center, Phnom Penh, Cambodia, e-mail: e.baran@cgiar.org*
[2]*Environmental Consultant, Phnom Penh, Cambodia*
[3]*Inland Fisheries Research and Development Institute, Phnom Penh, Cambodia*

Abstract

This chapter presents the methodological aspects of the development of a decision support system (DSS) based on Bayesian networks and aiming at assisting in the management of water-dependent resources (rice, fish, shrimp and crab). The principles of Bayesian networks are introduced, then the steps of model development are detailed and illustrated by the BayFish model being developed in the inland coastal zone of Bac Lieu Province (Vietnam). The particular feature of this DSS is that it is based on the contribution of local stakeholders. The process of building the model is detailed and illustrated by examples, in particular: the modalities of stakeholder consultations, the progressive arrangement of variables into a structured network, the justifications and weights defining each of the variables, the integration of databases in the model so that it combines quantitative and qualitative information, and the way outputs are calculated. The model, once completed, allows planners and decision-makers to visualize the trade-offs among various water management options. It also highlights the socio-political options and choices inherent to environmental management.

Introduction

Coastal zones are areas subject to high population pressure. In 1995, over 39% of the world's population lived within 100 km of a coast, whereas the coastal area accounts for only 20% of all land area (Burke *et al.*, 2000). The conflicts that result from this pressure are considerable, and technical as well as policy tools are needed to propose optimized land-use options and solutions to these conflicts.

On the technical side, in response to this need, decision support systems (DSS) have been developed in recent years to handle a large number of variables, their interactions and feedback loops. They have also been developed to integrate information originating from multiple disciplines, and to assess the outcomes of a given decision, which is almost always beyond the reach of individual experts. However, the specificity of land and water management is that it has implications for the livelihoods of millions of persons, and thus goes far beyond engineering and includes societal and political choices (Ostrom, 1990). In that regard, the contribution of civil society – either at the grass-roots level or through representatives – is to be considered as an intrinsic component of the management process and of the decision support tools (Kasemir *et al.*, 2000; Gregory *et al.*, 2001).

Among the various types of computer-based decision support tools, Bayesian networks are relatively unique in their ability to integrate quantitative information and data as well as qualitative expert knowledge and subjective choices (Varis and Fraboulet-Jussila, 2000; Cain 2001; Lynam *et al.*, 2002).

In this chapter, we introduce the principles of Bayesian networks applied to environmental management and to the role and modalities of stakeholder consultation, and use the ongoing development of a Bayesian support system in Bac Lieu Province (Vietnam) as an example of consultative integrated management of aquatic resources in inland coastal zones.

In Bac Lieu Province, saline water intrusion inland is partially controlled by sluice gates that create an artificial freshwater environment, which favours rice farming but is detrimental to brackish-water aquaculture and fish production. This leads to conflicts among various users whose water-quality requirements vary significantly (Hoanh *et al.*, 2003).

The model being developed to assist in water management in this province is named BayFish–Bac Lieu, following two similar models named BayFish–Mekong, focusing on Mekong fish production (Baran *et al.*, 2003), and BayFish–Tonle Sap, developed for Tonle Sap fisheries in Cambodia (Baran *et al.*, 2004). The model is intended mainly for the use of planners and decision-makers at the provincial level. The objectives of BayFish–Bac Lieu are:

- to help in optimizing the operations of the sluice gates in terms of fish production;
- to assist, through the production of scenarios, in making informed decisions about water management options;
- to identify and inform all stakeholders about the trade-offs inherent to water management options and production outputs; and
- to involve stakeholders in the management process.

In the following sections, we present an overview of the Bayesian approach, followed by the steps that have been achieved so far:

- a consultation with stakeholders for the identification of critical variables and zones,
- the creation of a conceptual framework that summarizes these interactions and
- the parameterization of the model with stakeholders and experts.

Two final steps will be achieved in 2005–2006 through the comprehensive integration of databases and results from other surveys and studies, and the production of scenarios and sensitivity analyses. Thus, this chapter focuses on the methodological aspects of the modelling approach.

Principles of Bayesian Networks

Bayesian networks consist of defining the system studied as a network of variables linked by probabilistic interactions (Jensen, 1996). Bayesian networks are also called Bayes' nets or Bayesian belief networks (BBN). These methods based on the calculation of dependent probabilities (Bayes' theorem) were originally developed in the mid-1990s as decision support systems (DSS) for medical diagnosis. Their principles and application to environmental management have been detailed in Charniak (1991), Ellison (1996), Cain (2001) and Reckhow (2002).

Variables representing the modelled environment can be either quantitative (e.g. 'number of fishers') or qualitative (e.g. 'fishing strategy'). For each variable, a small number of classes is specified. One of the challenges when building any kind of model is in defining enough, but not too many, variables. Probabilities based on the best available knowledge – either data-based or qualitative – quantify the link between two connected variables. Figure 16.1 shows an example.

In a driven variable, or child node, all the possible combinations of driving variables, or parent nodes, are integrated (see Fig. 16.2).

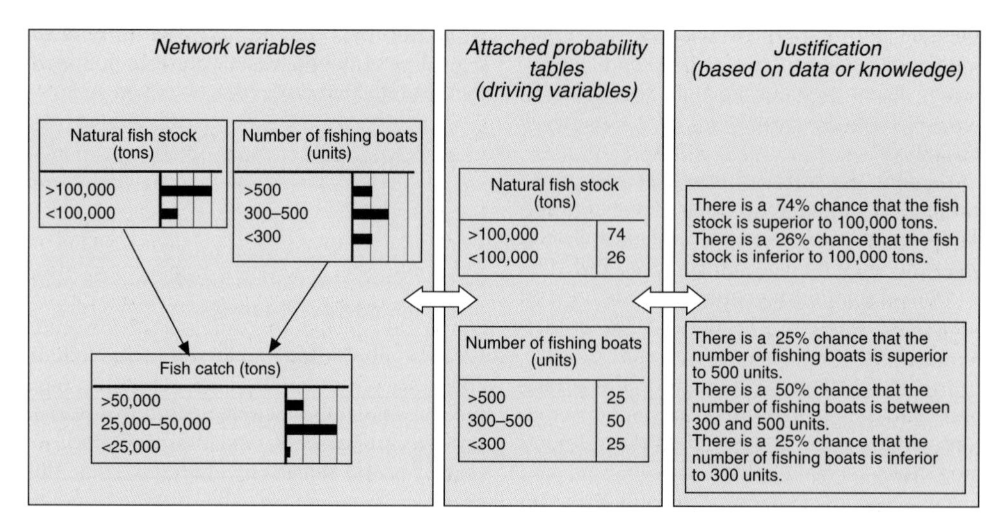

Fig. 16.1. Mini-network of three connected variables representing a hypothetical fishery (left). The probabilities of the first two driving variables are detailed in the middle section and the justification is detailed in the right section.

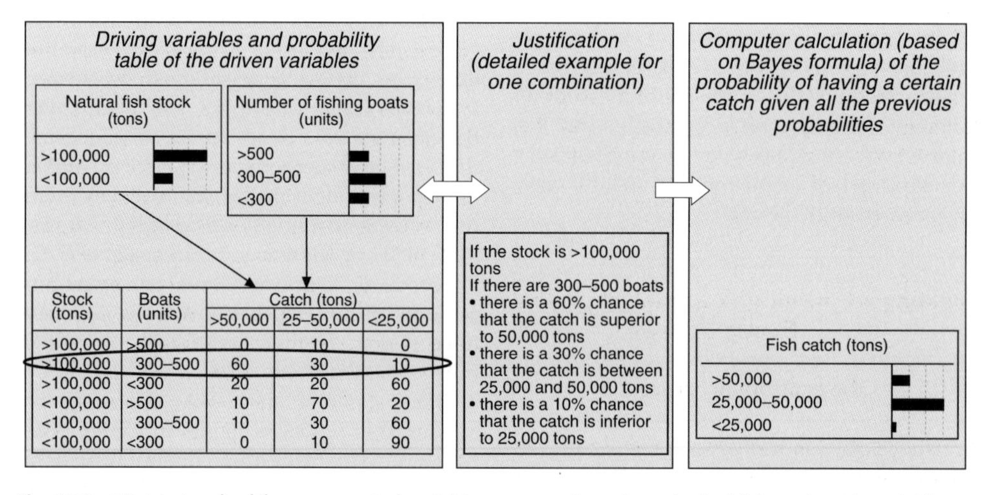

Fig. 16.2. Mini-network of three connected variables representing a hypothetical fishery (*continued*). The probability table of the driven variable is detailed in the middle section and the resulting probabilistic computation is given in the right section.

Thus, the sequence of tasks in model development is as follows:

- Building the model framework: (i) to identify the major variables of the system studied; and (ii) to arrange them into a meaningful network.
- Specifying the classes of each variable to specify a few relevant states for each variable.
- Parameterizing variables: (i) to define the probability of each state of each driving variable (a process called 'elicitation of prior probabilities'); and (ii) to define for each driven variable the probabilities of each combination of driving variables.

If data are available, the quantified relationship between two variables is automatically converted into probabilities. If data are not available, expert knowledge (an expert being any person having first-hand experience of the system studied) can be applied to express the known relationship between two variables in terms of probabilities. Ultimately, the software calculates, based on the Bayes formula of combined probabilities, the probability of having a certain output given the state of all the driving variables.

The possible integration of expert knowledge into a modelling framework contributed significantly to the success of the Bayesian approach, which is nowadays being more and more broadly used (e.g. McKendrick *et al.*, 2000; Hahn *et al.*, 2002; Soncini-Sessa *et al.*, 2002; Bertorelle *et al.*, 2004).

Different software is available to build and run Bayesian networks (review in Arthington *et al.*, 2004), although some modelling teams prefer to develop their own (e.g. Varis, 2003). For the development of this BayFish–Bac Lieu model, Netica software (<http://www.norsys.com>) was chosen. This package is intuitive and user-friendly, does not require the mastering of a computer language and is easily accessible on the Internet, where a freeware version allows the development of small models and the running of any large model.

Building a Consultative Model: Steps and Examples

The consultative process

In using Bayesian networks for environmental management, consultation with experts and stakeholders is acknowledged as being of critical importance (Borsuk *et al.*, 2001; Ravnborg and Westermann, 2002; Cain *et al.*, 2003). This consultation has been described with more or fewer details in almost all studies using Bayesian networks. However, for modelling applications dealing with societal issues such as natural resource management, studies focusing on the consultation process and its methodology are scarce (Reckhow, 2002). Some authors have addressed specific aspects of consultation, in particular the theoretical side (Gregory *et al.*, 2001; Beierle, 2002; Wilkins *et al.*, 2002; Seidel *et al.*, 2003), whereas others have highlighted the psychological pitfalls inherent in consultation with individuals or stakeholders (Anderson, 1998; De Bruin *et al.*, 2002; Fenton, 2004). On the more practical side, the guidelines provided by Cain (2001) and Ravnborg and Westermann (2002) for stakeholder consultations are among the most detailed. Acknowledging the lack of concise and pragmatic recommendations, Baran and Jantunen (2004) have proposed guidelines about stakeholder consultation for Bayesian modelling, detailing in particular possible options about the selection and number of stakeholders, group consultation options and pitfalls, and issues specific to Bayesian networks.

Example: the consultation for the BayFish–Bac Lieu model

Stakeholders were identified as follows: fishers, farmers and aquaculturists (for first-hand experience at the district level); technicians and extension officers in these disciplines (for the field experience at the provincial level); representatives of local organizations and of line agencies (for institutional roots); provincial representatives (for the administrative and political viewpoint); academe (for the research follow-up); and, in theory, equal proportions of men and women, even if achieving this balance proves impossible because of both professional and cultural reasons.

The stakeholders were informed in advance about the purposes of the exercise: (i) to identify the relevant environmental variables that drive aquatic food production in the area studied; (ii) to build a model framework that will reflect what users say about the interactions among these variables; and (iii) to give a weight to each variable and to the role it plays in food production or in the factors that drive food production.

Consultations were conducted on the following basis:

- Identification of the stakeholders matching the above criteria by the provincial authorities. Although other ways of identifying stakeholders do exist (Baran and Jantunen, 2004), this was the only possible one given the time frame and the centralized administrative structure in Vietnam.

- Ten to 15 participants per meeting; women were represented as well, but were a silent minority.
- A first round of half-day consultations at the commune level, to assess the contemporary state and constraints in different agro-ecological zones (mainly rice farming, aquaculture and brackish-water zone). Each zone was discussed separately. The consultations involved: (i) a facilitator familiar with the modelling approach to solicit contributions from the stakeholders and lead the debate; (ii) the modellers asking specific questions on variables and linkages, and checking the logic of the system; and (iii) a translator with field experience for the modellers to follow the debate in Vietnamese, who would not intervene too much and would be competent to cross-check the answers. The consultations were followed by conversion of the information gathered into a computer model by the modellers and the facilitator (another half day).
- A second round of consultations with the same stakeholders to verify model structure and variables identified during the first round.
- Additional discussions involving high-level technicians or scientists (e.g. agronomists from the Department of Agriculture and Rural Development, hydrodynamics modellers from the International Water Management Institute, etc.) and provincial managers (in particular, representatives of the Bac Lieu People's Committee) in order to: (i) check that variables and linkages significant at the provincial scale were not missing from the model developed at the district level; (ii) test the acceptability by provincial authorities of the model proposed; and (iii) identify the availability of data and statistics supplementing expert knowledge wherever possible.

Building the model framework

A preliminary model framework, introducing progressively the environmental variables and their links in the system modelled, is to be presented to stakeholders so that they can understand the approach and expectations. Then, stakeholders are requested to build from scratch a model of their own, starting with the main variables of interest, whose components and driving factors are progressively detailed.

Building the BayFish–Bac Lieu basic framework

Three half-day consultations were held in three different communes (one in the freshwater zone, one in the brackish-water zone and one in the saline zone outside the sluice gates) during the first round. Important variables were identified, which showed that total aquatic food production in the province is composed of shrimp, rice, fish and crab. The production of each commodity depended, to various degrees, on water supply and water quality. The three models developed in the three areas were then integrated into a single one by the modellers. This model was subsequently checked, modified and agreed upon during a second round of consultations, another half day for all the same stakeholders. The basic framework obtained is detailed in Fig. 16.3.

Each level was further detailed until a

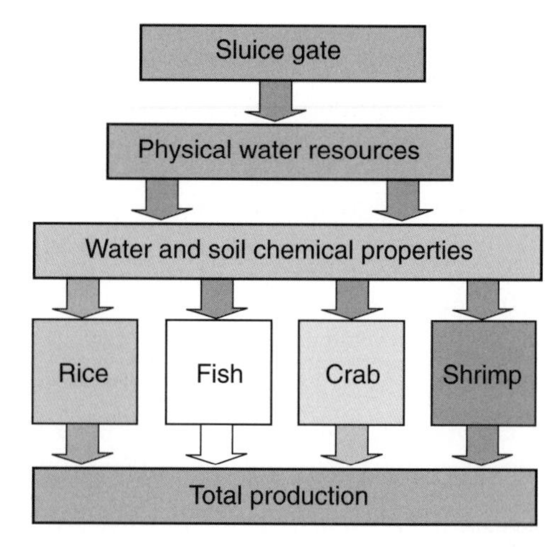

Fig. 16.3. Bac Lieu water and food model, overall structure flow diagram.

final framework was agreed upon by all stakeholders at the end of the consultations (Fig. 16.4).

Building the BayFish–Bac Lieu detailed framework

To show how variables are defined, links set and contributions weighted, we detail below an excerpt of the BayFish model, specifically the module detailing fish production *variables* (see Fig. 16.4). All probabilities assigned resulted from stakeholder consultations and averaging was used between different consultations only when stakeholders failed to agree upon one value.

F1. *Fish production* is the combination of *wild fish* and *aquaculture fish* production. Wild fish are caught in canals and aquaculture fish are cultivated with shrimp and crabs in ponds or in rice paddies. Overall, 30% of the fish produced are estimated to originate from aquaculture and 70% from wild fish capture. Fish production is season-specific, that is, fish grow more during the wet season, but at the moment this growth is not quantified in the model.

F1.1. *Wild fish* production depends on the combination of *estuarine fish* and *freshwater fish* production. Estuarine fish refer to marine fish that migrate to the province from the sea, whereas freshwater fish refer to fish living and migrating within the province and from the Mekong River. Estuarine fish are estimated to contribute 80% of the production and wild freshwater fish 20%.

F1.1.1. *Estuarine fish* production depends on *water quality for estuarine wild fish* and *marine inflow*. Stakeholders value the importance of local water quality vis-à-vis estuarine fish production at 70%, and that of marine inflow through the sluice gates at 30% (because a proportion of the marine fish comes from the Gulf of Thailand, outside the area controlled by sluice gates built along the South China Sea coastline).

F1.1.1.1. *Water quality for estuarine wild fish* depends on *organic water pollution* and *chemical water pollution*. Chemical pollution causes biological problems and organic pollution causes, among others, anoxia problems and fish kills. Stakeholders estimate that 70% of the problems are due to organic pollution (mainly from shrimp-processing factories), whereas 30% of the pollution problems relate to chemicals (mostly pesticides from rice farming).

F1.1.2. Freshwater fish production depends on *water quality for freshwater wild fish* and *upstream inflow*. Upstream inflow refers to fish larvae and food supply, originating mainly from the Mekong River. The weight given to water quality is 70% and that given to upstream inflow is 30%.

F1.1.2.1. *Water quality for freshwater wild fish* depends on *organic water pollution, chemical water pollution* and *water salinity*. Pollution problems have been detailed above and salinity is not acceptable to fish of freshwater origin. Subsequently, the weight given to salinity is 50%, whereas organic pollution and chemical pollution receive 35% and 15%, respectively.

F1.2. *Aquaculture fish* production results from the *semi-intensive aquaculture method, extensive aquaculture method, soil acidity* and *water quality for aquaculture fish*. Examples of species produced are *Oxyeleotris marmorata* (marble goby) and *Barbonymus gonionotus* (Java barb). Stakeholders consulted estimate that water quality contributes 40% to aquaculture fish production. Out of the 60% remaining, averaging between proportions given in different district consultations shows that 87.5% (i.e. 52.5% overall) is due to the aquaculture mode and 12.5% (i.e. 7.5% overall) to soil acidity. Extensive aquaculture farms make up 90% of the cases and semi-intensive ones 10%.

F1.2.1. The *semi-intensive fish aquaculture method* depends on stocking of *fish seeds*, application of *fish feed* as well as suitable *water quantity for fish*. Feed and seeds play an equal role, 80% overall, whereas the role of water supply is valued at 20%.

F1.2.1.1. *Fish seeds* are the basis for fish cultivation and only natural seeds are used. Some 40% of the farmers use few seeds (generally $3/m^2$) and 60% of the farmers use a lot of seeds (around $10/m^2$).

F1.2.1.2. *Fish feed* is usually caught locally from channels. Some 30% of the farmers use 'little' feed (i.e. around 10 tons/ha/year),

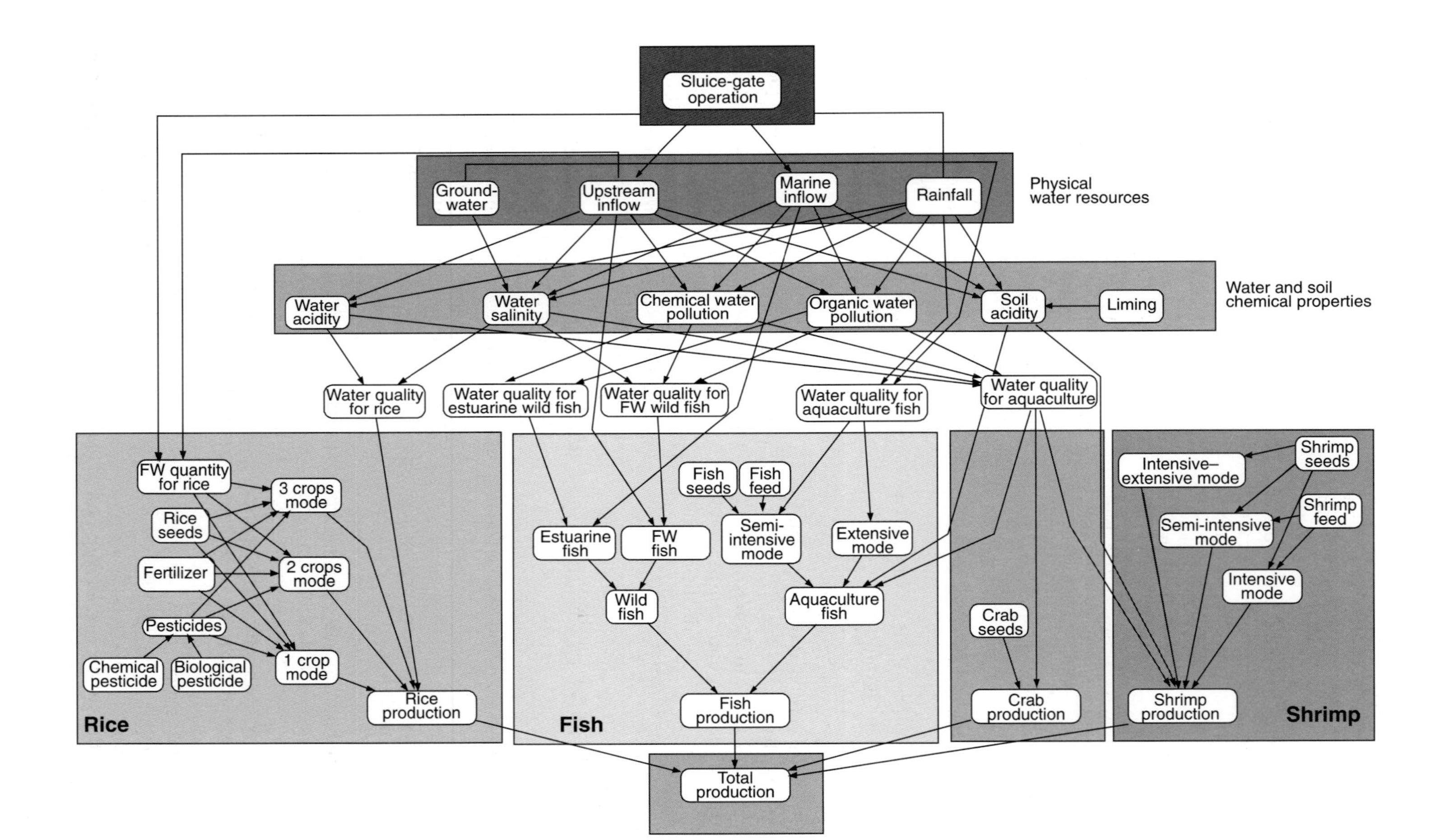

Fig. 16.4. Network of variables in the BayFish–Bac Lieu water-dependent food production model. FW, fresh water.

whereas 70% of them use 'a lot' of feed (about 14 tons/ha/year).

F1.2.1.3. *Water quantity for fish* results from *groundwater* and *rainfall*. Both groundwater and rainfall provide necessary unpolluted fresh water *in situ* for the ponds. The role of groundwater is dominant (weight estimated at 80% by stakeholders), while that of rainfall is only 20%.

F1.2.2. The *extensive fish aquaculture method* depends only on water supply (*water quantity for fish*). This latter result shows that, according to stakeholders, sluice-gate operations do not influence freshwater fish aquaculture (only groundwater and rainfall do).

Specifying the variable classes

The variables of a Bayesian model have to be defined and specified according to a number of states, such as 'Yes' or 'No', 'High' or 'Low', 'x<5', '5≤x≤7' or 'x>7', etc. Cain (2001) details the constraints of this operation; in particular, states have to be discrete (i.e. discontinuous) and limited in number in order to avoid overly complex and unmanageable probability tables. We provide below two examples of variable specification drawn from the BayFish–Bac Lieu model: variables *pesticides* (qualitative specification) and *rainfall* (fully quantitative specification).

Example 1: specifying the variable 'Pesticides' in the Rice module of the BayFish–Bac Lieu model

- *Biological pesticides* as well as *chemical pesticides* are used in the area. Biological pesticides such as DTN 32 and DTN 16 are of plant origin and are less toxic than chemical pesticides; however, they are also less efficient against pests. They are nevertheless preferred by some farmers, who can then grow shrimp or fish in the same location after the annual rice harvest.

  Chemical pesticides such as deltamethrin are very efficient and secure for rice production, but have a strong negative effect on the survival of shrimp and fish juveniles.
- For *biological pesticides*, a small dose is considered to be around 0.25 l/ha, according to farmers and extension officers, whereas a large dose consists of at least 0.5 l/ha.
- For *chemical pesticides*, a small dose is considered to be around 0.5 l/ha, according to farmers and extension officers, whereas a large dose consists of at least 1 l/ha. Most of the farmers who prefer large doses are in the rice-farming area and cannot grow shrimp anyway. One must note that those latter farmers used to spray about 3 l/ha a few years ago, thus the notion of 'a lot' has evolved over time.
- Having detailed in the parent nodes what products and doses were, the integrative child node *pesticides* is simply specified in terms of 'a little' or 'a lot' of pesticides (Fig. 16.5), depending on the combination amount of chemical and biological pesticides.

Example 2: specifying the variable 'Rainfall' in the Physical Water Resources module of the model

The *rainfall* variable is based on both long and short time-series data sets from the Provincial Departments of Meteorology and Hydrology: 1910–1990 (eight stations with gaps, from 1930 to 1950 and the mid-1970s) and 2002–2003 (three stations, no gaps). Frequency distribution graphs show that the rainy season occurs from July to November, with an average annual rainfall (in those 5 months) of 1325 mm. The dry season (from December to June) exhibits an average annual rainfall of 580 mm in those 7 months and the frequency distribution of dry-season rainfall data is nearly normal. The authors decided to use, as in the Vietnam River System and Plains (VRSAP) model (Hoanh, 1996), the dry-season months as an input in the model, with a threshold value of 580 mm to qualify either rainy or dry years. Distribution analysis shows that, in available data, years with more than average dry-season rainfall constitute 55.6% of the cases, whereas years with less than average dry-season rainfall constitute 44.4% of the cases (Fig. 16.6).

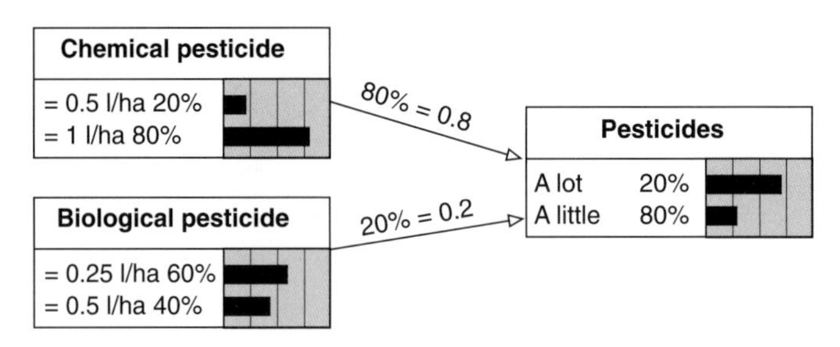

Fig. 16.5. Specification of the variable '*Pesticides*' in the Rice module of the model.

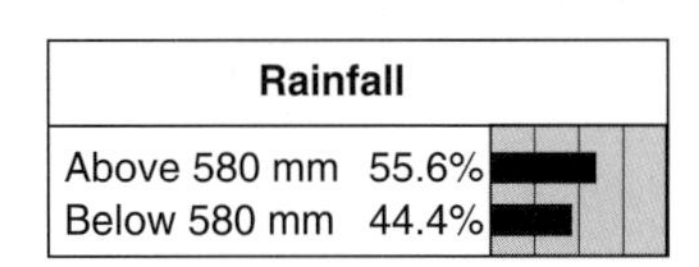

Fig. 16.6. Specification of the variable '*Rainfall*' in the Physical Water Resources module of the model.

Parameterizing the model variables

Parameterizing the variables of the model consists of attributing a probabilistic value to each possible combination of parent variables ('probability elicitation'). This is also to be done by stakeholders once all the variables of the framework have been identified and specified. An efficient way of eliciting probabilities is to ask stakeholders to give a weight to each variable, then to fill in all possible combinations of the weighted variables combined. We detail this process in two examples: *pesticides* in the Rice module and *total production* resulting from rice, fish, shrimp and crab production.

Example 1: parameterizing the variable 'Pesticides' in the Rice module of the model

- Around 40% of the farmers prefer to use small doses of *biological pesticides* to minimize the impact on fish or shrimp, whereas 60% of them use large doses.
- Around 20% of the farmers who use *chemical pesticides* prefer to use small doses, and 80% of them use large doses.
- Overall, 80% of the farmers use chemical pesticides and 20% use biological pesticides.

These statistics are illustrated in Fig. 16.7.

The software simply calculates resulting probabilities according to the Bayes formula, which can be expressed as

Probability P of event A knowing conditional event

$$E = P(A \mid E) = [P(E \mid A) \times P(A)] / P(E) \tag{16.1}$$

Example 2: parameterizing the variable 'Total production' of the BayFish–Bac Lieu model

Total production is usually understood in terms of tons of a commodity, but for most farmers consulted production means contribution to income. In the model developed, the contribution of each commodity to production is subsequently expressed in terms of contribution to income, and the estimates provided in different locations were simply averaged (as only one set of values could be input in the current model; Table 16.1).

Beyond the technicality of averaging or weighting answers lies a much more crucial question, that of units and thus viewpoints. In Bac Lieu Province, for instance, the 'contribution to total production' is dominantly perceived in terms of tons (e.g. contribution of the province to national rice production) and in terms of income (e.g. motivation for farmers to grow shrimp rather than rice when 1 kg of shrimp is worth about 50 kg of rice). In the BayFish–Bac Lieu model, we followed the dominant stakeholders' perception and expressed the share of each commodity in financial terms (Vietnamese dong or US dollars). However, other viewpoints might be preferred, such as food security (each commodity then being expressed in tons) or land use (number of hectares

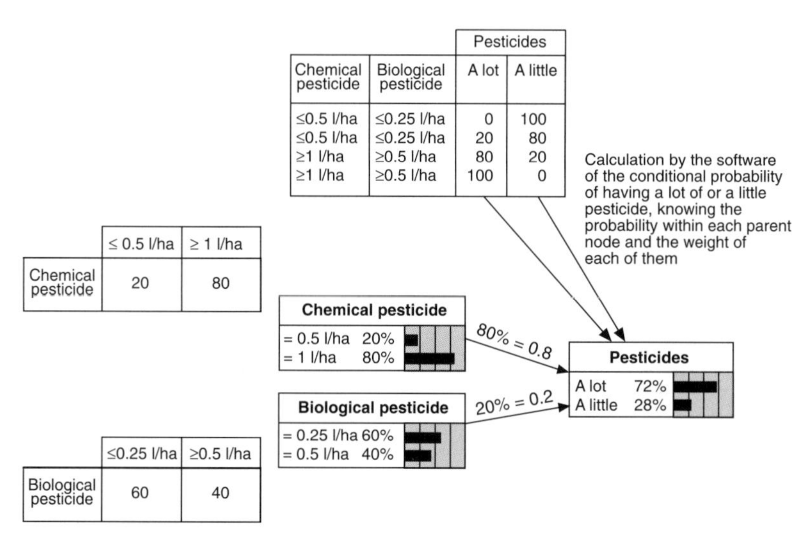

Fig. 16.7. Probabilities attached to pesticides and the resulting calculation of conditional probabilities.

needed to produce a given commodity), as detailed in the second part of Table 16.1.

It is to be noted that the management of sluice gates is to be done in response to a desired output, such as maximized income or maximum food security. The latter belongs to the socio-political realm, and the technical role of a decision support tool consists of identifying optimal ways to achieve an objective once this objective is clearly identified.

Conclusions

Bayesian networks have been developed in recent years in the fields of natural resource management (review in Cain, 2001), fisheries management (Varis and Kuikka, 1997; Kuikka *et al.*, 1999; Marcot *et al.*, 2001) and watershed management (Borsuk *et al.*, 2001; Varis, 2003). However, only a few of these models have been based on extensive stakeholder consultations.

In the field of environmental management, the strength of models based on Bayesian networks and stakeholder consultations is that they give stakeholders and managers a chance to visualize trade-offs between conflicting uses and needs, thus contributing to conflict resolution and management optimization. The consultation-based approach also permits us to overcome

Table 16.1. Contribution of each commodity to Production, and meaning of 'Production'. In italics, the values kept in the model.

	Contribution (%) of each commodity to 'Total production'				
Location	Rice	Fish	Crab	Shrimp	Production understood as
Phong Thanh	20	5	5	70	Income
Phong Thanh Tay	5	15	15	65	Income
Ho Phong	10	15	15	60	Income
Vinh Loi	20	5	5	70	Income
Mean of 4 locations	*13.8*	*10*	*10*	*66.3*	*Income*
Vinh Loi	50	10	0	40	Area of land necessary year-round
Ho Phong	0	5	15	80	Area of land necessary in dry season
Ho Phong	10	15	15	60	Area of land necessary in wet season
Vinh Loi	90	5	2.5	2.5	Food security

the lack of data in certain fields, and results in integration of both quantitative data and qualitative information based on experience. However, it remains possible to replace qualitative, consultation-based information by data when the latter become available (Fig. 16.8).

The BayFish–Bac Lieu model is based on extensive stakeholder consultations, but it also integrates existing environmental databases (in particular, rainfall records from the Provincial Departments of Meteorology and Hydrology, and outputs of the VRSAP hydrodynamic model; Hoanh *et al.*, 2001). The Bayesian structure allows the prediction (in qualitative terms so far, i.e. a percentage between 'good' or 'bad') of changes in water management (in particular, through sluice-gate operations) for each of the four main food commodities.

Although not dynamic, the BayFish model includes spatial as well as temporal dimensions (e.g. quality of estuarine *versus* freshwater; effect of liming on soil acidity, etc.). It is not detailed by geographic area but integrates the specificities and requirements of the main shrimp and rice production zones.

The model is intuitive and user-friendly, as clearly confirmed by the straightforward contribution of farmers and fishers. Running and modifying it require computer conversance, but no knowledge of any specific programming language. Netica, the software required to run it, is freely available on the Internet. Lastly, sources are open (i.e. the model structure and contents are fully modifiable) and of very small size (a few hundred kilobytes).

A sensitivity analysis (not detailed here) allows for the identification of the variables that contribute most to the overall variability of the system modelled; this analysis, based on entropy calculation (Coupé *et al.*, 1999, 2000), classifies variables according to their degree of influence on the output of interest.

The model produced, although still under development, already provides a picture of the different and conflicting water quality requirements of rice, shrimp, crab and fish that all contribute to the economy and food supply of Bac Lieu Province. There is currently a conflict between the government-run policy of expanding rice culture area, which requires fresh water, and lucrative shrimp farming, which requires brackish water. Pushing further the questioning initiated during the stakeholder consultations, a policy targeting economic return would drive sluice-gate operations favourable to shrimp culture, whereas a policy targeting food security for the poor would command sluice-gate operations favourable to fish and rice, and one favouring biodiversity conservation would place weight on fish and crab natural production.

A model like BayFish allows the quantification of trade-offs and the production of scenario analyses, but its technical requirements also highlight the need for a clear identification of the socio-political options that drive environmental management. Thus, consultative modelling can efficiently assist in the management of inland coastal zones provided that the question 'What for?' is addressed and tentatively answered.

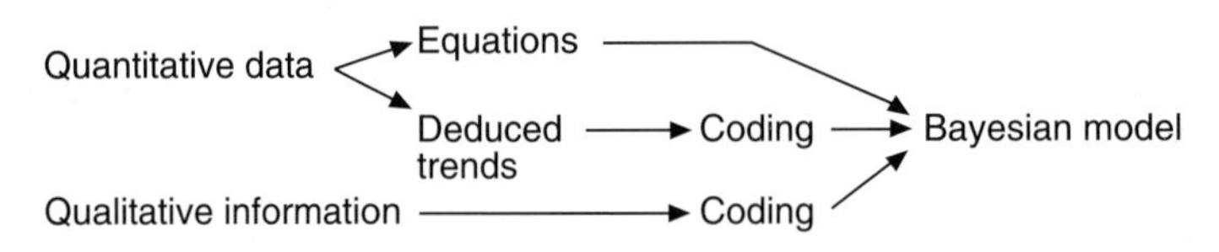

Fig. 16.8. Possible inputs for a Bayesian network.

References

Anderson, J. (1998) Embracing uncertainty: the interface of Bayesian statistics and cognitive psychology. *Conservation Ecology*. <http:// www.consecol.org/vol2/iss1/art2>

Arthington, A.H., Baran, E., Brown, C.A., Dugan, P., Halls, A.S., King, J.M., Minte-Vera, C.V., Tharme, R.E. and Welcomme, R.L. (2004) *Water Requirements of Floodplain Rivers and Fisheries: Existing Decision-support Tools and Pathways for Development*. Report for the Comprehensive Assessment of Water Management in Agriculture. International Water Management Institute (in press). 59 pp.

Baran, E. and Jantunen, T. (2004) Stakeholder consultation for Bayesian Decision Support Systems in environmental management. *Proceedings of the Regional Conference on Ecological and Environmental Modeling (ECOMOD 2004)*. Universiti Sains Malaysia, 15–16 September 2004, Penang, Malaysia (in press).

Baran, E., Makin, I. and Baird, I.G. (2003) *BayFish: a Model of Environmental Factors Driving Fish Production in the Lower Mekong Basin*. Contribution to the Second International Symposium on Large Rivers for Fisheries, 11–14 February 2003, Phnom Penh, Cambodia.

Baran, E., Jantunen, T., Hort, S. and Chheng, P. (2004) *Building 'BayFish-Tonle Sap', a Model of the Tonle Sap Fish Resource*. ADB/WorldFish Center project 'Technical Assistance for capacity building of IFReDI'; WorldFish Center and Inland Fisheries Research and Development Institute, Department of Fisheries, Phnom Penh, Cambodia, 26 pp.

Beierle, T. (2002) The quality of stakeholder-based decisions. *Risk Analysis* 22(4), 739–749.

Bertorelle, G., Bruford, M., Chemini, C., Vernesi, C. and Hauffe, H.C. (2004) New, flexible Bayesian approaches to revolutionize conservation genetics. *Conservation Biology* 18(2), 584.

Borsuk, M.E., Clemen, R.T., Maguire, L.A. and Reckhow, K.H. (2001) *Stakeholder Values and Scientific Modeling in the Neuse River Watershed*. Group Decision and Negotiation 10, 355–373. <http://www2.ncsu.edu/ncsu/CIL/WRRI/ken%27s_page.html>

Burke, L., Kura, Y., Kassem, K., Revenga, C., Spalding, M. and McAllister, D. (2000) *Pilot Analysis of Global Ecosystem: Coastal Ecosystems*. World Resources Institute, Washington, DC.

Cain, J. (2001) *Planning Improvements in Natural Resources Management: Guidelines for Using Bayesian Networks to Support the Planning and Management of Development Programmes in the Water Sector and Beyond*. Centre for Ecology and Hydrology (CEH), Wallingford, UK, 124 pp.

Cain, J., Jinapala, K., Makin, I., Somaratna, P., Ariyaratna, B. and Perera, L. (2003) Participatory decision support for agricultural management: a case study from Sri Lanka. *Agricultural Systems* 76, 457–482.

Charniak, E. (1991) Bayesian networks without tears. *Artificial Intelligence* 12(4), 50–63. <http://www-psych.stanford.edu/~jbt/224/Charniak_91.pdf>

Coupé, V.M.H., Peek, N.B., Ottenkamp, J. and Habbema, J.D.F. (1999) Using sensitivity analysis for efficient quantification of a belief network. *Artificial Intelligence in Medicine* 17, 223–247.

Coupé, V.M.H., van der Gaag, L.C. and Habbema, J.D.F. (2000) Sensitivity analysis: an aid for belief-network quantification. *Knowledge Engineering Review* 15, 1–18.

De Bruin, B., Fischbeck, P., Stiber, N. and Fischhoff, B. (2002) What number is 'fifty-fifty'? Redistributing excessive 50% responses in elicitated probabilities. *Risk Analysis* 22(4), 713–723.

Ellison, A.M. (1996) An introduction to Bayesian inference for ecological research and environmental decision-making. *Ecological Applications* 6(4), 1036–1046.

Fenton, N. (2004) Biases and Fallacies in Reasoning about Probability. <http://www.dcs.qmw.ac.uk/~norman/BBNs/Biases_and_fallacies_in_reasoning_about_probability__about_this_section.htm>

Gregory, R., McDaniels, T. and Fields, D. (2001) Decision aiding, not dispute resolution: creating insights through structured environmental decisions. *Journal of Policy Analysis and Management* 20(3), 415–432.

Hahn, M.A., Palmer, R.N., Merrill, M.S. and Lukas, A.B. (2002) Expert system for prioritizing the inspection of sewers: knowledge base formulation and evaluation. *Journal of Water Resources Planning and Management* March/April 2002, 121–129.

Hoanh, C.T. (1996) Development of a Computerized Aid to Integrated Land Use Planning (CAILUP) at regional level in irrigated areas: a case study for the Quan Lo Phung Hiep region, Mekong Delta, Vietnam. PhD thesis. ITC Publication Number 38, International Institute for Aerospace Survey and Earth Sciences, Enschede, Netherlands.

Hoanh, C.T., Tuong, T.P., Kam, S.P., Phong, N.D., Ngoc, N.V. and Lehmann, E. (2001) Using GIS-linked hydraulic model for managing water quality conflict for shrimp and rice production in the Mekong

River Delta, Vietnam. In: Ghassemi, F., Post, D., Sivapalan, M. and Vertessy, R. (eds) *Proceedings of MODSIM 2001, International Congress on Modelling and Simulation, Vol. 1: Natural Systems (part one)*, 10–13 December 2001, Canberra, Australia, pp. 221–226.

Hoanh, C.T., Tuong, T.P., Gallop, K.M., Gowing, J.W., Kam, S.P., Khiem, N.T. and Phong, N.D. (2003) Livelihood impacts of water policy changes: evidence from a coastal area of the Mekong River Delta. *Water Policy: Official Journal of the World Water Council* 5(5/6), 475–488.

Jensen, F.V. (1996) An introduction to Bayesian networks. UCL Press, London, 178 pp.

Kasemir, B., Dahinden, U., Gerger Swartling, Å., Schüle, R., Tabara, D. and Jaeger, C.C. (2000) Citizens' perspectives on climate change and energy use. *Global Environmental Change* 10, 169–184.

Kuikka, S., Hildén, M., Gislason, H., Hansson, S., Sparholt, H. and Varis, O. (1999) Modeling environmentally driven uncertainties in Baltic cod (*Gadus morhua*) management by Bayesian influence diagrams. *Canadian Journal of Fisheries and Aquatic Sciences* 56(4), 629–641.

Lynam, T., Bousquet, F., D'Aquino, P., Barreteau, O., Le Page, C., Chinembiri, F. and Mombeshora, B. (2002) Adapting science to adaptive managers: spidergrams, belief models, and multi-agent systems modeling. *Conservation Ecology* 5(2), 24. <http://www.consecol.org/vol5/iss2/art24>

Marcot, B.G., Holthausen, R.S., Raphael, M.G., Rowland, M. and Wisdom, M. (2001) Using Bayesian belief networks to evaluate fish and wildlife population viability under land management alternatives from an environmental impact statement. *Forest Ecology and Management* 153(1–3), 29–42.

McKendrick, I., Gettinby, G., Gu, Y., Reid, S. and Revie, C. (2000) Using a Bayesian belief network to aid differential diagnosis of tropical bovine diseases. *Preventive Veterinary Medicine* 47, 141–156.

Ostrom, E. (1990) *Governing the Commons*. Cambridge University Press, New York.

Ravnborg, H. and Westermann, O. (2002) Understanding interdependencies: stakeholder identification and negotiation for collective natural resource management. *Agricultural Systems* 73, 41–56.

Reckhow, K.H. (2002) Bayesian approaches in ecological analysis and modeling. In: Canham, C.D., Cole, J.J. and Lauenroth, W.K. (eds) *The Role of Models in Ecosystem Science*. Princeton University Press. <http:/www2.ncsu.edu/ncsu/CIL/WRRI/ ken's_page.html >

Seidel, M., Breslin, C., Christley, R., Gettinby, G., Reidc, S. and Revie, C. (2003) Comparing diagnoses from expert systems and human experts. *Agricultural Systems* 76, 527–538.

Soncini-Sessa, R., Castelletti, A. and Weber, E. (2002) Participatory decision making in reservoir planning. IEMSS conference proceedings, Lugano, Switzerland. <http://www.iemss.org/iemss2002/>

Varis, O. (2003) *WUP-FIN Policy Model: Finding Ways to Economic Growth, Poverty Reduction and Sustainable Environment*. WUP-FIN Socio-economic Studies on Tonle Sap no. 10, MRCS/WUP-FIN, Phnom Penh, Cambodia, 38 pp. <http://www.eia.fi/ wup-fin/>

Varis, O. and Fraboulet-Jussila, S. (2000) Causal Bayesian network approach to integrated watershed planning and management. In: Al-Soufi (ed.) *Proceedings of the International Workshop on Hydrologic and Environmental Modelling in Mekong Basin*, Phnom Penh, 11–12 September 2000. MRC Publications, Phnom Penh, Cambodia, pp. 204–214.

Varis, O. and Kuikka, S. (1997) Joint use of multiple environmental assessment models by a Bayesian meta-model: the Baltic salmon case. *Ecological Modelling* 102(2–3), 341–351.

Wilkins, D., Mengshoel, O., Chernyshenko, O., Jones, P., Hayes, C. and Bargar, R. (2002) *Collaborative Decision Making and Intelligent Reasoning in Judge Advisor Systems*. <http://www.computer.org/proceedings/hicss/ 0001/00011/00011061.PDF>

17 Aquatic Food Production in the Coastal Zone: Data-based Perceptions on the Trade-off between Mariculture and Fisheries Production of the Mahakam Delta and Estuary, East Kalimantan, Indonesia

P.A.M. van Zwieten,[1] A.S. Sidik,[2] Noryadi,[2] I. Suyatna[2] and Abdunnur[2]
[1]*Aquaculture and Fisheries Group, Wageningen University, Wageningen, The Netherlands, e-mail: paul.vanzwieten@wur.nl*
[2]*Faculty of Fisheries and Marine Sciences, University of Mulawarman, Samarinda, East Kalimantan, Indonesia*

Abstract

In less than two decades, from 45,000 to 70,000 ha, or up to 70%, of the mangroves in the Mahakam Delta, East Kalimantan, were converted to shrimp ponds. This is expected to affect the productive and buffering function of intact mangroves, observable as shifts in composition and a possible reduction in productivity of the coastal fisheries. The trade-off between mariculture and fisheries is explored with data from fisheries statistics, surveys and reported information. Analysis of trends and developments in total catch, catch by species category, catch rate, fishing effort, pond production and productivity gave no direct quantitative evidence of reduced coastal production and productivity. Shrimp-pond productivity (125 kg/ha/year) is low, stable and highly variable (CV = 69%) at the aggregated level of the delta. Since 1989, fishing efforts have increased and patterns diversified, but aggregated catch rates did not decrease. Catches of rays and sharks decreased and the pelagic/demersal catch ratio increased. A shift towards more resilient species categories with a high turnover rate took place from 1993 to 1999, 4 to 10 years after the boom in pond construction. However, these clear shifts are not self-evidently related to mangrove conversion. Reasons for this are discussed. The potential for detection of changes in resource outcome and assessment of the trade-off between mariculture and fisheries, at both the local level and through aggregated fisheries statistics, is limited because of the high variability in outcome. This implies a limited capacity for resolution of resource-use conflicts when evaluating competing claims informed by existing data and information on resource change.

Introduction

Can coastal aquaculture increase the total aquatic production of estuarine and coastal marine areas of the Mahakam Delta in East Kalimantan in a sustainable way? Mangrove forests have important functions in the life histories of fish and crustaceans: as a corridor

 Environment and Livelihoods in Tropical Coastal Zones (eds C.T. Hoanh, T.P. Tuong, J.W. Gowing and B. Hardy)

for diadromous species and as nursery, spawning and feeding areas for a host of maritime species. Annually, nearly 3000 km^2, or 2%, of the world mangrove area is converted, for instance, for pond mariculture. Conversion to ponds changes (essential) fish and crustacean habitats by closing off areas to the dynamics of sea–river–land interactions. This results in increased cultured production, but there is evidence that it also leads to decreased coastal productivity and diversity (Turner, 1977; Turner and Boesch, 1987; Blaber, 2002; Mumby *et al.*, 2004). Thus, pond culture competes with other natural resource-use claims, such as fisheries and biodiversity conservation, in both mangroves and adjacent estuarine and coastal habitats. Informed decision-making in the management of estuarine systems requires a quantified view of the ecological impacts of pond production on adjacent marine ecosystems that depend on well-functioning mangroves. The capacity of stakeholders using and managing aquatic resources to develop an informed view on the trade-offs in marine resources is hampered by the high diversity and variability of the estuarine environment as well as by the high diversity and adaptability of its uses. A first requisite for enabling the development of evaluative capacity is a thorough examination of trends, shifts and variabilities in relevant existing data and information, for instance, from statistical monitoring of fisheries.

Essential in the development of a shared view by users and managers on the state of aquatic resources and the assessment of trade-offs in resource use is the capacity to detect trends and shifts in coastal aquatic production, productivity and diversity. This capacity is dependent on the predictability and size of the daily, seasonal and inter-annual variability in outcome of the various resources. Variability in outcome obscures the perception of changes and trends, while aggregation of data or observations reduces temporal and spatial detail. This inherently means that different stakeholders using and managing a resource have differing capacities to evaluate trends depending on their position in the decision-making process with regard to the flow of information from the resource (van Densen, 2001). A high variability in resource outcome leads to a limited capacity to: (i) perceive trends and changes as they occur; (ii) attribute changes to causes; and (iii) relate individual experience to aggregated information, and, with that, to agree on the state of a resource and on the causes of change by stakeholders who have differential access to information. This in turn results in difficulties in communication between and among stakeholders. If informed agreement on states (and causes) of changes cannot be reached, the capacity to resolve conflicts will be hampered.

Within 20 years, up to 70% of the mangrove forest of the Mahakam Delta (Fig. 17.1) was converted to shrimp ponds, resulting in the destruction of large tracts of essential fish habitat. We can reasonably hypothesize that these rapid changes will be reflected in the daily and annual resource outcome of coastal fishermen and pond farmers as well as in the existing time-series of fisheries monitoring data. We explore the capacity for perception of changes in resource outcome of pond farmers, fishermen and fisheries managers in the Mahakam Delta by examining existing data and information on fisheries and brackish-water mariculture production and productivity. While shrimp culture in the 'mining mode' leads to habitat destruction, resulting (potential) shifts in diversity and productivity of natural resources that could represent a long-term threat to the resource base and use options will not appear to be evident.

Time-series of fisheries-dependent data collected by fisheries management institutions through monitoring of catch, fishing effort and other production statistics, and existing data and information collected through specific surveys on fisheries and shrimp-pond production, were used to address the following questions:

1. What are the developments in fisheries (effort, catch and catch rates) and mariculture (area, production and productivity)?
2. Is there evidence of a decreased productivity of shrimp ponds (= *tambaks*) over time for cultured shrimp (*Udang windu*, *Peneaus monodon*) and wild shrimp (*U. bintik*, *Metapeneaus monoceros* and *M. brevirostris*)?

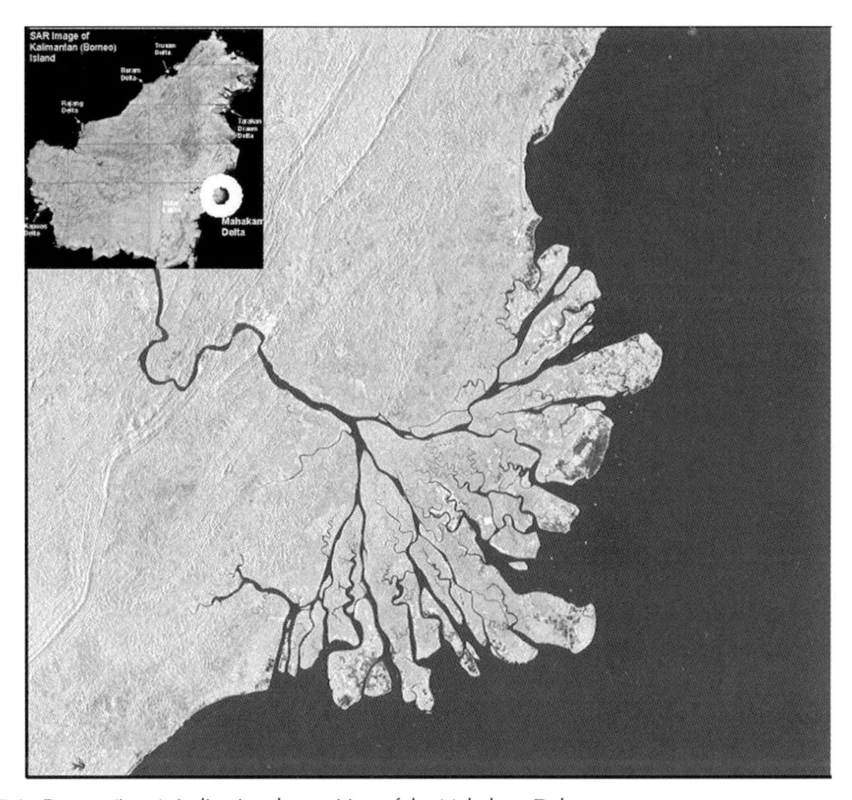

Fig. 17.1. Borneo (inset), indicating the position of the Mahakam Delta.

3. What is the potential for detection of changes in productivity by fishermen and pond farmers given their daily catches and pond harvests?
4. What is the potential for detection of changes in marine aquatic production and productivity based on aggregated statistical fisheries information?
5. What are the quantitative indications of a trade-off between the two marine production activities?

Study Site: Development of Shrimp Culture and Fisheries

The Mahakam Delta is about 150,000 ha, of which around 110,000 ha are categorized as salt- or brackish-water mangrove. Administratively, it is located in the regency of Kutai Kartanegara. The landscape is a deltaic plain, with the meanders of the Mahakam River ending in a complex fan-shaped network of multiple distribution channels and small rivers subject to tidal influence. Before 1950, the natural vegetation was a typical mangrove forest in an upstream to seashore succession of mixed-freshwater forest to dense *Rhizophora* spp., with some degraded forest where oil palm plantations had been present. *Nypa* stands dominated, with 55% of the area covered, ahead of freshwater mangroves (17%) and dense *Avicennia* stands.

With the onset of the 20th century, the delta had few settlements: fishermen and coconut planters who had settled in its early decades largely left during World War II. This situation remained until the early 1970s, when oil exploration and production caused an influx of people – mainly from Sarawak – attracted by the labour opportunities. In their wake, fisheries became important as well. With limited possibilities for preservation and transport, local fishermen did not consider shrimp to be valuable – small-sized shrimp were sun-dried and sold as *ebi* (salted dried shrimp) – until two cold-storage facilities built in the early 1970s provided capital to fishermen for modernization of their boats and fishing equipment. Around 1975, the introduction of trawls increased shrimp catches considerably. In 1980, a decision by the Indonesian government put an end to all trawl fishing, a ruling carried out over the next few years. A few milkfish ponds, developed with methods from South Sulawesi, were already established in the delta as early as 1932: trials for the introduction of shrimp to substitute for the loss of trawl production and to cater to increased international demand achieved success in the early 1980s. This initiated a new immigration wave from Sulawesi. The pond opening rate increased after the introduction of excavators, which replaced manual labour around 1989, and again after the monetary crisis of 1997 (Fig. 17.2). The resulting boom in shrimp production turned the Mahakam Delta into one of the wealthiest areas in Indonesia.

Most stakeholders – pond owners, fishermen, oil companies, government and NGOs – are pessimistic about the future of the delta, which, in their view, will increasingly become 'an ecological and economic desert under the combined assault of mangrove clearing, erosion, pond development, diseases, salinity, and productivity and income decreases'. The most important issues are related to land conflicts and erosion. Furthermore, all stakeholders agreed that maintaining the sustainability of shrimp-pond production and protecting or improving mangrove areas for the reproduction of fish and crustaceans have a high priority, indicating local awareness of potential trade-offs in natural resource use. Nevertheless, when weighing ecological and economic issues, it appeared that mangroves were seen mainly as valuable if cleared for ponds. Only fishermen did not appear to be much troubled by the concerns expressed by other stakeholders, except on the issue of protection of nursery and breeding areas, indicating local awareness of potentially conflicting resource use (Bourgeois *et al.*, 2002).

Data and Methods of Analysis

Data

Fisheries monitoring data available from 1980 to 2004 in statistical yearbooks were obtained from the Dinas Kelautan dan Perikanan in Samarinda (provincial office) and Tenggarong (regency office). They were used to examine developments in fisheries and mariculture (question 1), trends in overall productivity of ponds (question 2), the potential to detect changes in marine production (question 4) and evidence of trade-offs between the two activities (question 5). The two offices monitor catch by species category and gear type, effort by gear type and pond production by species category by district and by fresh and marine waters. All data presented and analysed refer to the Regency of Kutai that encompasses the Mahakam Delta.

Detection of trends in pond productivity (question 2) and in changes by pond farmers (question 3) was examined with data obtained from a survey on the management and productivity of the tambak culture conducted by the Faculty of Fisheries and Marine Science, Mulawarman University, Samarinda (UNMUL-FPIK), commissioned by Total Indonesia. The survey encompassed 18 locations distributed over the delta in different mangrove habitats, indicated by salinity. Between July and September 2003, 125 farmers were interviewed. Physical and chemical parameters were measured from a selected pond managed by the people interviewed. We used the following data:

- Production of the last harvest (kg/ha) of stocked *P. monodon* and wild *Metapenaeus* spp. ($n = 96$)
- Total area and number of ponds ($n = 125$),
- Year of pond construction (pond age) ($n = 124$); and
- Management factors: (i) salinity ($n = 125$); pH ($n = 124$), % water exchange ($n = 120$); (ii) stocking density of tiger shrimp ($n = 118$); and (iii) input of nitrogen, phosphate and potassium (NPK) fertilizers ($n = 59$), triple superphosphate (TSP) ($n = 49$) and lime ($n = 15$) (all kg/ha).

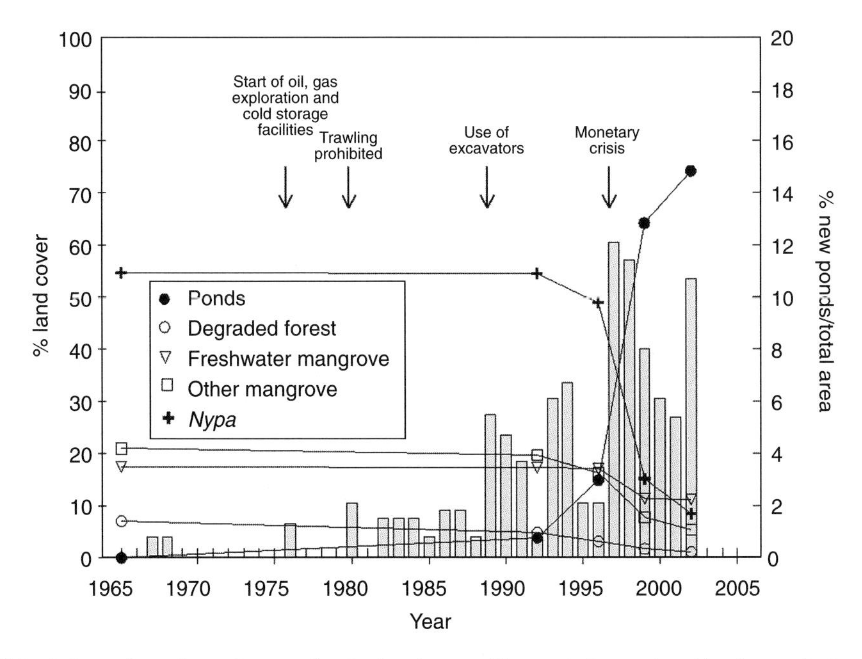

Fig. 17.2. Estimated pond opening rate (bars) related to specific events (arrows) and land cover in the Mahakam Delta (lines with markers). Based on a GIS database constructed from historic settlement analysis, maps, aerial photographs and satellite photographs (adapted from Bourgeois *et al*., 2002).

Shrimp ponds are stocked and harvested two to three times a year. All parameters in the survey referred to the last harvest period, except for salinity and pH, which were measured *in situ*. For each factor with n <125, the value of the non-recorded observations = 0, except for stocking density, which was not known when not recorded, as often multiple stockings take place over a harvest period because of the scarcity of shrimp postlarvae. Analysis of variance was used to examine the significance of difference ($\alpha \leq 0.05$) in the mean harvest of ponds with non-zero values for a parameter vis-à-vis those with observations = 0 or not known. This was never the case.

Daily catch variability (question 4) was examined for four small shrimp trawlers from the village of Muara Badak, located on the north coast of the delta, based on daily log-book recordings over a period of 6 months.

Estimates of total pond area development, taken from four studies (Table 17.1), are based on interpretations of satellite imagery. Dutrieux (2001) and Bourgeois *et al.* (2002) also based their estimates on aerial photography and interpretations of old maps, including historical settlement analysis. These are the most consistent time-series as the interpretations are performed by the same team. However, all studies suffer from a lack of (access to) methodological descriptions and/or limited ground truthing, hampering the evaluation of the discrepancies between the estimates. Despite this, it can be concluded that in 2004 at least 50–70% of the total brackish- and saltwater mangrove had been converted. We have used the estimates of Dutrieux (2001), Bourgeois *et al.* (2002) and Bappeda Kukar (2003) to interpolate and extrapolate the total pond area from 1980 to 2004 through

$$\text{Tambak area (ha)} = 6006 \times 10^{0.17\,(\text{year 1993})} \tag{17.1}$$

and checked the results with information on the pond opening rate reconstructed by Bourgeois *et al.* (2002) (Fig. 17.2).

Table 17.1. Area of the Mahakam Delta covered by mangrove, settlements or shrimp ponds (tambak). All analyses are based on interpretation of satellite imagery and GIS maps (see text for an evaluation of the estimates).

Item	Year								
	1984[1]	1991[2]	1992[1]	1996[2]	1998[1]	1999[1]	2001[2]	2001[3]	2004[4]
Mangrove (ha)		105,999		95,096			40,219	–	60,818
Settlements		74		125			129		–
Tambak (ha)	420	3,629	3,687	14,480	52,300	67,000	69,354	50,000 (31,000, in prep.)	45,297 (2,036, in prep.)
Total (ha)	110,000	109,702	110,000	109,702	111,000	110,000	109,702	110,000	108,153

[1] Dutrieux (2001).
[2] Bappeda Kukar (2003).
[3] Bourgeois *et al.* (2002).
[4] From Fakultas Perikanan dan Ilmu Kelautan, 2005 (FPIK) (2005). Produktivitas tambak di kawasan Delta Mahakam (Shrimp pond productivity on Mahakam Delta). Draft Report to Total Indonesia, Balikpapan, March 2005. pag. var.)

Methods of analysis

Data treatment

The 24 statistical yearbooks contain catch information on 60 species (groups); for the regency of Kutai, 39 contain time-series of data. Trends analysed were by: (i) total catch; (ii) catch of demersal/benthopelagic and pelagic species (Froese and Pauly, 2005); and (iii) catch of 38 categories of fish, crustaceans and molluscs (two categories combined). Catch information is collected for 25 fishing methods, but we aggregated the numbers of gears and total catch into the six main categories recorded before 1988: large trawls, small trawls and seines (*Pukat kantong*: shrimp trawls, lampara and Danish, beach and purse seines), gill nets (*Jaring insang*: drift nets, encircling nets, shrimp gill nets, stationary gill nets, trammel nets); lift nets (*Jaring angkat*: boat/raft nets, *bagan*, scoop nets, other lift nets) and hooks and lines (*Pancing*: tuna longlines, longlines, set longlines, other poles and lines, troll lines). A category 'traps' (*Perangkap*: guiding barriers, stow nets, portable traps, other traps) was also aggregated with a category 'others' (Anadara shell fish, muro-ami and others) recorded only occasionally with low levels of output. Catch rate by gear group was calculated by dividing the annual catch per gear group by the associated number of gears.

The productivity of shrimp ponds was calculated by dividing the total annual production and the production by species category by the estimated area (ha) of ponds in that year.

Data analysis

Trends were analysed with the polynomial regression model:

$$G(m)_i = a + b^*year_i + c^*year_i{}^*year_i + \epsilon_i \quad (17.2)$$

where $G(m)_i$ is the time-series of catch (fisheries) or production per hectare (tambaks), a is the intercept, b and c are the slope, $year_i$ is the linear regression term; $year_i{}^*year_i$ are the quadratic terms; and ϵ_i is the residual error.

Only significant parts of the model ($\alpha \leq 0.05$) were retained, resulting in four possible models: mean, linear, quadratic and polynomial. Concavity and explained variance of the quadratic and linear terms of significant polynomial models were evaluated to obtain an indication of direction, timing and strength of the reversal in long-term trends. Trends were tested over the whole period and over the period from 1989 onward, when pond opening rates increased dramatically. An indication of recent change relative to the long-term trend was obtained by examining the short-term trend over the last

5 years. Serial correlation, producing additional variability obscuring long-term trends, was examined in time-series of shrimp-pond productivity by cross-correlation at lag = 1 (year) of the residuals of the linear trend analysis.

The effect of age on stocked and wild shrimp productivity was assessed by a two-stage regression analysis using general linear modelling (SAS-Institute, 1989). The first variance reduction was achieved with all significant ($\alpha \leqslant 0.05$) factors and sensible interactions that could have an effect on the production of each species and for which data were available. In the next step, the residuals were regressed on age as an explanatory factor. All time-series were orthogonalized before analysis. All data were ^{10}log-transformed to fulfil the assumption of normal distribution of residuals.

Variability, expressed as coefficient of variation (CV is standard deviation scaled by the mean), is used here to scale (basic) uncertainties in fisheries and aquaculture outcomes. CV is a powerful indicator of the capacity to perceive trends and shifts. For instance, low to medium variabilities (CVs of 23–60%) in daily catches are found in trawl fisheries, medium CVs of 50–75% in gill-net fisheries and high to extreme CVs of 80–500% in light fisheries (van Densen, 2001; Oostenbrugge *et al.*, 2002; van Zwieten *et al.*, 2002; Jul-Larsen *et al.*, 2003b). After variance reduction of ^{10}log-transformed data by known (spatial, temporal, management) factors, the residual variance from the models, also called random or basic uncertainty (van Zwieten *et al.*, 2002), can be expressed as CV through (Aitchison and Brown, 1957)

$$CV = 100 * \sqrt{e^{2.303*\sigma^2} - 1} \qquad (17.3)$$

where σ is the standard deviation in the residuals of the ^{10}log-transformed data.

All CVs are directly comparable. Variabilities at higher scales of aggregation are estimated by dividing the variability at a lower scale by $\sqrt{n}$, with n as the number of units aggregated. Similarly, disaggregation is done by multiplying by $\sqrt{n}$. Both procedures assume independence of observations.

Trend perception can be approximated statistically through power analysis. The number of years needed to significantly detect a trend b in the time-series can be calculated iteratively given decision limits α and β, and the trend-to-noise (b/s) ratio following the inequality (van Zwieten *et al.*, 2002):

$$\left|\frac{b}{s}\right|\sqrt{\frac{n(n-1)(n+1)}{12}} \geq (t_{\alpha/2} + t_{\beta}) \qquad (17.4)$$

where b is the trend parameter (slope) in the linear regression, n is the number of observations (year), and $t_{\alpha/2}$ and t_{β} are the decision rules of a t-distribution, where α is the probability of a type I error (a trend is rejected falsely) and β the probability of a type II error (a trend is accepted falsely). In our analysis, $\alpha = \beta = 0.1$; s is the standard deviation of the residuals.

Data-based Perceptions on Changes in Aquatic Food Production

Over 24 years, the total fishery and mariculture production in Kutai has increased by 3.3% per year from 5700 to 27,200 t. Total fish catch increased at a slower rate of 2.9% per year from 5400 to 17,100 t. After a drop in levels in the late 1980s, crustacean catches reached 1300 t in 2001, the same level as the average catch of 1980 to 1987. The crustacean fishery decreased in importance from around 20% of the total production in the early 1980s to 4–5% at present. From 1980 onward, shrimp culture increased from 5% (285 t) of the total production to 29% (7803 t) in 2004 (Fig. 17.3).

Mariculture

Total annual pond production increased linearly with the area under cultivation, while the shrimp production increased by a factor of 1.24 (Fig. 17.4). Over a 24-year period, fisheries statistics indicate a shift in mariculture from fish to prawns around 1995 (Fig. 17.5). Initially, stocked *P. monodon* dominated shrimp production. From 1990 to 2000, during the boom in pond construction, wild

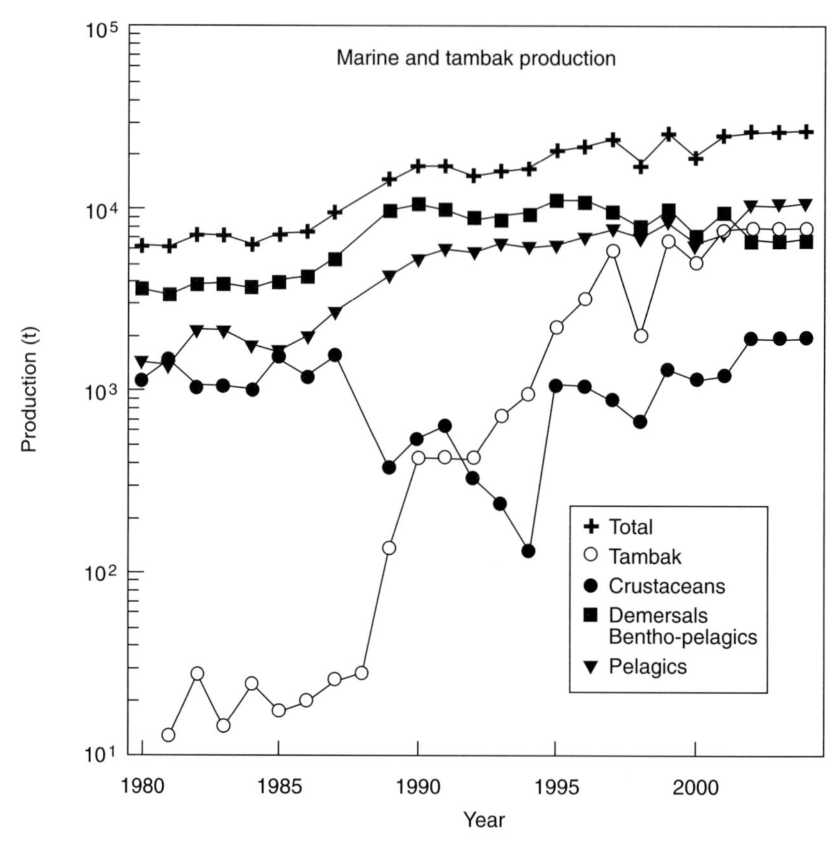

Fig. 17.3. The development of the total marine and brackish-water fisheries and tambak culture production of the regency in Kutai, East Kalimantan. Closed markers, capture fisheries; total, aggregated fisheries and culture production (data from statistical yearbooks, Fisheries Service).

Metapeneaus spp. formed 63–81% of the total production. Presently, about 50% of the production again is stocked *P. monodon*. The production of mullet and milkfish increased concurrently, until 1997, to 1357 and 1059 t, respectively. Mullet decreased to about 300 t and milkfish remained stable. The production of other fish species (wild stocks of, for instance, *Lates calcarifer*), never important, is now zero. Inter-annual variability in production is high for all species, ranging from a CV of 93% (*P. monodon*) to 179% (mullet). Over the last 5 years, the production of all categories has stabilized.

Tambaks are cultured extensively: with 125.4 kg/ha, long-term productivity is stable and low, but with a high inter-annual variability (CV of 69%) at the aggregated level of the delta. Since 1980, productivity of both stocked and wild shrimp has increased by a factor of 1.09/year, and both have a residual inter-annual CV of, respectively, 76% and 125%. Stocked shrimp productivity has stabilized at 41 kg/ha over the last 5 years, but wild shrimp decreased to a stable level of 44 kg/ha after a peak in 1997. In both cases, only the linear trends were significant ($\alpha \leq 0.05$), but with low power: the observed trends could be detected only with 20–26 years of data ($\alpha = \beta = 0.1$) (Table 17.2). No evidence of serial correlation was found in the residual time-series of stocked shrimp productivity. But the residual productivity of wild shrimp of subsequent years is strongly correlated ($\rho = 0.55$, $r^2 = 0.30$, $P < 0.01$), indicating that natural processes regulating the overall productivity of wild shrimp in the delta are important for cultured productivity.

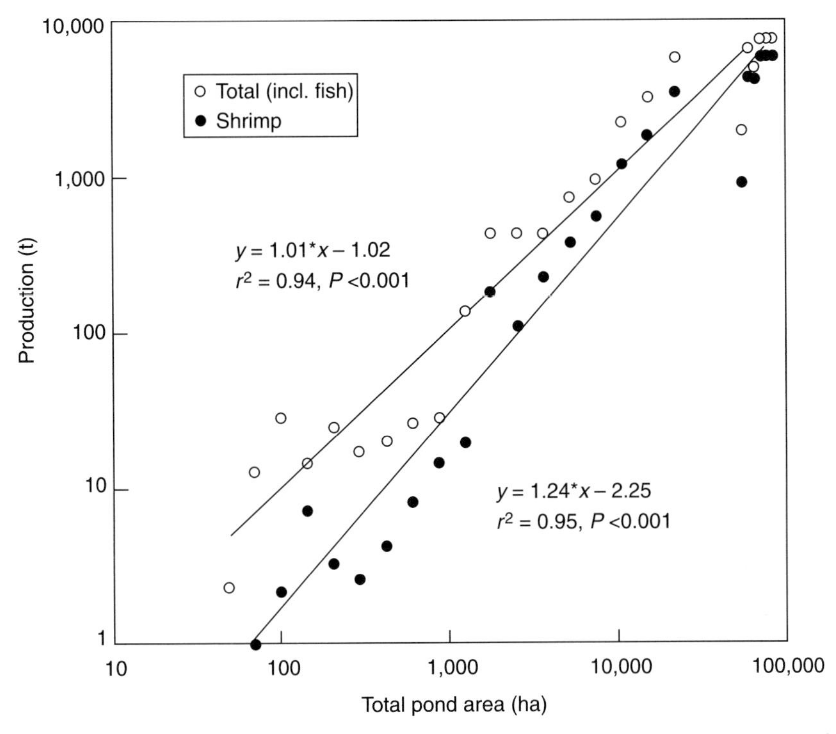

Fig. 17.4. Total and shrimp production in the mariculture of the Mahakam Delta related to area under cultivation (total pond area). See text for the estimation procedure of total pond area.

Average shrimp productivity of one harvest calculated from the UNMUL-FPIK survey is 66 kg/ha ($CV_{harvest} = 119$). With two to three harvests per year, this amounts to an annual productivity of 130–190 kg/ha, with a $CV_{year} = CV_{harvest}/\sqrt{n} = 84\% - 68\%$. The average productivity per harvest of stocked *P. monodon* is 42.5 kg/ha ($CV_{harvest} = 117\%$; $CV_{year} = 83\% - 68\%$) and of wild *Metapeneaus* spp. is 25 kg/ha ($CV_{harvest} = 163\%$; $CV_{year} = 115\% - 94\%$). Thus, the annual average productivity is higher than the estimate from fisheries statistics. However, the variabilities are in the same order of magnitude, which is unexpected as aggregation of pond production over the whole delta scaled by the total pond area should lower variability. The high variability in fisheries statistics could be due to data collection methods, to our estimation procedure of the area under cultivation or to high co-variance in the production per pond inducing increased variability. The latter may result, for example, from overall harvest failure induced by weather or through variable availability of shrimp larval production.

Larger-sized ponds had lower productivity, also noted by Bourgeois *et al.* (2002), but pond size explained more variation in wild shrimp (13%) than in stocked shrimp (4%) (Table 17.3). None of the management factors had a significant effect on stocked *P. monodon* productivity. Fertilizer application had a significant positive effect on wild *Metapenaeus* spp. (Table 17.3). However, as the mean harvest of fertilized ponds did not differ significantly from that of non-fertilized ponds, the effect was cancelled in the overall variability. Lowered productivity with increasing age was observed with stocked shrimp, but the age effect explained only 5% of the residual variance after correction for size (Table 17.3).

Fisheries

The fisheries around the delta are small-scale and multigear and have a limited spatial range: at present, up to 50% of the vessels are

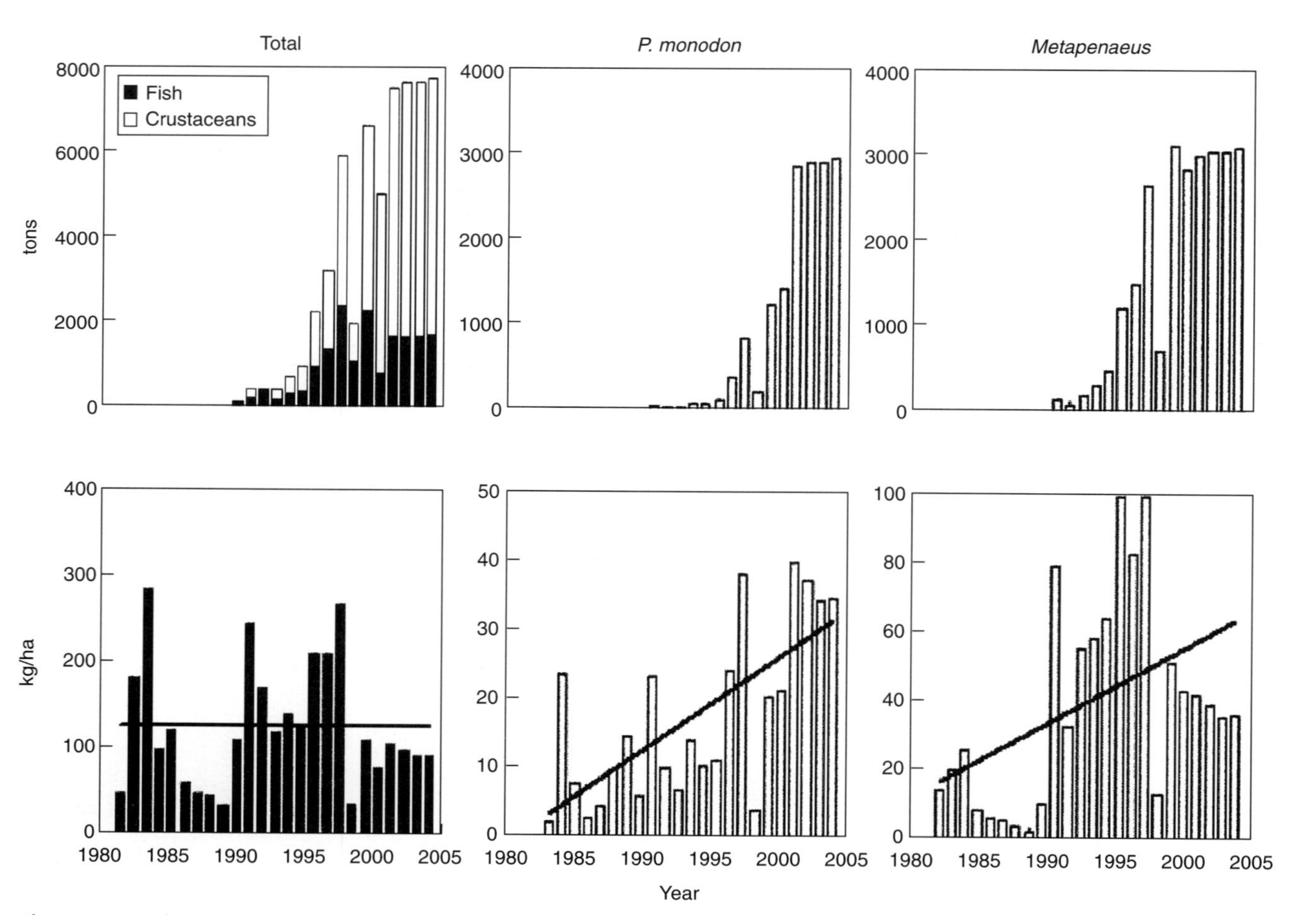

Fig. 17.5. Development of the total, and shrimp, production (top panels) and productivity (bottom panels) in the Mahakam Delta, 1980–2004. The mean of the time-series is displayed if the trend is not significant ($\alpha \leq 0.05$) (data from statistical yearbooks, Fisheries Service).

Table 17.2. Trend and power analysis of the marine catch (1989–2004) and tambak productivity (1980–2004).

Species	PDT	df	CV	B	B:S	*n*	Model	r^2	Year of trend reversal	Slope/concavity
Marine catch (tons)										
Rays (Rajiformes) – 10	H	15	70	−0.14	0.49	9	Lin	0.85		−
Sharks (Elasmobranchii) – 10	H	15	83	−0.06	0.19	16	Lin	0.46		−
Croakers, drums (Sciaenidae) – 9	L	15	5	−0.02	0.82	6	Pol	0.67	1995	−
Grunters, sweetlips (Haemulidae) – 5	M	15	45	0.02	0.12	21	Pol	0.62	1998	−
Yellow-tail fusiliers (Caesionidae) – 9	M	14	51	−0.03	0.13	21	Pol	0.86	1996	−
Skipjack tuna (Scombridae) – 4	M	14	54	0.03	0.15	18	Pol	0.57	1999	−
Big eyes (Holocentridae) – 9	L	15	84				Pol	0.82	1996	−
Trevallies (Carangidae) – 2	M	14	32				Qua	0.32	1996/7	−
Goatfish (Mullidae) – 5	L	15	69				Qua	0.59	1996/7	−
Scads (Carangidae) – 2	L	15	117				Qua	0.54	1996/7	−
I. halibut (Pleuronectiform) – 5	M	15	130				Qua	0.37	1996/7	−
Rainbow sardine (Clupeidae) – 3	L	14	44	0.04	0.21	15	Lin	0.51		+
Common squid (Cephalopoda)	L	9	46	0.08	0.43	10	Lin	0.61		+
Pony f./slipmouths (Leiognathidae) – 5	L	14	87	0.08	0.24	14	Lin	0.55		+
Mud crabs (Portunidae) – 12	L	15	93	0.10	0.28	12	Lin	0.66		+
Metapenaeus shrimp (Penaeidae) – 15	L	14	95	0.08	0.21	15	Lin	0.50		+
Giant tiger prawn (Penaeidae) – 13	L	13	96	0.10	0.29	12	Lin	0.62		+
Sea cat-fishes (Ariidae) – 5	M	15	37				Pol	0.68	1995	+
Threadfins (Polynemidae) – 11	L	15	39	0.02	0.12	22	Pol	0.53	1994	+
Anchovies (Engraulidae) – 3	L	15	60	0.05	0.20	16	Pol	0.68	1993	+
Fringescale sardinella (Clupeidae) – 3	L	15	62	0.03	0.11	23	Pol	0.66	1995	+
Mullets (Mugilidae) – 11	M	15	81				Pol	0.63	1998	+
Black pomfrets (Carangidae) – 6	M	15	63				Qua	0.40	1996/7	+
Eastern little tuna (Scombridae) – 4	L	15	71				Qua	0.29	1996/7	+
Barramundi (Centropomidae) – 7	H	15	81				Qua	0.56	1996/7	+
Narrow band king mackerel (Scombridae) – 4	M	15	114				Qua	0.59	1996/7	+
Indopac king mackerel (Scombridae) – 4	M	14	123				Qua	0.60	1996/7	+
Other fish		12	185	−0.14	0.27	13	Lin	0.54		−
Other shrimp – 16		14	45	0.05	0.28	12	Lin	0.67		+
Total catch – 1		15	10	0.006	0.14	19	Lin	0.34		+
Tambak productivity (kg/ha)										
Penaeus monodon		23	76	0.04	0.13	20	Lin	0.44		+
Metapenaeus spp.		23	126	0.04	0.09	26	Lin	0.29		+
Total tambak productivity		23	69				Mean	ns		

Mean models were obtained with red snapper (Lutjanidae) – 8 (CV = 29, PDT = M); Spiny lobster (Palinuridae) – 12 (36, L); Indian mackerel (Scombridae) – 4 (38, L); Jack trevallies (Carangidae) – 2 (42, M); Groupers (Serranidae) – 8 (44, M); Silver pomfret (Stromateidae) – 6 (44, M); Wolf herrings (Chirocentridae) – 3 (51, L); Banana prawns (Penaeidae) – 14 (76, L); and other marine catch (207).

PDT, population doubling time (H, high, > 4.4–14 years; M, medium, > 1.4–4.4 years; L, low, < 15 months); df, degrees of freedom; CV, coefficient of variation; b, slope of the linear trend (significance $\alpha \leqslant 0.05$); b/s, trend-to-noise ratio; *n*, number of years of data needed to detect trend b ($\alpha = \beta = 0.1$); Lin, linear; Pol, polynomial; Qua, quadratic; r^2, coefficient of determination; ns, non-significant; −, negative slope/downward concavity; +, positive slope/upward concavity. Numbers associated with names refer to the species aggregations in Fig. 17.8.

Table 17.3. Variance reduction in productivity (kg/ha) of wild (*Metapenaeus monoceros*) and stocked (*Penaeus monodon*) shrimp in four models with pond size, age and management factors as explanatory factors.

	Wild shrimp				Stocked shrimp				
Model	df	MSE	r^2	b	df	MSE	r^2	b	Trend to noise ratio (b:s)
Total	85	0.296			95	0.213			
Res size	84	0.262[c]	0.13	−0.579	94	0.207[a]	0.04	−0.308	
Res size age	83	n.s.	–	–	93	0.198[a]	0.09	−0.021	0.048
Total	33	0.379							
Res TSP	32	0.296[b]	0.24	0.448					
TSP size	31	0.274[b]	0.32	−0.653					
Total	29	0.515							
TSP*NPK	28	0.300[b]	0.30	0.235					
TSP*NPK size	27	0.224[b]	0.49	−1.069					

[a] Significance ≤ 0.05.
[b] Significance ≤ 0.01.
[c] Significance ≤ 0.001; n.s., non-significant.
df, degrees of freedom; MSE, mean square error; r^2, coefficient of determination; b, slope; s, standard deviation; Res, residuals; size, pond size; age, number of years since pond construction; TSP, triple superphosphate; NPK, nitrogen, phosphate, potassium; TSP*NPK, interaction term (from FPIK-UNMUL survey on 125 tambaks in the Mahakam Delta).

non-motorized; the remainder have diesel engines of < 5 hp or outboard engines. Most gears are more or less stationary: gill nets and traps/barriers each account for about 30% of the gears and lift nets/bagans account for another 18%. The active fishing gears, trolls and lines, and trawls each account for about 10% of the gears (Fig. 17.6). The annual catch of 1000–3500 kg/gear/year is comparable with that of any small-scale fishery (Jul-Larsen *et al.*, 2003a). Three gear categories show variable but increasing catch rates over the past 15 years (gill nets, trawls and lift nets/bagans), whereas those of lines and traps are variable but stable (Fig. 17.7). This is contrary to expectation as increased effort generally leads to decreased catch rates. Changes in gear, catchability resulting from increased gear efficiency, target species, spatial allocation of effort or changes in carrying capacity due to environmental forcing could each result in the observed developments. As yet, none of these explanations can be excluded.

Since 1980, the marine fisheries showed an increasing and (in recent years) stabilizing catch of pelagic and demersal/benthopelagic species, but with a shift in importance to pelagics (the ratio of pelagics/demersals increased from 0.4 to 1.6). A long-term overall stable but highly variable catch of shrimp dropped to almost zero around 1988 (Fig. 17.8). Trends reveal a clear shift from shrimp to fish species after 1987, possibly as a result of the trawl ban. After that, shrimp and fish catches, respectively, initially decreased and increased rapidly in concurrence for most species (Fig. 17.8, Table 17.2). Since 1989, rays and sharks, highly vulnerable to fishing pressure (Myers and Worm, 2003) and with high population doubling times (PDT) (Froese and Pauly, 2005), are the only species categories with a significant ($\alpha \leq 0.05$) long-term linear decreasing trend. All species with a long-term linear increasing trend had low PDT. From 1995 to 1999, catches in nine categories started decreasing or stabilizing, whereas from 1993 to 1998 ten categories showed a reversal to increasing trend. These reversals indicate a shift from species that are less resilient to a changing environment, including increased fishing effort (high–medium PDT), to species highly resilient to changes but also with highly dynamic population sizes (low PDT). The distribution of high, medium and low PDT

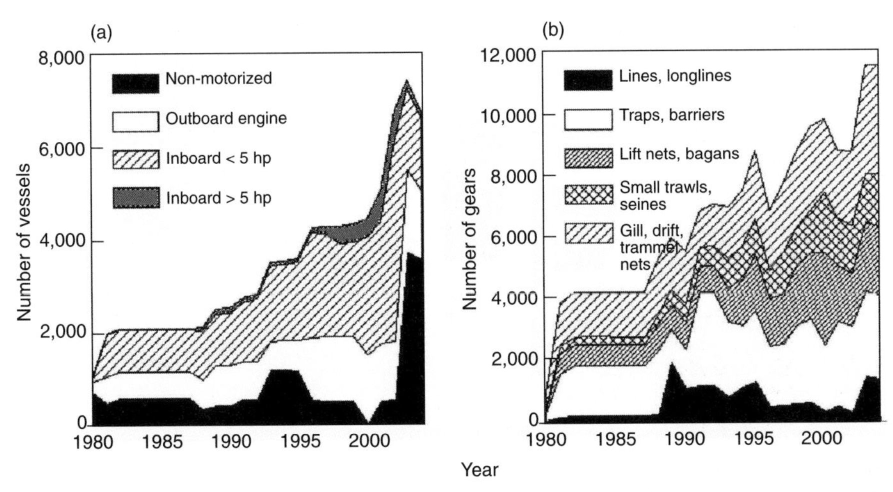

Fig. 17.6. The development in fishing effort around the Mahakam Delta (a) the number of vessels and motorization of vessels; (b) the number of gears by type (data from statistical yearbooks of the Fisheries Service). The number of non-motorized vessels was underestimated before 2003 data collection procedures were changed (pers. info., Fisheries Service, Samarinda).

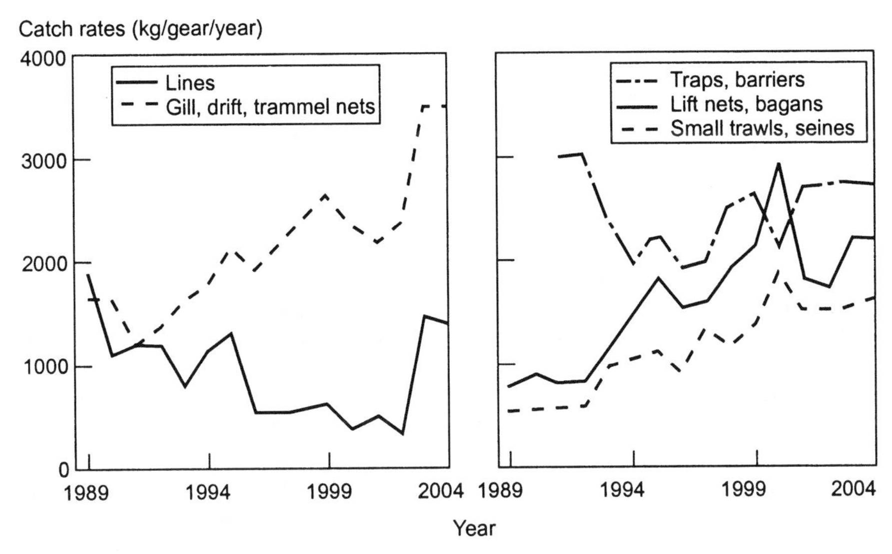

Fig. 17.7. The development in catch per unit of effort (= catch rate) for the main gear groups of the fisheries around the Mahakam Delta.

is, respectively, 2, 9 and 8 for categories that have a long-term, a recently decreasing or no trend, and 1, 5 and 10 for those with increasing trends (Table 17.2).

Inter-annual variability for all categories is extremely high. For instance, the residual inter-annual CV of *Metapenaeus* spp. of 95% is as high as the daily variability experienced by four small shrimp trawler fisheries in the estuary (van Zwieten, unpublished data). Though a shrimp fisher will experience high inter-annual variability because of high variability in annual local shrimp biomass, it can be demonstrated that the CV from the fisheries statistics is outside any ordinary experience of a fisherman. Dis-aggregation over about 1600 (see Fig. 17.6) small-trawler fishermen presently fishing in the estuary

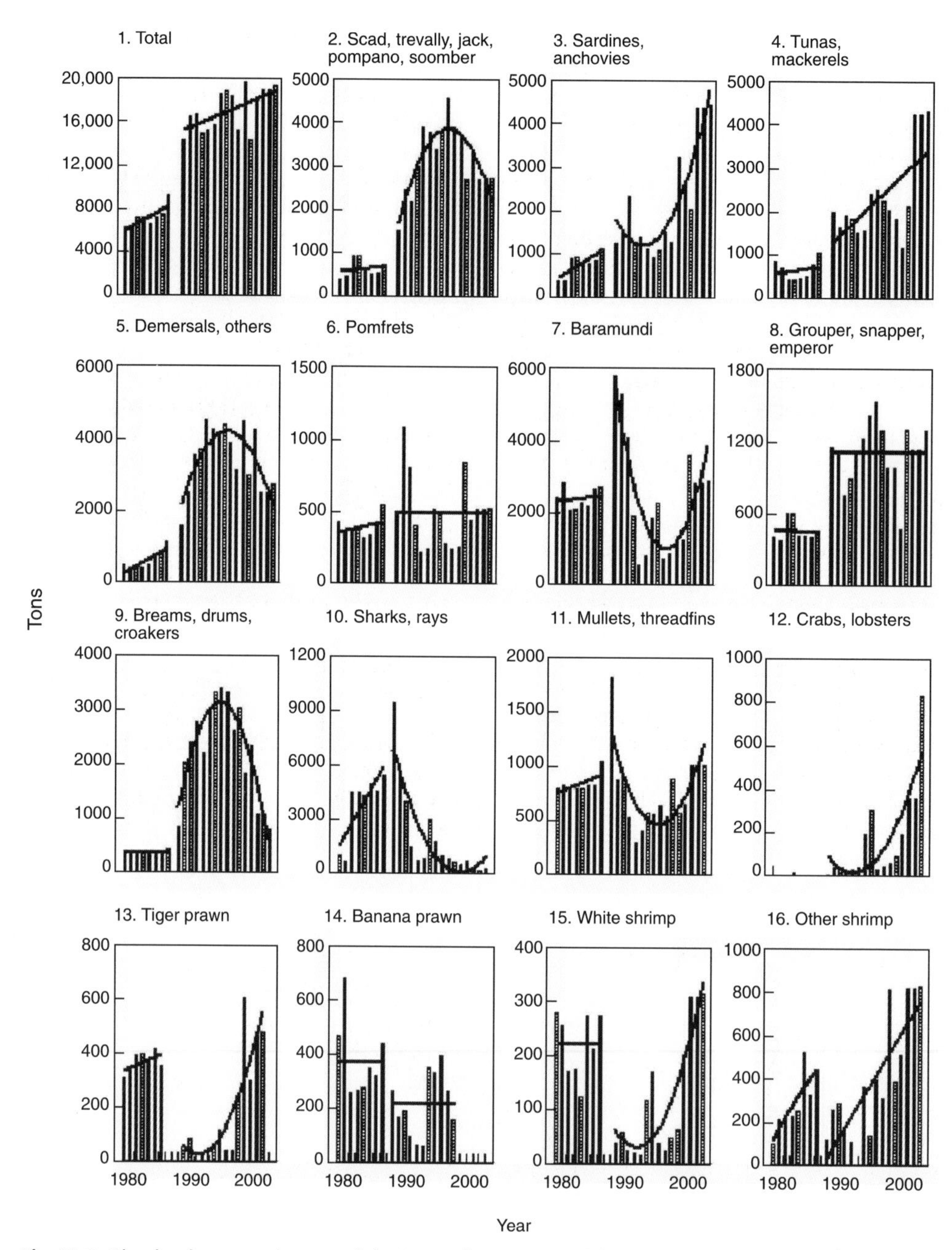

Fig. 17.8. The development of capture fisheries production around the Mahakam Delta, 1980–2004. Top panel, total and pelagic species catch; middle panels, demersal species; lower panel, shrimp species. The aggregation of 39 species categories in the 16 categories presented here is explained in Table 17.2. The mean of the time-series is displayed if the trend is not significant ($\alpha \leq 0.05$) (data from statistical yearbooks of the Fisheries Service).

yields a $CV_{dis} = 95*\sqrt{1600} = 3800\%$. The cause of this high variability may be due largely to the data collection methods used. With this caveat on data quality, all categories with long-term linear trends exhibited inter-annual catch variabilities of CV < 80, except crustaceans, pony fish/slipmouths and sharks and rays, which showed higher inter-annual variability. To better compare the capacity for trend perception in all species, we used the highest (b = 0.14) and lowest (b = 0.02) absolute slopes in long-term linear trends to calculate the number of years of data needed for detection according to Equation 17.4. Trend detection for a 38% decrease or increase per year required from 2 (croakers/drums) to 11 (Indian halibut) years of data (Fig. 17.9). Trend detection for a 5% increase or decrease per year requires from 5 to 50 years of data for these two extremes. The four most variable species represent around 20% of the total catch. Of these, the two mackerels (*Scomberomorus commerson* and *S. guttatus*) and the scad (*Decapterus* spp.) are coastal schooling and long-shore migratory pelagics whose populations are probably not local. Two important shrimp categories, known to have high inter-annual fluctuations in population densities, are extremely variable as well.

Discussion

We can now answer our five questions: after the boom in pond development from 1989 onward, changes were observed in several indicators of fisheries and mariculture production and productivity. Shrimp pond area and fishing effort in all gear categories increased, resulting in an overall increase in production. Productivity for both activities either increased or stabilized.

Management factors and size and age of the pond explained only a limited amount of variability in shrimp productivity, and no evidence of decreased productivity of ponds can be found over the period examined. In the past 5 years, productivity stabilized around 85 kg/ha. The observed decreased productivity over time of wild *Metapenaeus* spp. should be treated with caution and could be the result of chance, as wild shrimp stocks are regulated by natural processes, and influx in ponds is a chance process depending on local availability of shrimp larvae.

A comparison with other production systems reveals that the Mahakam Delta pond productivity and production are low and extremely uncertain, and can best be described as a boom-and-bust operation. The

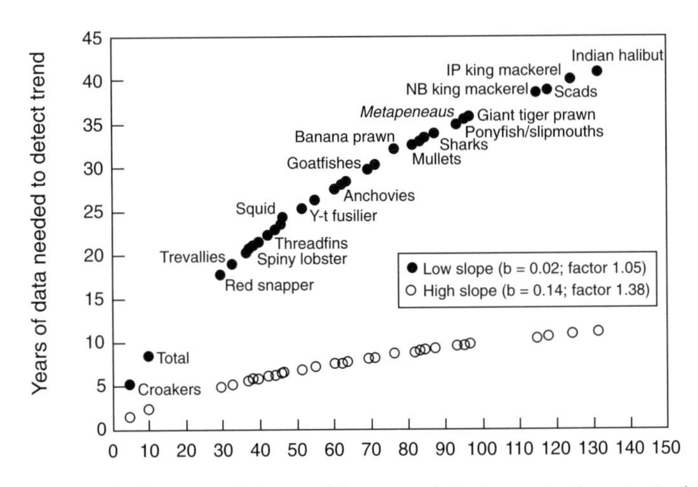

Fig. 17.9. Power analysis for the range of observed linear trends in the catch of species in the Mahakam estuary, expressed as number of years of data needed related to the observed variability in the statistical monitoring data. The data points of the high slopes (closed circles) and low slopes refer to the same species categories (see Table 17.2 and text). CV, coefficient of variation.

average harvest in the extensive shrimp culture in Probolingo, East Java, was 300 kg/ha/crop (Hariati *et al.*, 1998). With an average of 125 kg/ha/year (shrimp: 85 (statistical monitoring) to 130–190 (survey) kg/ha/year) and a mode in farm size in the Mahakam of 4–7 ha (Bourgeois *et al.*, 2002), the annual production per farm was 500–875 kg. The annual shrimp output, fished in around 180 days/year, of four small-scale shrimp trawler fishermen in the estuary is around 900 kg (van Zwieten, unpublished data). The overall CV in their total daily multispecies catch is 75%; taken separately, the four shrimp species in their catch (range 0.15–1.3 kg/day) had a high CV of 100–155%. This CV *per species per day* is comparable to the CV *per harvest* for individual pond owners of 119%. Inter-annual CV in catches from a wide range of fish stocks range from 9% to 100%; inter-annual CV in crop production ranges from 10% (intensively managed agriculture) to 70% (marginal rainfed agriculture) (van Densen, 2001). The extreme variability in the Mahakam pond productivity means that the potential for detecting changes, for instance, as a result of better management practices, is low both on an aggregated level of fisheries statistics and for individual pond owners. This is corroborated by Bourgeois *et al.* (2002): while causes of variability are readily discussed by stakeholders, there is little consensus about them, except for the effect of rainfall. The importance of better management practices and mangrove protection is high on the list of everyone involved, but experience is limited on what this could achieve. It will not be easy to gain this experience barring a long-term commitment to better management practices.

Notwithstanding data aggregation, lowering variability, limited data quality resulted in high variability in fishery statistics, obscuring the potential for detecting changes in marine production and productivity. Nevertheless, the observed trends, variabilities and shifts in fishery catches do have biological realism: the described developments in relative species composition are general trends appearing in many fish communities under stress. Although a significant shift in catch towards shorter-lived species took place from 1993 to 1999, 4 to 10 years after the start of the boom in pond construction, fishery statistics gave no direct evidence of a trade-off between the pond production and fisheries on the aggregated level of the whole estuary. Catch rates, at their aggregated level, show no indication that the fishery in its current operational set-up is affected by the rapidly decreasing mangrove coverage.

The high variabilities in fishery statistics alone already imply that the observability of a trade-off is limited. Multiple causes and non-linearity of the changes taking place will exacerbate the problem. First, the observed 6–9 years' monotonic rise and fall in annual catch of most categories is more likely due to changes in the structure of the fish community through environmental forcing or density-dependent processes (Bjørnstad *et al.*, 1999). Year-class variation in stocks would produce short-term trends of 3–4 years or shorter given the residence time of any year class in an exploited stock of most of the species examined. Whether these changes are cyclical or represent a more irreversible change as a result of mangrove habitat destruction or upstream processes in the Mahakam River cannot be concluded. Next, fishing patterns – gears used, spatial effort allocation – changed with increased effort throughout the 1990s as well. Changing fishing patterns in small-scale fisheries are often an adaptation to changing fish community structures (Jul-Larsen *et al.*, 2003a). Whether the fisheries are adapting to community changes as a result of environmental forcing or are themselves a major cause of the community changes depends on their relative impact. Lastly, while a certain minimum area of mangrove is necessary for the maintenance of coastal productivity, the mangrove area/yield relationship is non-linear (Baran and Hambrey, 1998). Many fish and crustacean species are transients and are not critically dependent on mangroves (Krumme and Saint-Paul, 2003; Mumby *et al.*, 2004). Where forests extend along a coastline and are connected by coastal currents, impacts on coastal fisheries may only be noticeable when most of the functional part of the forest has disappeared. Type, spatial arrangements and connectivity of mangrove forest are important and patterns of mangrove–water interface, tree communities and age of mangrove stands have been suggested as affecting fish and crustacean community structure (Primavera, 1997; Ronnback *et al.*, 1999; Manson *et al.*, 2003).

In the Mahakam Delta, the potential for detection of changes in relevant indicators as a result of management action is limited, while the high variability in individual outcome also limits the utility of local observations of fishermen and shrimp-pond farmers. The acceptability of taking measures to protect mangrove habitat could be reduced as any positive impact will be difficult to observe in the short term on a local level as well as in aggregate fishery statistics, limiting their use for evaluation of impacts. Over the long term, the utility of monitoring data appears to be high, but improved data quality and capacity to evaluate information will be required to increase diagnostic power. The challenge is to devise tools for informed decision-making with limited data in a spatially and temporally complex environment with a high fish and crustacean diversity, starting from what is already available. Better use of existing data, information and knowledge will aid in improving the evaluative capacity of those involved in managing coastal resources.

Acknowledgements

This study was carried out as part of the WOTRO/KNAW/ICOMAR East Kalimantan Pilot Research. The Netherlands Foundation for the Advancement of Tropical Research (WOTRO) is thanked for its financial support. LIPI and the University of Mulawarman in Samarinda are thanked for their logistical support. Earlier sections of the work were commented upon by Marc Verdegem and Leo Nagelkerke. R. Bourgeois and TotalFinaElf are thanked for making available their socio-economic study. The Dinas Kelautan dan Perikanan (Fisheries Service) in Samarinda is thanked for its assistance in disclosing the wealth of information present in its statistical yearbooks.

References

Aitchison, J. and Brown, J.A.C. (1957) *The Lognormal Distribution: With Special Reference to Its Uses in Economics.* Monographs, Department of Applied Economics, University of Cambridge 5, Cambridge University Press, Cambridge, UK, 167 pp.

Bappeda Kukar (2003) *Rencana Detail Tata Ruang Kawasan Delta Mahakam. Kerjasama antara Bappeda Kutai Kartanegara dengan LAPI Institut Teknologi Bandung, Tenggarong.* (Detailed spatial planning of Mahakam delta area. Cooperation between Bappeda Kutai Kartanegara and LAPI Bandung Institute of Technology, Tenggarong.) Internal report. 196 pp.

Baran, E. and Hambrey, J. (1998) Mangrove conservation and coastal management in South-east Asia: what impact on fishery resources? *Marine Pollution Bulletin* 37(8–12), 431–440.

Blaber, S.J.M. (2002) Fish in hot water: the challenges facing fish and fisheries research in tropical estuaries. *Journal of Fish Biology* 61(1), 1–20.

Bjørnstad, O.N., Fromentin, J.-M., Stenseth, N.C. and Gjøsæter, J. (1999) Cycles and trends in cod populations. *Proceedings of the National Academy of Sciences US* 96, 5066–5071.

Bourgeois, R., Gouyon, A., Jésus, F., Levang, P., Langeraar, W., Rahmadani, F., Sudiono, E. and Sulistani, B. (2002) *Socioeconomic and Institutional Analysis of Mahakam Delta Stakeholders.* Final Report to TotalFinaElf. Contract No. 501125/DKI/204. 108 pp.

Dutrieux, E. (2001) The Mahakam Delta Environment from the 80's up to now: a synthesis of a 15-year investigation. In: Kusumanstanto, T., Bengen, D.G., Widigdo, B. and Soesono, I. (eds) *Optimising Development and Environmental Issues at Coastal Areas. Problems and Solutions for Sustainable Management of Mahakam Delta.* Proceedings of international workshop held in Djakarta, 4–5 April 2001. Centre for Coastal and Marine Resources Studies, Bogor Agricultural University, Bogor, Indonesia, in association with the Ministry of Marine Affairs and Fisheries and TotalFinaElf, pp. 69–72.

Froese, R. and Pauly, D. (eds) (2005) *FishBase*. World Wide Web electronic publication. <http://www.fishbase.org> (accessed February 2005).
Hariati, A.M., Wiadnya, D.G.R., Prajitno, A., Sukkel, M., Boon, J.H. and Verdegem, M.C.J. (1998) Recent developments of shrimp, *Penaeus monodon* (Fabricius) and *Penaeus merguiensis* (de Man), culture in East Java. *Aquaculture Research* 26, 819–829.
Jul-Larsen, E., Kolding, J., Overå, R., Nielsen, J.R. and van Zwieten, P.A.M. (2003a) *Management, Co-management or No Management? Major Dilemmas in Southern African Freshwater Fisheries* 1. Synthesis report, FAO Fisheries Technical Paper No. 426/1, FAO, Rome, 127 pp.
Jul-Larsen, E., Kolding, J., Overå, R., Nielsen, J.R. and van Zwieten, P.A.M. (2003b) *Management, Co-management or No Management? Major Dilemmas in Southern African Freshwater Fisheries* 2. Case studies, FAO Fisheries Technical Paper No. 426/2, Rome, 266 pp.
Krumme, U. and Saint-Paul, U. (2003) Observations of fish migration in a macrotidal mangrove channel in Northern Brazil using a 200-kHz split-beam sonar. *Aquatic Living Resources* 16(3), 175–184.
Manson, F.J., Loneragan, N.R. and Phinn, S.R. (2003) Spatial and temporal variation in distribution of mangroves in Moreton Bay, subtropical Australia: a comparison of pattern metrics and change detection analyses based on aerial photographs. *Estuarine, Coastal and Shelf Science* 574, 653–666.
Mumby, P.J., Edwards, A.J., Arlas-González, J.E., Lindeman, K.C., Blackwell, P.G., Gall, A., Gorczynska, M.I., Harborne, A.R., Pescod, C.L., Renken, H., Wabnitz, C.C.C. and Llewellyn, G. (2004) Mangroves enhance the biomass of coral reef fish communities in the Caribbean. *Nature* 427, 533–536.
Myers, R.A. and Worm, B. (2003) Rapid worldwide depletion of predatory fish communities. *Nature* 423, 280–283.
Oostenbrugge, J.A.E. v., Bakker, E.J., van Densen, W.L.T., Machiels, M.A.M. and van Zwieten, P.A.M. (2002) Characterising catch variability in a multispecies fishery: implications for fishery management. *Canadian Journal of Fisheries and Aquatic Science* 59, 1032–1043.
Primavera, J.H. (1997) Fish predation on mangrove-associated penaeids: the role of structures and substrate. *Journal of Experimental Marine Biology and Ecology* 215(2), 205–216.
Ronnback, P., Troell, M., Kautsky, N. and Primavera, J.H. (1999) Distribution pattern of shrimps and fish among Avicennia and Rhizophora microhabitats in the Pagbilao mangroves, Philippines. *Estuarine, Coastal and Shelf Science* 48(2), 223–234.
SAS Institute (1989) SAS/STAT® User's Guide, Version 6, Vol. 2, 4th edn. SAS Institute Inc., Cary, North Carolina.
Turner, R.E. (1977) Intertidal vegetation and commercial yields of penaeid shrimp. *Transactions of the American Fisheries Society* 106, 411–416.
Turner, R.E. and Boesch, D.F. (1987) Aquatic animal production and wetland relationships: insights gleaned following wetland loss or gain. In: Hook, B. (ed.) *Ecology and Management of Wetlands*. Croon Helms Ltd., Beckenham, UK, 592 pp.
van Densen, W.L.T. (2001) On the perception of time trends in resource outcome: its importance in fisheries co-management, agriculture and whaling. PhD thesis, Twente University, Enschede, Netherlands, 299 pp.
van Zwieten, P.A.M., Roest, F.C., Machiels, M.A.M. and van Densen, W.L.T. (2002) Effects of inter-annual variability, seasonality and persistence on the perception of long-term trends in catch rates of the industrial pelagic purse-seine fisheries of Northern Lake Tanganyika (Burundi). *Fisheries Research* 54, 329–348.

18 Managing Diverse Land Uses in Coastal Bangladesh: Institutional Approaches

M. Rafiqul Islam
Program Development Office for Integrated Coastal Zone Management, Dhaka, Bangladesh, e-mail: rafiq@iczmpbd.org.

Abstract

Land use in coastal Bangladesh is diverse, competitive and conflicting. Agriculture, shrimp farming, salt production, forestry, ship-breaking yards, ports, industry, settlements and wetlands are some of the uses. Land uses have gone through major changes. Land use in the 1950s had been mainly for paddy cultivation, but salinity intrusion and tidal flooding prevented further intensification. Hence, in the 1960s–1980s, the World Bank and others helped with large-scale polderization in order to boost rice production. A decade later, drainage congestion inside and heavy siltation outside the polders made the southwestern area unsuitable both for agriculture, and, in extreme cases, even for human habitation. However, as the region has a history of traditional shrimp farming, polders provided an opportunity for intensive shrimp farming. Crop land and mangroves were transformed to shrimp farming. This created social conflict. Planned management of diverse land use, including zoning, has been recommended since. This chapter focuses on the complexities of land use in Bangladesh and the adopted institutional approaches.

Introduction

Land uses in coastal Bangladesh[1] (Fig. 18.1) have gone through major changes over the last half century. The land is intensively used for agriculture, settlements, forests, shrimp ponds (known locally as *ghers*), water bodies and fisheries, salt production, industrial and infrastructure developments, tourism and preservation and management of environmentally important and special areas. With the continually increasing population, the following features emerge:

- demand for expansion in all land uses (urban area, settlement, shrimp, etc.),
- increasing demand for new uses (tourism, export-processing zones and others), and
- encroachment and conversion of land from one use to another.

The above-mentioned circumstances call for planned management of land resources, including zoning. Though land zoning and regulations of land use have been advocated for a long time, actions and/or steps in this regard are almost totally lacking (Brammer, 2002). A start was made recently to bring

[1] The coastal zone of Bangladesh consists of 19 southern zilas and the EEZ (Exclusive Economic Zone).

 Environment and Livelihoods in Tropical Coastal Zones (eds C.T. Hoanh, T.P. Tuong, J.W. Gowing and B. Hardy)

relevant agencies together to discuss and develop a proposal for land zoning (PDO-ICZMP, 2004a). The complexities of land use and the institutional approaches adopted to define indicative land zoning are described in this chapter.

Land and Land Use

Land – a declining resource

Land is the basic natural resource that provides habitat and sustenance for living organisms, as well as being a major focus of economic and livelihood activities. Bangladesh has a population of 123 million living on a land area of 147,000 km^2 (PDO-ICZMP, 2004b). The population is increasing and the land is being converted from directly productive purposes, such as crop cultivation, to other uses such as housing, roads and urban development, and this trend is expected to continue. Some of the statistics provide an alarming picture:

- Some 220 ha of arable land is being lost daily to uses such as road construction, industry, houses, etc. (Islam *et al.*, 2004).
- At least 86,000 ha of land was lost to river/estuarine erosion between 1973 and 2000 (MES, 2001), though this is compensated by land generated through accretion.
- Some 70% of the land of Barisal and Khulna divisions is affected by different degrees of salinity, which reduces agricultural productivity (Rahman and Ahsan, 2001).
- Some 50% of the coastal lands face different degrees of inundation, thus limiting their effective use. This situation is expected to worsen further because of the effects of climate change.

In the coastal zone also, the population is expected to increase from 36.8 million in 2001 to 43.9 in 2015, and to 60.8 million by 2050 (PDO-ICZMP, 2005a). Present per capita agricultural land of 0.056 ha will decrease to 0.025 ha by 2050. On top of this, about 54% of the people of coastal Bangladesh are functionally landless and more than 30% are absolutely landless. Among the landholders, 80% are small farmers, 18% are medium farmers and only 2% are large farmers (PDO-ICZMP, 2004b). These have decisive impacts on major economic and livelihood activities, on land use and subsequently on the quality of land.

Present land use

Land use in Bangladesh is generally determined by physiography, climate and land height in relation to water level (Brammer, 2002). These together make a highly complex environment characterized by five main land types related to depth of seasonal flooding: 30 or more agro-ecological zones encompassing differences in soils, climate and hydrology; and areas with varying degrees of risk of disastrous floods, drought and cyclones. About 60% of the lands are inundated to a depth of 30 cm or more. The Bangladesh Bureau of Statistics publishes land-use statistics regularly. Emphasis is mainly on agriculture. Land uses are classified as net cropped area, current fallow, current waste, forest and area not available for cultivation. Along this line, SRDI (Soil Resources Development Institute) produces agricultural land-use maps for the country identifying many different types of agricultural land use.

In 2003, an estimate was made (Table 18.1) capturing a broader perception of land use and recognizing seasonal variations (ASB, 2003). Two complications were identified: areas under river and water bodies increased greatly in the wet season and estuarine/riverine wetlands (known locally as *chars*) cultivated during the dry season went under water in the wet season.

In coastal Bangladesh, agriculture, shrimp farming, salt production, forestry, ship-breaking yards, ports, industry, human habitation and wetlands are some of the uses in an area of only 47,000 km^2 inhabited by 36.8 million people. Land use in the coastal zone is diverse, competitive and conflicting.

Early 20th century land use

A vivid description of how modifications were being made to the natural levees beside some rivers was given by Mukerjee (1938):

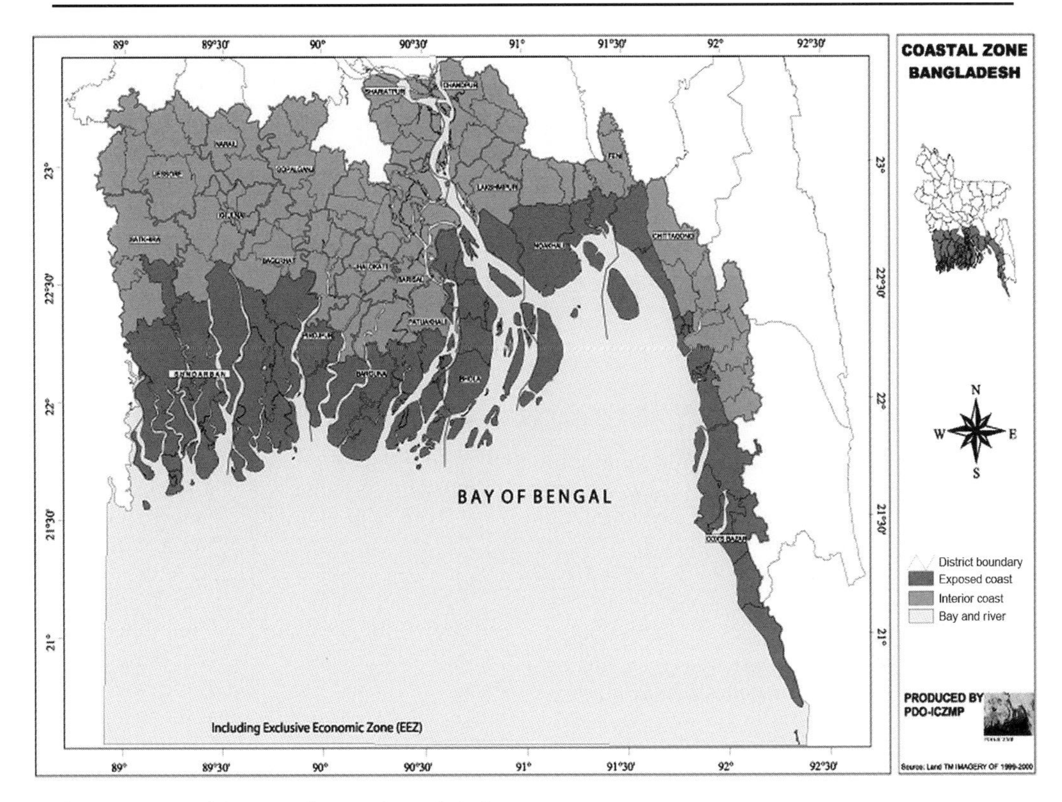

Fig. 18.1. Map of the coastal zone of Bangladesh.

> Small embankments were constructed and gaps in them were closed at high tide and opened at low tide so as to achieve partial drainage of the land behind, along with reduced salinity. Thus the human intervention of embankment construction has interacted with the delicately balanced natural processes, and has drastically hastened the raising of riverbed levels.

Local landlords, since the 17th century, constructed small dykes or embankments around individual land to limit saline water overflow and prevent crop damage. This traditional mechanism of construction of embankments through local efforts practically ceased in 1947. Land use, at that stage, was for paddy cultivation, especially low-yielding locally adapted varieties. In very limited areas of the southwest, traditional shrimp culture was practised.

Land use in the 1950s–1960s

In the 1950s–1960s, existing embankments deteriorated for lack of proper maintenance so that salinity intrusion and tidal surges caused routine crop damage. Crop failures caused by saline inundation or monsoon flooding were reported in most areas once every 3 years (Nishat, 1988). The Green Revolution at that time called for more intensive rice cultivation. Hence, the government recognized the need for protection of the coastal areas and construction and development of embankments started in 1961. The Coastal Embankment Project (CEP) was established, with assistance from the World Bank, in 1967. The Dutch term 'polder' was used to designate areas that are surrounded by dykes or embankments, separating them hydrologically from the main river system and offering protection against tidal floods, salinity intrusion and sedimentation. The embankments include regulators and other structures to control water intake and drainage of the empoldered area.

The primary purpose of empolderment was to increase agricultural production. During the first phase, 92 polders were constructed with 4022 km of embankments and 780 drainage sluices. The gross polder area

Table 18.1. Land use in Bangladesh showing seasonal variation (from ASB, 2003).

Classification	Area (km^2)	
	Dry season	Wet season
Rivers	6,400	7,700
Main rivers	2,860	3,940
Rivers in Sundarban	1,660	1,660
Other rivers	1,880	2,100
Standing water bodies	4,245	9,500
Haors	450	3,700
Beels	177	1,500
Baors	55	60
Ponds, tanks, ditches	3,000	3,500
Kaptai Lake	563	740
Forest	19,610	19,610
Cultivated	77,600	73,500
Field crops	51,000	17,140
Tree crops	4,900	4,900
Seasonal fallow	17,000	16,760
Current fallow	4,100	4,100
Seedbed only	600	600
Brackish-water aquaculture	1,900	1,900
Salt beds	50	0
Rural build-up	7,000	7,000
Non-cropped village land	8,400	8,400
Urban	7,000	7,000
Infrastructure	2,100	2,100
Estuarine area	8,600	8,600
Total	147,570	147,570

protected by June 1971 was 1.01 million ha (Talukder, 1991). Though no evaluation of the impact of the CEP on agricultural production has been performed, it became apparent that empoldering has increased the scale of production. Crops were saved from salinity and flooding; some yields increased by 200–300% (Nishat, 1988). The dominant land use during this period was paddy cultivation, primarily traditional local varieties. Modern paddy varieties and technological packages were introduced. Other uses remained the same: salt production, mangrove forest and traditional shrimp farming.

Land use in the 1970s–1980s

In the 1970s–1980s, the World Bank and other donors helped to continue large-scale polderization of coastal Bangladesh. Polders became part of the natural setting of coastal Bangladesh (Fig. 18.2), with a total of 123 coastal polders implemented (PDO-ICZMP, 2004b). To further enhance agricultural production, it was soon realized that internal water management had to be established within these polders. Changes in land use occurred because of the intensification of paddy cultivation with the attempted expansion of modern varieties and conversion of agricultural land to non-agricultural use (Sereno, 1981). During this time, coastal afforestation started with the objective of protection from cyclones and foreshore erosion.

Land use in the 1990s

In the 1990s, southwestern coastal Bangladesh experienced drainage congestion inside and heavy siltation outside the pold-

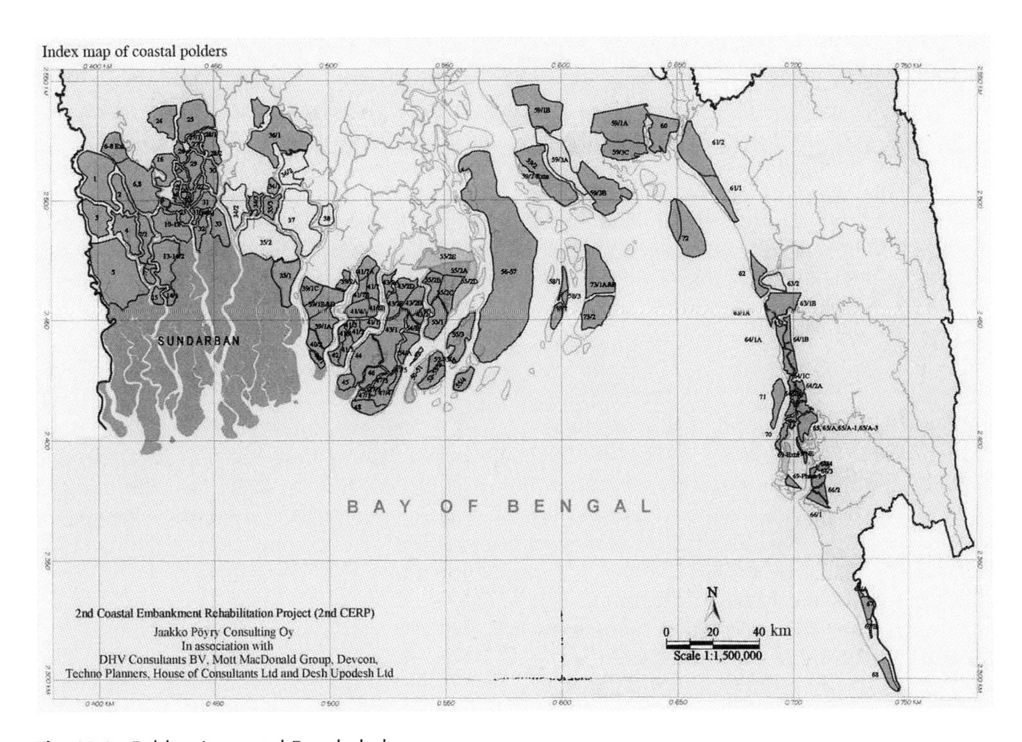

Fig. 18.2. Polders in coastal Bangladesh.

ers because of extensive polderization of the hydro-dynamically active delta, and subsequently the achievements from polderization gradually evaporated. The area became unsuitable both for agriculture and, in extreme cases, even for human habitation. This was termed a 'man-made disaster' (Rahman, 1995). Poverty and out-migration from the area occurred. The Khulna–Jessore Drainage Rehabilitation Project was conceived, with emphasis on structural solutions, including the construction of large regulators. Local people did not support a structural solution and resorted to wide-scale protest. The traditional system of allowing natural siltation under the concept of 'tidal river management' was adopted and land became suitable again for cultivation.

Emergence of commercial shrimp farming

Around the same time, increased demand and a high price for shrimp on the international market occurred. As the southwest had a history of limited-scale traditional shrimp farming, polders provided an opportunity for intensive shrimp farming. Many coastal polders constructed to protect agricultural land from inundation of salt water were turned into large shrimp *ghers*. The priority was reversed and salt water was willingly being allowed in the *ghers* to raise shrimp. Land previously used for agriculture and mangroves was transformed, often forcibly, to shrimp farming. Wide-scale land-use conflict emerged and created social unrest. Shrimp farming is now established as an important industry, contributing 5.2% to GDP, and the second-highest foreign exchange earner of the country. Shrimp areas expanded from 51,812 ha in 1983 to 137,996 ha in 1994 and to 141,353 ha in 2002 (DoF, 1995, 2003).

Land-use Conflicts

Most coastal lands are suitable for more than one use. Hence, the many diverse uses of limited land have created conflict. Many studies have highlighted these conflicts, especially between shrimp farming and other uses (Nuruzzaman, 1979; Karim and Stellwagen, 1998). In addition, one land use or another has manifold implications for socio-environmental conditions. The introduction of shrimp farming has gradually changed the land-use patterns of the surrounding farms, transforming agriculture and mangrove areas into shrimp-farming areas (Haque, 2004 and Fig. 18.3). Several studies reported a reduction in land for cattle grazing (Maniruzzaman, 1998), death of trees and other vegetation (Alauddin and Tisdell, 1998), increased salinity of soil and water and a reduction in the drinking-water supply because of the introduction of shrimp farming.

Firoze (2003) and Majid and Gupta (1997) elaborated upon the social and environmental impacts of commercial shrimp culture. As agricultural lands were turned into shrimp polders, the share-croppers and landless wage labourers found themselves losing their livelihoods, and began movements to resist the introduction of shrimp in their areas. This often resulted in violence. During the last two decades, more than 150 people have been killed and thousands injured in shrimp-related violence (Firoze, 2003). Influential and rich shrimp farmers, to harass leaders of the anti-shrimp movements, also initiated thousands of court cases, many of which are still pending.

Brackish-water shrimp cultivation, on a commercial scale, has brought large-scale environmental degradation. Shrimp polders retain saline water for months at a time, and the salinity seeps on to adjacent farms and spreads soil salinity. The loss of mangrove areas to aquaculture is a common feature, with Chakoria Sunderban being the classic example (Chowdhury *et al.*, 1994; Brown, 1997). From 1967 to 1988, the total area of Chakoria Sundarban mangroves decreased from 7500 ha to only 973 ha (Fig. 18.4) (Chowdhury *et al.*, 1994).

Management Approaches

To accommodate diverse land uses, changed patterns of land use and land suitability, zoning has been proposed as a management approach. Hossain and Lin (2002) suggested that, to reduce social conflicts and promote effective and sustainable resource use, land

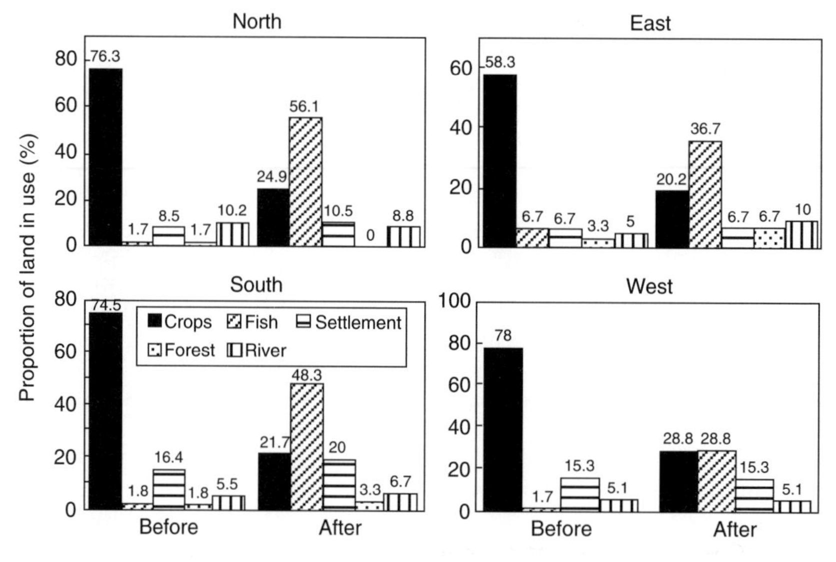

Fig. 18.3. Induced changes in land use around a farm, before and after the introduction of shrimp (from Haque, 2004).

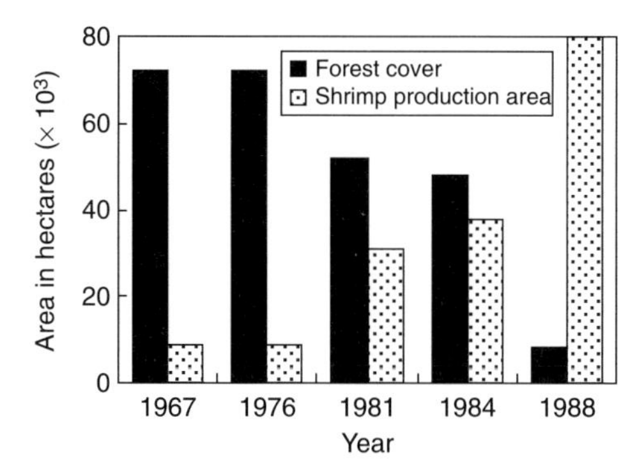

Fig. 18.4. Changes in area under forest and shrimp production in the Chakoria Sundarbans, 1967–1988 (after Choudhury *et al.*, 1994).

should be zoned on the basis of suitability: the most suitable zone, a moderately suitable zone and an unsuitable zone. Zoning can also be coupled with time-sharing: for example, shrimp farming during monsoon months and salt production during summer months. This will minimize the unplanned horizontal expansion of any land use, particularly shrimp farming or salt production, and maximize productivity from smaller areas through vertical integration. In the southwest, the pattern emerging is integrated rice–shrimp/fish farming. This will improve the socio-economic condition of the people as well as maintain the ecological balance in the coastal region.

Land suitability: specific research results

Several studies have dealt with comparative land suitability analyses on present land uses. Hossain and Lin (2002) analysed remote-sensing data of Landsat TM and thematic information of Cox's Bazar to identify suitable areas for mangrove afforestation and shrimp farm and salt bed development. Most of the suitable areas identified in the study actually coincided with the existing land use, which is important for appropriate zoning to optimize resource allocation and minimize conflicts between user groups. Further, the land suitability maps for mangrove afforestation, shrimp farms and salt beds were overlaid together to distinguish the combined land suitability categories as well as land-use conflicts. The results for potential land-use conflicts among salt, shrimp and mangrove areas were not prominent but indicated possible conflicts, mainly between mangrove and salt, as well as between mangrove and shrimp. Combining the present land-use map of the study area with the identified suitable land-use map revealed that most areas of existing shrimp farms and salt beds were developed within the suitable area, except for mangrove afforestation. Similar observations were made by Alam *et al.* (1990), Shahid *et al.* (1992) and Islam *et al.* (1997).

Ibrahim (2004) prepared land suitability maps for both *bagda* (brackish-water shrimp) and *T. aman* (transplanted paddy) crops in three administrative zilas: Khulna, Bagerhat and Satkhira. The land suitability factors for both *T. aman* and *bagda* are similar, except for soil salinity. High salinity increases the yield of *bagda* but reduces the yield of the *T. aman* crop. Based on physical suitability, areas used exclusively for *bagda* could be clearly delineated. Similarly, *T. aman* and other agricultural crops could be separated from *bagda* practices. Present land use for these two indicated mismatched areas in these three districts. Quader *et al.* (2004), using remote-sensing techniques, analysed the suitability of different land uses and found that the Khulna–Satkhira area was more suitable for shrimp farming than Cox's Bazar.

Institutional and other approaches

To reduce conflicting land use, maximize potential land use and facilitate integration of different uses in the coastal zone, several projects began: Coastal Embankment Rehabilitation Project (CERP), Systems Rehabilitation Project (SRP), Delta Development Project (DDP), Khulna–Jessore Drainage Rehabilitation Project (KJDRP) and 3rd and 4th Fisheries Projects. For example, under the Delta Development Project, it was concluded that the combination of rice cultivation and shrimp cultivation on the same land (in rotation) was possible without direct negative effects on rice yields (Nishat, 1988).

These projects have shown that planned management of land uses could offer both economic and environmental benefits.

Recognizing the need for integrated and comprehensive planning and other socio-economic support, the government of Bangladesh established the Program Development Office (PDO) for Integrated Coastal Zone Management. It is a multiministerial and multisectoral initiative led by the Ministry of Water Resources. The PDO is composed of representatives from different ministries and a small number of national experts. The office is supported by focal points, established in 34 government organizations, universities and research organizations, and by three task forces dealing with livelihoods, policy and strategy, and knowledge management issues. The whole set-up is steered by an interministerial technical committee composed of heads of relevant government departments, universities, NGOs, civil and chamber bodies and an interministerial steering committee consisting of secretary-level representatives of different ministries.The PDO, in its preparatory phase, plans to deliver three key outputs: a Coastal Zone Policy, Coastal Development Strategy and a 'priority investment programme', which are backed by supporting outputs in relation to coastal livelihoods, thus enabling the institutional environment and knowledge base. Land use is core to these outputs.

Several government policy documents have highlighted the importance of optimizing land use and land zoning for integrated planning of resource management. These are the National Fish Policy (1998), National Water Policy (1999), National Agricultural Policy (1999), Draft Shrimp Strategy (2004) and Coastal Zone Policy (2005). The National Fish Policy states that 'coastal areas will be demarcated for shrimp farming'. The Draft Shrimp Strategy states further that 'areas suitable for shrimp cultivation will be identified using a land-zoning process which will limit brackish-water shrimp aquaculture to coastal areas' and 'the objective of land zoning is to optimize land use'. The Coastal Zone Policy states that 'actions shall be initiated to develop land-use planning as an instrument of control of unplanned and indiscriminate use of land resources' and 'zoning regulations would be formulated and enforced in due course'.

However, the key policy is the National Land-Use Policy (2001), which describes

- zoning based on land use,
- ensuring the best use of land through zoning, and
- enactment of a zoning law to allow local government institutions to prepare zoning maps.

The National Land Use Policy places special emphasis on coastal areas. Recognizing the complexities of coastal land use, the policy makes provisions for an inter-agency task force to prepare an outline of coastal zoning.

No concrete and effective steps had been taken in the country towards restricting or regulating the conversion of agricultural land to non-agricultural uses (Nuruzzaman, 1979; Brammer, 2002). The following laws, however, are aimed at managing public land and water bodies, allocating and managing public and private shrimp land, and conserving the environment:

- Embankment and Drainage Act (1952),
- Bangladesh Water and Power Development Board Ordinance (1972),
- Manual for Land Management (Jalmohal) (1990),
- Shrimp Mohal Management Policy (1992), and
- Shrimp Farm Taxation Law (1992).

The emerging land-zoning concept

As shown above, land zoning has been advocated since the 1980s. Several broad zoning studies have been or are being carried out, notably the agro-ecological zoning of Bangladesh (FAO, 1988) and the SRDI land-zone mapping. Other information bases are organized on a narrower basis, such as SRDI's land suitability assessment for different crops. However, zoning, along sectoral lines, does not provide a basis for choices between often conflicting sectoral objectives. Karim and Stellwagen (1998) emphasized that 'no efforts have so far been initiated to classify the coastal land into various economic zones and develop them according to their development potential'. Integrated development is the outcome of such choices and multisector zoning should provide a tool for achieving the best choices for economic land use in an area, on the basis of its needs and potentials. Many agencies in Bangladesh already recognize this need for integrated zoning in support of planning for 'best possible' economic land use, while preventing land degradation and protecting the environment. DoF (2002) stated that 'coastal zoning would improve land-use planning, minimize conflicts over land tenure and identify appropriate areas for shrimp farming and areas that need to be protected [for grazing of livestock, common access, etc.]'.

The emerging concept is to formulate land zoning, with administrative boundaries as the unit, in accordance with the (dominant) land use and economic activities, as well as their potentials and vulnerabilities. Hence, this zoning has to be more than just a description of the current situation and must account for major underlying ecological and socio-economic factors and processes that have led to the current situation and that may be important for future trends and hazards. The approach should therefore take into account important ecological and socio-economic factors.

Zoning as a tool for area development is, of course, an ambitious goal that can only be attained in stages. Mutsaers and Miah (2004) have outlined the following conceptual basis for the first stage in the process to delineate an indicative coastal land zoning:

- a stepwise approach with clear intermediate results;
- use of an administrative boundary[2] as the unit for zoning, such as the upazila as the unit for indicative land zoning. Further detailed versions using union and later field blocks will be developed;
- use of only existing data. Field information to be collected for the purpose of validating the zoning (ground-truthing);
- proactive interaction with relevant agencies at different stages of the elaboration process; and
- support and backing of a structured technical support group involving government and non-government agencies. The Ministry of Land will be involved as an implementing agency of the Land-Use Policy.

The expected outputs will be

- zoning of the coastal area within broader zones, such as 'agriculture zone,' 'shrimp zone', etc
- each upazila will be associated preferably with one and only one of the land zones
- a typology will be given for each zone explaining its characteristics in some detail.

Land Zoning – Consensus Building

For the zoning to acquire the status of an accepted planning tool, there has to be a high level of consensus and ownership among relevant agencies. The need for discussion among different line agencies was identified. The Program Development Office (PDO), as a multi-sectoral, multi-agency set-up, provided a platform for relevant stakeholders for such a discussion. A 'Technical Discussion on Coastal Land Zoning' was arranged in August 2004 (PDO-ICZMP, 2004a). The approach and the outputs were discussed and agreed upon.

Moreover, to prepare the proposal of coastal land zoning, it was considered essential to draw on the expertise available among different agencies. A ten-member technical support group was formed, notably with representatives from the following agencies:

- Soil Resources Development Institute;
- Department of Fisheries,
- Forestry Department,
- Bangladesh Agricultural Research Council,
- Bangladesh Shrimp Foundation.

[2] Administratively, Bangladesh is divided into six divisions, 64 zilas, 507 upazilas and 4484 unions. The coastal zone encompasses 19 zilas, 147 upazilas and 1913 unions.

The group met several times during the process to contribute through information and data, participate and review ongoing work. Some of the group members also participated in the ground-truthing visits in the field.

Indicative Land Zones

An indicative land zoning has emerged (Fig. 18.5, PDO-ICZMP, 2005b), identifying the following eight zones:

- shrimp (brackish-water) zone,
- shrimp (sweet-water) zone,
- salt–shrimp zone,
- forest zone,
- mangrove (including Sundarban) zone,
- urban and commercial zone (industrial, port, export-processing zones and ship-breaking yards),
- tourism zone, and
- agricultural zone.

Results of the indicative land zones have been presented to field-level stakeholders at regional workshops and to policy planners at national workshops. There is now a national consensus on indicative land zoning.

Conclusions

Even with agreement on indicative land zoning among many agencies, only a start has been made. This version of land zoning is expected to be used as a basis for detailed land zoning, as elaborated in the Land-Use Policy (MoL, 2001). The challenge is to give a legal status to this broad zoning. However, the strength of this exercise is that it has brought relevant agencies together on an institutional platform. A consensus has been

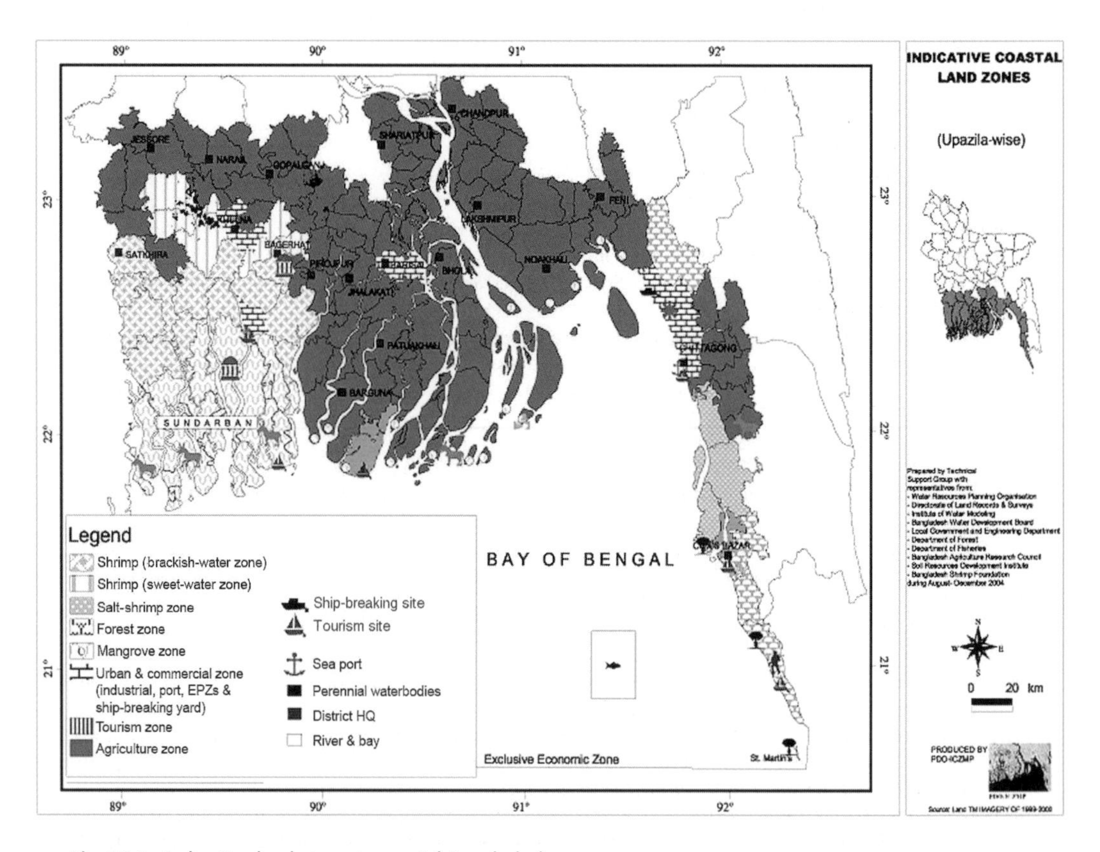

Fig. 18.5. Indicative land zones in coastal Bangladesh.

reached to aim for further detailed land zoning, but taking one step at a time. Land zoning, complemented by policy and investment support, can be instrumental in managing diverse land uses in the coastal zone. Continued research on land use will support optimum, sustainable and environmentally friendly land use and the subsequent modern management of land use through land zoning.

References

Alam, M.S., Shamsuddin, S.D. and Sikder, S. (1990) Application of remote sensing for monitoring shrimp culture development in a coastal mangrove ecosystem in Bangladesh. In: *Proceedings of the Twenty-third International Symposium on Remote Sensing of Environment*, 18–25 April 1990, Bangkok, Thailand, pp. 109–119.

Alauddin, M. and Tisdell, C. (1998) *The Environment and Economic Development in South Asia: an Overview Concentrating on Bangladesh*. Macmillan, London.

ASB (Asiatic Society of Bangladesh) (2003) Land use. In: *Banglapedia: National Encyclopedia of Bangladesh*. Asiatic Society of Bangladesh, Dhaka, 6, 235–239.

Brammer, H. (2002) *Land Use and Land Use Planning in Bangladesh*. The University Press Limited, Dhaka, Bangladesh, 554 pp.

Brown, B.E. (1997) The conversion of mangroves for shrimp aquaculture development. In: *Integrated Coastal Management: South Asia*. University of Newcastle, Newcastle, UK, 4, pp. 1–19.

Chowdhury, A.M., Quadir, D.A. and Islam, M.J. (1994) Study of Chokoria Sundarbans using remote sensing techniques. ISME (International Society of Mangrove Ecosystems), Japan. Technical Report 4, pp. 1–33.

DoF (Department of Fisheries) (1995) *Shrimp Resources Statistics*. Central Shrimp Cell, Department of Fisheries, Dhaka, Bangladesh.

DoF (Department of Fisheries) (2002) *Shrimp Aquaculture in Bangladesh: a Vision for the Future*. Department of Fisheries, Dhaka, Bangladesh, 7 pp.

DoF (Department of Fisheries) (2003) *Fishery Statistical Yearbook of Bangladesh: 2002–2003*. Department of Fisheries, Dhaka, Bangladesh, 41 pp.

FAO (Food and Agriculture Organization) (1988) *Land Resources Appraisal for Agricultural Development in Bangladesh*. Report 2, Agroecological regions. Food & Agriculture Organization of the UN, Rome.

Firoze, A. (2003) The southwest coastal region: problems and potentials. *The Daily Star*, XIV, Issue 215, Dhaka, Bangladesh.

Haque, A.K.E. (2004) *Sanitary and Phyto-Sanitary Barriers to Trade and Its Impact on the Environment: the Case of Shrimp Farming in Bangladesh*. IUCN Bangladesh Country Office, Dhaka, Bangladesh, 63 pp.

Hossain, M.S. and Lin, C.K. (2002) Land suitability analysis for integrated coastal zone management in Cox's Bazar, Bangladesh: a remote sensing and GIS approach. In: *Proceedings of the Coastal Zone Asia Pacific Conference*, May 2002, Bangkok, Thailand.

Ibrahim, A.M. (2004) Use of GIS, RS and ALES in coastal land use zoning. In: *Proceedings of Technical Discussion on Coastal Land Zoning*, 2 August 2004. Program Development Office for Integrated Coastal Zone Management Plan, Dhaka, Bangladesh, pp. 47–50.

Islam, M.J., Alam, M.S. and Elahi, K.M. (1997) Remote sensing for change detection in the Sunderban, Bangladesh. *Geocarto International* 12, 91–100.

Islam, M.S., Razzaque, M.A., Rahman, M.M. and Karim, N.H. (2004) Present and future of agricultural research in Bangladesh (in Bangla). In: *Agriculture in the 21st Century: Challenges and Possibilities*. Ministry of Agriculture, Dhaka, Bangladesh, pp. 20–27.

Karim, M. and Stellwagen, J. (1998) *Shrimp Aquaculture*. Final Report, Vol. 6, Fourth Fisheries Project, Department of Fisheries, Dhaka, Bangladesh, 101 pp.

Majid, M.A. and Gupta, M.V. (1997) Research and information needs for fisheries and development and management. In: *Proceedings of National Workshop on Fisheries Resources Development and Management in Bangladesh*, 29 October–1 November 1995. MOFL/BOBP/FAO/ODA, pp. 160–177.

Maniruzzaman, M. (1998) Intrusion of commercial shrimp farming in three rice-growing villages of southern Bangladesh: its effects on poverty, environment and selected aspects of culture. PhD thesis. University of the Philippines, Quezon City, Philippines.

MES (Meghna Estuary Study) (2001) *Hydro-Morphological Dynamics of the Meghna* Estuary. MES Project, Bangladesh Water Development Board, Dhaka, Bangladesh.

MoL (Ministry of Land) (2001) *Jatiyo Bhumi Babohar Niti* (National Land Use Policy). MoL, Government of Bangladesh, Dhaka, Bangladesh.

Mukerjee, R. (1938) *The Changing Face of Bengal: A Study in Riverine Economy.* University of Calcutta, Calcutta, India, 293 pp.

Mutsaers, H.J.W. and Miah, A.H. (2004) Land use zoning: concepts and methodology. In: *Proceedings of Technical Discussion on Coastal Land Zoning.* Program Development Office for Integrated Coastal Zone Management Plan, Dhaka, Bangladesh, pp. 55–70.

Nishat, A. (1988) Review of present activities and state of art of the coastal areas of Bangladesh. In: *Coastal Area Resource Development and Management.* Part II, Coastal Area Resource Development and Management Association (CARDMA), Dhaka, Bangladesh, pp. 23–35.

Nuruzzaman, K.M. (1979) Physical planning legislation in Bangladesh: a study of proper legislation needs. Master's degree thesis, University of Sheffield, UK.

PDO-ICZMP (Program Development Office for Integrated Coastal Zone Management Plan) (2004a) *Proceedings of Technical Discussion on Coastal Land Zoning.* PDO-ICZMP, Water Resources Planning Organisation, Dhaka, Bangladesh, 80 pp.

PDO-ICZMP (Program Development Office for Integrated Coastal Zone Management Plan) (2004b) *Where Land Meets the Sea: A Profile of the Coastal Zone of Bangladesh.* The University Press Limited, Dhaka, Bangladesh, 317 pp.

PDO-ICZMP (Program Development Office for Integrated Coastal Zone Management Plan) (2005a) *Living in the Coast: Urbanization.* PDO-ICZMP, Water Resources Planning Organisation, Dhaka, Bangladesh, 36 pp.

PDO-ICZMP (Program Development Office for Integrated Coastal Zone Management Plan) (2005b) *Coastal Land Uses and Indicative Land Zones.* PDO-ICZMP, Water Resources Planning Organisation, Dhaka, Bangladesh, 64 pp.

Quader, O., Islam, Z., Rahman, H., Sarkar, M.H. and Khan, A.S. (2004) Suitable site selection of shrimp farming in the coastal areas of Bangladesh using remote sensing techniques (4 S Model). *Proceedings of the XXth ISPRS (International Society for Photogrammetry and Remote Sensing) Congress,* 12–23 July 2004, Istanbul, Turkey. ISPRS, Istanbul, Turkey.

Rahman, A. (1995) *Beel Dakatia: Environmental Consequences of a Development Disaster.* The University Press Limited, Dhaka, Bangladesh.

Rahman, M. and Ahsan, M. (2001) Salinity constraints and agricultural productivity in coastal saline area of Bangladesh. In: *Soil Resources in Bangladesh: Assessment and Utilization. Proceedings of the Annual Workshop on Soil Resources,* 14–15 February 2001. Soil Resources Development Institute, Dhaka, Bangladesh, pp. 1–14.

Sereno, G. (1981) *A Reconnaissance Study of Changes in Settlement and Related Non-Agricultural Land Use in Bangladesh.* FAO/UNDP Land Use Policy Project. Ministry of Agriculture, Dhaka, Bangladesh.

Shahid, M.A., Pramanik, M.A.H., Jabbar, M.A. and Ali, S. (1992) Remote sensing application to study the coastal shrimp farming area in Bangladesh. *Geocarto International* 2, 5–13.

Talukder, B.M.A. (1991) Current status of land reclamation and polder development in coastal lowlands of Bangladesh. In: *Polders in Asia: Atlas of Urban Geology,* Vol. 6. Economic and Social Commission for Asia and the Pacific (ESCAP), United Nations, New York.

19 Widening Coastal Managers' Perceptions of Stakeholders through Capacity Building

M. Le Tissier and J.M. Hills
Envision Partners LLP, University of Newcastle, Newcastle upon Tyne, United Kingdom, e-mail: m.le-tissier@envision.uk.com

Abstract

Environmental integration and mainstreaming is the process that ensures consideration of environmental sustainability in development projects. The agenda for promoting environmental sustainability has been firmly set at the local, regional and national levels through a variety of conventions and legislation. However, mechanisms for developing the capability and capacity to realize these goals remain elusive. Natural scientists have one sectoral view of coastal resources, embedded in a numeric and reductionist framework, whereas social scientists take the opposite approach. Policymakers and other stakeholders will have their own perspectives. A more holistic view of stakeholders' perceptions of the coast from those charged with determining coastal policy and implementation can permit sources of conflict to be identified by managers and appropriate action to be taken. Capacity building is one tool that can lead to an increase in the perceptions of stakeholders' needs and coastal resource issues in coastal managers. However, much training in integrated coastal management (ICM) focuses on increasing scientific knowledge rather than providing a robust framework for management. Recent work in developing capacity in state-level Indian coastal managers has had a more holistic approach, encompassing not just science, but socio-economic and governance issues also. In addition, this capacity-building approach uses a 'virtual scenario' approach, in which groups of delegates are required to develop a strategic ICM plan for a local 20–40 km stretch of coast. An ICM matrix is used to provide a framework for understanding the coast and the impact of management interventions. This 'virtual scenario' approach, coupled with experience in conflict reduction matrices, has been shown to provide coastal managers with a wider appreciation of stakeholder conflict in the coastal zone.

Introduction

Capacity building is one tool that can lead to an enhanced awareness of the diversity of perceptions of stakeholders and lead to a re-evaluation of the nature of knowledge and understanding required by coastal managers (Chircop, 1998; Harvey *et al.*, 2002). However, training in integrated coastal management (ICM) often focuses on capacity building of individuals to increase their scientific knowledge rather than providing a robust framework for management. This leads to negligible impact as science-based solutions are rarely practical, socially acceptable, applicable or sustainable and they lack appropriate cultural context (UNESCO, 1988; Belfiore, 1999; Chircop, 2000). The need for

 Environment and Livelihoods in Tropical Coastal Zones (eds C.T. Hoanh, T.P. Tuong, J.W. Gowing and B. Hardy)

multidisciplinarity is consistently advocated, but this usually occurs within a setting of sophisticated, high-tech specialist approaches rather than within one of a generalist approach (UNESCO, 1988).

A recently completed project working with officers from the federal and individual state government of India found that the greatest barrier to developing ICM capacity was an unwillingness to work with, and a mistrust of, other stakeholders, preventing a multistakeholder consensus approach to the management of the coastal zone. In common with the findings of Poitras *et al.* (2003), this problem largely arose from: (i) the novelty of consensus building as an approach to determining management solutions for the coastal zone; (ii) the lack of incentives within the workplace to seek a compromise; (iii) the apprehension of having to negotiate; and (iv) the uncertainty of the outcome and control of the resulting management process.

This chapter describes the development of a training programme to overcome this barrier that required a re-evaluation of the training requirements for ICM capacity building and training methodologies. It also illustrates a process for developing organizational capacity for ICM that seeks to avoid the barriers of specific sectoral and disciplinary approaches.

Background to Integrated Coastal Management

The coastal zone often becomes a zone of conflict, with multiple users competing for limited space and resources. Although attention is often focused on primary stakeholders who directly use space and resources, conflict also exists between secondary stakeholders who are involved in managing the space and resources (e.g. government departments, NGOs, aid agencies, etc.). Much of this conflict between secondary stakeholders arises from competition for ownership of space and resources, but also because different organizations have different institutional arrangements for implementing their policy.

Coastal zone management was introduced in the 1970s (Nichols, 1999; Olsen, 2000). By the 1990s, coastal zone management (CZM) had evolved into *integrated coastal management* (ICM), conceived as a holistic management tool working across sectoral, disciplinary and institutional boundaries (Burbridge, 1997; Ducrotoy and Pullen, 1999; Nichols, 1999; Turner and Bower, 1999; Olsen, 2000), although maybe it has not lived up to its holistic ambitions (Nichols, 1999; Sudara, 1999). Subsequent projects and policy approaches have tended to reflect the particular interests of the particular proponents of the analysis. This can lead to a narrow problem-solving exercise rather than to a holistic management process (Olsen, 1996, 2000).

Sophisticated scientific understanding of the coastal zone cannot in itself achieve ICM. For example, fisheries modelling and quota setting do not deliver solutions to unsustainable fishing-based livelihoods in areas of high poverty. Perceptions of coastal resources among groups can be varied, diverse and conflicting (Fig. 19.1). Natural scientists seek to predict changes in coastal resources embedded in a numeric and reductionist framework (Olsen, 2000; Vallega, 2000). In contrast, social scientists seek to describe the patterns of interactions of people in networks of social relations, their maintenance and the conflict that arises from competing interests (Knight, 1992; Ostrom *et al.*, 1992; Wilson and Jentoft, 1999). ICM should seek to determine a holistic view of stakeholders' perceptions of the coast in order to identify sources of conflict where appropriate knowledge from all disciplines can be employed to better understand the linkages and interdependencies of socio-economic and coastal environmental dynamics and arrive at more robust solutions (Vallega, 2000; Bowen and Riley, 2003).

The Integrated Coastal Zone Management and Training Project

This project explored capacity building for ICM from the perspectives of: (i) skills in training; and (ii) the course requirement for developing capacity in ICM. The project sought to develop capacity in state-level

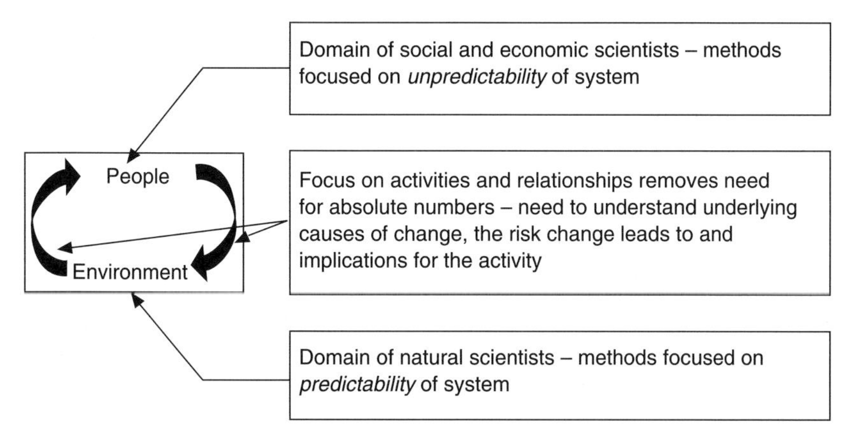

Fig. 19.1. The source of conflict between sectoral groups. Conflict arises because of a focus on either the provider (environment) or the user (people). A focus on the interactions and activities that link the two creates greater understanding of the dynamics of the coastal zone.

Indian government environmental officers for writing and implementing coastal management plans under their 1991 Coastal Zone Notification (<http://www.envfor.nic.in/legis/legis.html>). The project goal was to develop a programme that would promote a holistic approach to environmental management, encompassing socio-economic and governance issues as well as natural sciences, within a tradition of specialist, single-sector management. The training process was designed to lead trainees through a 'virtual scenario' case study approach, in which groups of trainees are required to develop a strategic ICM plan for a local 20–40 km stretch of coast during the duration of the course (Fig. 19.2). Case studies permit role-playing in an environment that simulates the work situation found in an ICM programme. They are safe and do not impose penalties for 'wrong' answers, and they can help improve decision-making skills under conditions of scientific uncertainty and competing interests (Suman, 2001). Central to the project goal was the development of training ability within India so that capacity building could continue beyond the life of the project. Training teams were established at two universities – Anna University at Chennai, Tamil Nadu, and Jadavpur University at Kolkata, West Bengal.

The following section outlines the training of local trainers, the overall training framework, specific training tools and the course design.

Training of trainers

The project required trainers to develop new skills in order to support a sustainable training approach. A major objective of training in ICM must be to remove existing discipline-biased perspectives in favour of approaches that promote an open and inclusive process to contextualize the various, and often conflicting, values and perceptions of the many stakeholders in the coastal zone.

Traditional teaching techniques such as class lecturing and research assignments cannot attain such learning objectives (Grant, 1998; Chircop, 2000; Fletcher, 2001). Furthermore, it is unlikely that a coastal manager can ever be an 'expert' in the many disciplines and sectors that have inputs into ICM. Indeed, one might argue that the role of a coastal manager is as an executive, coordinating and managing knowledge inputs rather than being the source of the knowledge itself. Thus, ICM is a team effort requiring individual inputs from a wide variety of sectors and disciplines.

Our approach was to develop training teams whose composition included expertise in the range of natural, social and economic disciplines and from the range of sectors

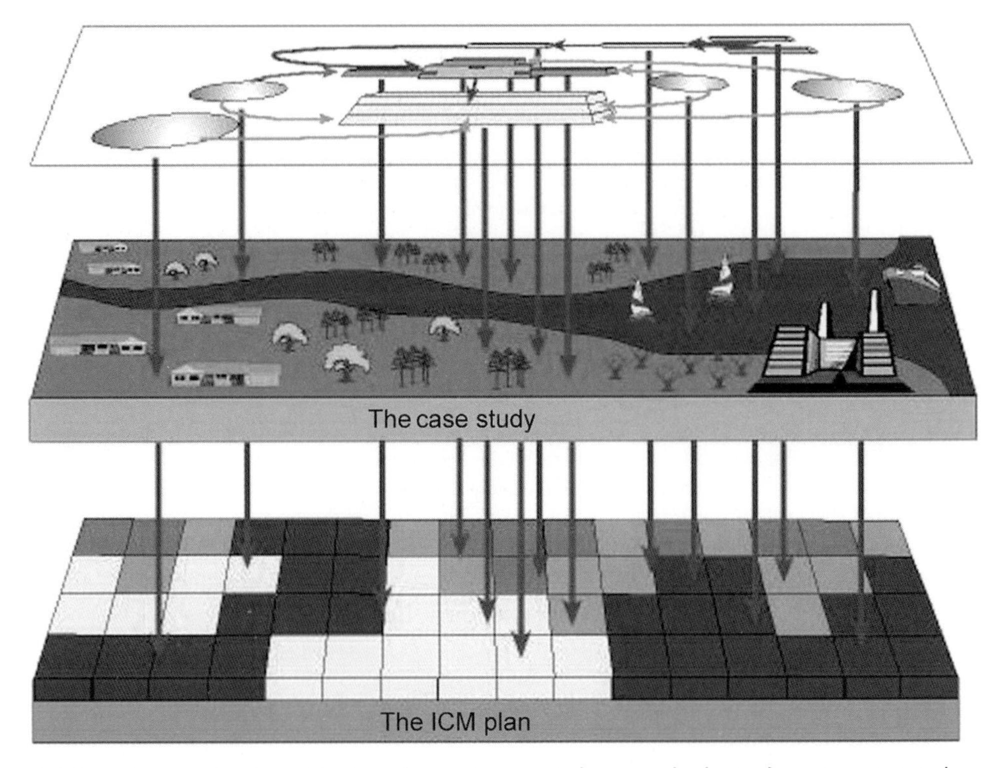

Fig. 19.2. Conceptual outline of the 'virtual scenario' case study approach. The top layer is a conceptual model of how the various dynamics of integrated coastal management commonly interface (see Fig. 19.3). The middle layer represents the various case studies, tasks and exercises that illustrate or codify the various dynamics of the top layer, in real life. The bottom layer represents the outputs of the second layer by populating an ICM plan, as traditionally conceived and implemented (with resources, responsibilities, institutional elements, time scales, etc.). The arrows represent the process of embellishment of each dynamic of the ICM model, with relevant case study information, to form elements of the ICM plan.

having a role in an ICM initiative (e.g. university, NGO, institute and government). The training teams were given a course based on the Certificate of Learning and Teaching in Higher Education given to new lecturers at Newcastle University, which included models of learning; training needs analysis; content, structure, format and materials for training; assessment and evaluation; course organization; and experiential training. This programme provided the necessary skills to conduct case study-based training.

A training framework

To guide trainers and trainees through the process of ICM encompassed within the course structure, a formulaic framework was constructed (Fig. 19.3) that is designed to address four fundamental stages of the ICM process:

- identification of the knowledge, information and inputs required from each discipline and sector to support the ICM process;
- integration of sectoral information on physical, biological and human dimensions of the coastal system;
- identification of significant interactions among processes operating within, and between, the three dimensions; and
- analysis of these interactions to identify the key issues and a range of applicable management options.

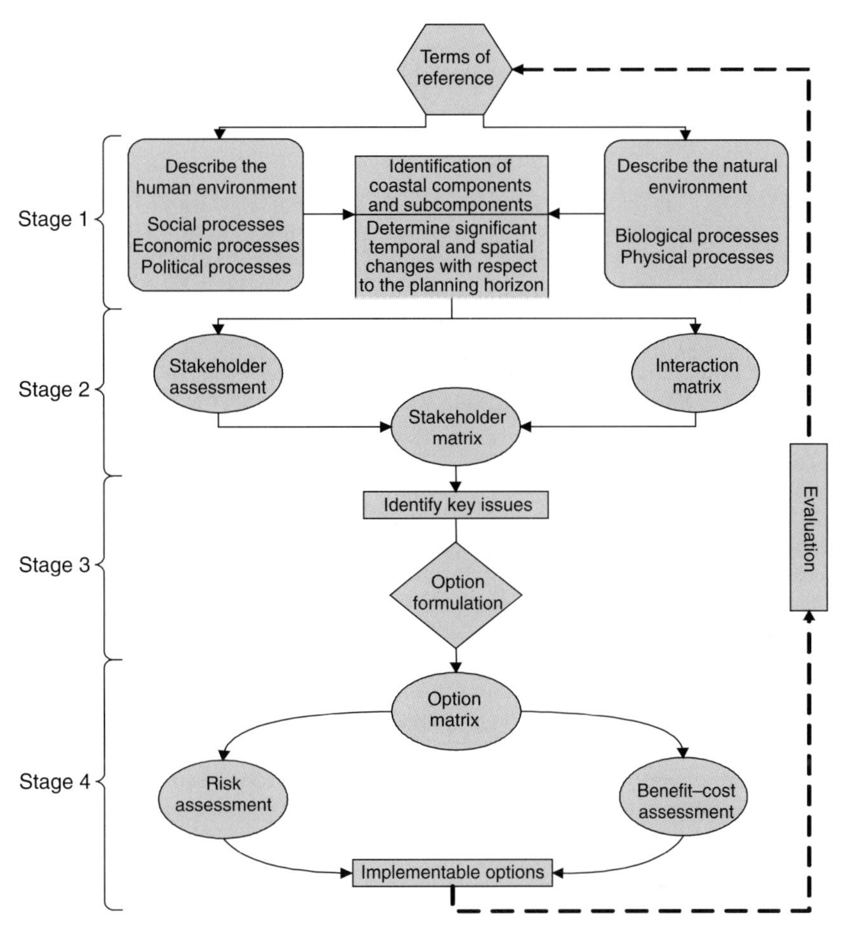

Fig. 19.3. Training framework for ICM. The framework describes four stages that gather information (round-cornered boxes), integrate information (square-cornered boxes), apply analytical tools (ovals) and negotiate priorities and solutions (diamond box). For explanation, see text.

The training framework is designed to be applied using a case study that allows the training process to take place within a virtual scenario of ICM, presenting to the delegates an experience that is as real as possible for the experiences they might expect to encounter in their workplace. The framework provides a structured and sequential guide to the process of ICM that can be used by both trainers and delegates to underpin a training course.

To achieve this process, the framework provides a means to assimilate information that can be interrogated using simplified tools accessible to non-specialists. These tools can synthesize and analyse a diverse array of interactions in order to guide coastal managers towards the determination of applicable options. Each stage in the framework incrementally filters, refines and reviews the outputs from the previous stage. This leads users from an essentially knowledge-led foundation derived from a wide range of sectoral and disciplinary sources, which is then analysed to identify the key issues/problems that are affecting the plan area, through to the formulation of potentially applicable integrated management options. The ICM process described by the training framework incorporates an integrated approach to ICM because it includes the spatial characteristics (physical, chemical, biological, ecological) of land and marine forms of coastal regions, temporal aspects of dynamic processes occurring within the plan area and the planning horizon of the intended ICM plan, and the interrelationships between the various human uses of coastal areas and resources, as well as associated socio-economic interests and values.

The framework is used throughout a short course in ICM as a means of helping trainers ensure that all training activities contribute to the ICM approach, provide a common structure for delegates to validate their development and progression and provide a common reference between the trainer and delegate.

The framework is 'entered' through a *terms of reference* (ToR) that sets out the goals and objectives of the training exercise in the form of a statement that covers the background to the plan area and the management objectives as well as other information, including the geographic boundaries and timescale of the plan.

The ToR provides a benchmark against which the relevance of information and activity can be evaluated. Progression through the framework then takes place as four discrete stages that develop sequentially.

Stage 1 develops a knowledge base focused on understanding the plan area and the changes that are taking place within it. Stage 2 provides an evaluation of how changes are affecting the plan area and the stakeholders that use the available resources, using stakeholder and matrix-based tools. Stage 3 leads to an identification of the key issues for management focus to fulfil the ToR, and to the design of a range of options for amelioration of identified issues. Both issues and options are attained through negotiation and consensus among the delegates, drawing on their own experiences and expertise supplemented by 'outside' expertise as necessary. Finally, Stage 4 provides a means to evaluate the likely success of each option in reducing the risk to people and property, while assessing its likely cost and appraising the outcomes against the original plan objectives set out in the ToR.

Training tools

The framework is supported by four types of tools that facilitate the interrogation and analysis of information. Tool 1 is a stakeholder assessment that provides insight into the major socio-economic drivers, and also allows an assessment to be made of the relationship between the relative influence and importance of stakeholder groups and the intended outcomes identified in the ToR. Tool 2 involves matrices that provide a structure for prioritizing information and for ensuring that discussions become clearly directed and non-sectorally entrenched (Fig. 19.4).

Three forms of the matrix are used:

- The interaction matrix provides a means of exploring the interactions between the main components of the biological, physical and human environment and their expected changes.
- The stakeholder matrix provides an overview of the impacts on people of future changes in the coastal area.
- The options matrix can be used to check for stakeholder benefits from proposed management options, to filter out options that have strong negative impacts on stakeholders and also to enhance management options to maximize stakeholder benefits.

The ICM matrix is used to provide a framework for understanding the coast and the impact of management interventions (Fig. 19.4). However, it can also provide a model for the conflicts of interventions on stakeholder groups to be determined by the delegates. Using this matrix, delegates experience iterative searching for low-conflict management solutions for the target coastline. This virtual scenario approach, coupled with experience with conflict reduction matrices, has been shown to provide coastal managers with a wider appreciation of stakeholder conflict in the coastal zone.

Tool 3 is a risk assessment that can be used to document the evaluation of risk before and after an intervention takes place. By considering a range of alternative management options, an evaluation of the relative risk can be made and management options compared. Tool 4 involves the evaluation of benefits and costs using a simple framework to enable a basic, and subjective,

		COASTAL ENV.					LAND USE		PORTS AND HARBOURS			INDUSTRY			LIVELIHOODS		
		Open shore	Estuaries	Old dunes	Frontal dunes	Wet hinterland	Plantations	Agriculture	Creek harbours	Beach	Constructed ports	Tourism	Fisheries	Aquaculture	Natural material	Semi-Pukka	Concrete houses
COASTAL ENV.	Sea level change	✓	✓	H	✓	✓	✓		F	✓	F	H		✓	✓		
	Climate change				✓		✓	✓			✓		✓	✓	F	F	
	Local erosion	✓	✓		✓					✓					F		
	Local accretion	✓	✓						F	✓	F						
LAND USE	Conversion of wetlands		✓			✓								✓			
	Construction – tourist			H	✓							H					
	Construction – general			✓								✓					
	Railway			✓								H					
PORTS AND HARBOURS																	
INDUSTRY	Pollution – aquaculture		✓			✓								✓			
	Pollution – sewage											✓					
	Pollution – litter											H					
	Failing aquaculture													✓			
	Increasing tourists											H					
	Declining CPUE												F				
LIVELIHOODS	Urban expansion														✓	✓	✓

Fig. 19.4. An example of a completed matrix. The columns indicate system components categorized into functional groupings of the natural coastal and human-built environment. Rows identify forces of change originating from each of the functional groups. Where a change will interact with a component of the system, it is checkmarked. No attempt is made to qualify whether the impact is positive or negative, or its magnitude. Two primary stakeholders are identified: hoteliers and fishermen allowing changes that are threats to users and uses of the system to be revealed. The matrix identifies the principal foci of management needs against which the goals and objectives of the plan should be directed. CPUE, catch per unit effort; F, fishing; H, hotels.

assessment and comparison of the benefits and costs associated. This can be used to evaluate the various management options for a particular issue.

Course design

The framework formed an underpinning guide that supported a course structure built around the virtual scenario case study approach organized into three modules. The first module involves the role of information and knowledge in ICM, whereby delegates sort through information on the designated area and try to build cross-sectoral linkages. The second module involves field visits to key sites identified from the previous stage, as well as meeting with a range of stakeholders to provide a closer appreciation of the main issues and problems within the designated area. The final module involves writing an outline ICM plan. The task of the coastal management plan is to advise on ways in which coastal resource development can be integrated into a coastal system without any loss of the resource or functional integrity of the environmental system in order to reduce risks to people, their livelihoods and their property.

Discussion and Conclusions

The training framework described here addresses a concern that training for development should be focused on strengthening the capacity to practise (Franks, 1999; Mann, 1999). This is particularly important in many countries where coastal problems are both so acute and persistent (Olsen *et al.*, 1997) that there is not time to engage in long-term research programmes in order to obtain the perfect management solution.

Previous models for ICM, based on GESAMP (1996) and Olsen *et al.* (1997), have largely focused on identifying the stages in the ICM process with little guidance on the nature of information required to support such a process. The framework described here aims to provide guidance in a training scenario that will allow coastal practitioners to engage in the ICM process.

The approach to training facilitates integrative thinking, active *versus* passive learning and unambiguous communication (Grant, 1998), as well as re-orienting attitudes towards a cross-sectoral and multidisciplinary approach that supports concepts of sustainability (Hopkins *et al.*, 1996; Pooley and O'Connor, 2000). This ensures an inclusive process for all sectors and disciplines, avoiding polarization of different groups within the management process or the development of a 'closed' language with minimal multisectoral ownership (Endter-Wada *et al.*, 1998; Norton, 2000).

Training programmes, using the framework described here, have been given to cadres of officers from the federal and state government in India in 2000 and 2001, and personnel associated with the Char Development and Settlement Project II in Bangladesh in 2001. Immediately after the course, all participants were asked to evaluate the training provided using a scoring and comment format. Results showed that participants scored at least 70%, and as high as 93%, in the categories of attainment of course objective, relevance, structure of course and presentation, and their comments suggested that the courses would be beneficial to their professional roles in delivering ICM within a range of coastal areas in southern Asia.

References

Belfiore, S. (ed.) (1999) *Education and Training in Integrated Coastal Management: The Mediterranean Prospect.* FrancoAngeli, Milan, Italy.

Bowen, R. and Riley, C. (2003) Socio-economic indicators and integrated coastal management. *Ocean and Coastal Management* 46, 299–312.

Burbridge, P.R. (1997) A generic framework for measuring success in integrated coastal management. *Ocean and Coastal Management* 37, 175–189.

Chircop, A. (1998) Introduction to capacity-building section. *Ocean and Coastal Management* 38, 67–68.

Chircop, A. (2000) Teaching integrated coastal management: lessons from the learning arena. *Ocean and Coastal Management* 43, 343–359.

Ducrotoy, J.P. and Pullen, S. (1999) Integrated coastal zone management: commitments and developments from an international, European, and United Kingdom perspective. *Ocean and Coastal Management* 42, 1–18.

Endter-Wada, J., Blahna, D., Krannich, R. and Brunsonet, M. (1998) A framework for understanding social science contributions to ecosystem management. *Ecological Applications* 8, 891–904.

Fletcher, S. (2001) Empowering learning through design: a comparative analysis of ICM course development for web-based and classroom delivery. *Marine Policy* 25, 457–466.

Franks, T. (1999) Capacity building and institutional development: reflections on water. *Public Administration and Development* 19, 51–61.

GESAMP (IMO/FAO/UNESCO-IOC/WMO/WHO/IAEA/UN/UNEP Joint Group of Experts on the Scientific Aspects of Marine Environmental Protection) (1996) *The Contributions of Science to Coastal Zone Management.* Reported studies, GESAMP, 72 pp.

Grant, W.E. (1998) Ecology and natural resource management: reflections from a systems perspective. *Ecological Modelling* 108, 67–76.
Harvey, N., Clarke B.D., and von Baumgarten, P. (2002) Coastal management training needs in Australia. *Ocean and Coastal Management* 45, 1–18.
Hopkins, C., Damlamian, J. and Ospina, G.L. (1996) Evolving towards education for sustainable development: an international perspective. *Nature and Resources* 32, 2–11.
Knight, J. (1992) *Institutions and Social Conflict*. Cambridge University Press, New York.
Mann, P. (1999) Knowledge, training and development: an overview. *Public Administration and Development* 19, 1–3.
Nichols, K. (1999) Coming to terms with 'integrated coastal management': problems of meaning and method in a new arena of resource regulation. *Professional Geographer* 51, 388–399.
Norton, B.G. (2000) Biodiversity and environmental values: in search of a universal earth ethic. *Biodiversity and Conservation* 9, 1029–1044.
Olsen, S.B. (1996) *Increasing the Efficiency of Integrated Coastal Management*. Coastal management report No. 2220, Coastal Resources Centre, University of Rhode Island, Rhode Island, 10 pp.
Olsen, S.B. (2000) Educating for the governance of coastal ecosystems: the dimensions of the challenge. *Ocean and Coastal Management* 43, 331–341.
Olsen, S., Tobey, J. and Kerr, M. (1997) A common framework for learning from ICM experience. *Ocean and Coastal Management* 37, 155–174.
Ostrom, E., Walker, J. and Gardner, R. (1992) Covenants without a sword: self-governance is possible. *American Political Science Review* 86, 404–417.
Poitras, J., Bowen, R. and Wiggin, J. (2003) Challenges to the use of consensus building in integrated coastal management. *Ocean and Coastal Management* 46, 391–405.
Pooley, J.A. and O'Connor, M. (2000) Environmental education and attitudes: emotions and beliefs are what is needed. *Environment and Behavior* 32, 711–723.
Sudara, S. (1999) Who and what is to be involved in successful coastal zone management: a Thailand example. *Ocean and Coastal Management* 42, 39–47.
Suman, D. (2001) Case studies of coastal conflicts: comparative US/European experiences. *Ocean and Coastal Management* 44, 1–13.
Turner, R.K. and Bower, B.T. (1999) Principles and benefits of integrated coastal zone management (ICZM). In: Salomons, W. *et al.* (eds) *Perspectives on Integrated Coastal Zone Management*. Springer-Verlag, Berlin, pp. 13–34.
UNESCO (1988) Year 2000 Challenges for Marine Science Training and Education Worldwide. *UNESCO Reports in Marine Science*.
Vallega, A. (2000) Introduction: coastal education – a multifaceted challenge. *Ocean and Coastal Management* 43, 277–290.
Wilson, D. and Jentoft, S. (1999) Structure, agency and embeddedness: sociological approaches to fisheries management institutions. In: Symes, D. (ed.), *Alternative Management Systems for Fisheries*. Blackwell Science Ltd., Oxford, UK.

20 Can Integrated Coastal Management Solve Agriculture–Fisheries–Aquaculture Conflicts at the Land–Water Interface? A Perspective from New Institutional Economics

C. Brugere[1]

Fishery Policy and Planning Division, Food and Agriculture Organization, Rome, Italy, e-mail: cecile.brugere@fao.org

Abstract

The principles of integrated coastal area management have been widely adopted and advocated by the international community. However, integrated coastal management has been less successful in practice and, in many areas, conflicts over resource use still prevail. This chapter explores the causes for such conflicts from the perspective of New Institutional Economics (NIE). It argues that conflicts are not only the result of competition for resources, but predominantly the outcome of institutional failures, that is, the ability of institutions in place (if not their simple absence) to address coastal zone issues. The case of shrimp culture development in India is used as an illustration of the NIE concepts presented. Decentralization and devolution, when meeting specific institutional requirements identified with NIE as a framework of analysis, are suggested as a suitable reform process to stimulate environmentally and socially sustainable coastal zone development. In this respect, sectoral capacity building of user groups followed by strengthening of local government capacity to integrate and adequately address sectoral concerns are considered as practical measures for improving the efficiency of current coastal zone management schemes.

Introduction

The aim of this chapter is to demonstrate that the lack of results achieved through integrated coastal management is not directly related to the occurrence of competitive conflicts for resources at the land–water interface,[2] but to a failure of institutions to provide a suitable environment to address coastal issues. Limited achievements of integrated coastal manage-

[1] Views expressed in this chapter are the author's and do not necessarily reflect those of the Food and Agriculture Organization. The author is very grateful for the valuable comments and suggestions provided by her colleagues in the Fisheries Department (Serge Garcia, Rebecca Metzner, Florence Poulain, Grimur Valdimarsson, Ulf Wijkstrom and Rolf Willmann) for improving an earlier draft.

[2] Unless specified otherwise, 'coastal zone' refers specifically to the geographical area where marine and terrestrial environments meet (by opposition to freshwater-land environments). It is understood to have the same meaning as the perhaps more narrowly defined 'land–water interface'.

ment schemes and reasons for conflicts are explored from the perspective of New Institutional Economics (NIE). Much time has been devoted in the past to studying physical phenomena in coastal zones. Today, however, the problem has shifted from understanding environmental processes to allocating limited, yet renewable, coastal resources to often-competing economic uses. In most instances, such uses are traditional and not supported by legal rights, as is the case with coastal fisheries. In addition, land–water interfaces are ecologically sensitive zones. Modification of the ecosystem resulting from resource use tends to affect other uses through the creation of externalities.[3] As coastal resources tend to support high-value economic activities (e.g. shrimp farming) as well as the livelihoods of many poor people (e.g. small-scale fishing or farming), managing and allocating coastal resources while minimizing interferences involve addressing simultaneously growth and equity issues. Efficient and socially acceptable distribution is a key to the sustainable development, in both economic and environmental terms, of coastal areas. Conflicts are an inevitable outcome of a distribution perceived as unfair by specific groups of resource users.

The paradigm of integrated coastal management has been to identify conflicts occurring at the land–water interface and to provide means of resolving them to ensure sound management of coastal environments. Concepts from NIE can shed light on the actual *causes* of conflicts and suggest necessary steps towards institutional reform to limit their occurrence and reconcile multiple resource uses for successful coastal development.

This chapter begins by setting the context and principles of integrated coastal zone management. A second section focuses on the reasons for conflicts in coastal zones, which are then placed in the perspective of NIE to better understand their causes (third section). A fourth section presents the requirements for successful decentralization, devolution and institutional strengthening towards improved coastal conflict management and sustainable resource use. Despite the range of economic activities occurring in coastal areas, such as transport, energy generation, military, tourism development, etc., the scope of the chatper has been narrowed to deal specifically with agriculture, aquaculture and fisheries interactions.

Integrated Coastal Management in Context

Uptake of integrated coastal management at the international level

The importance of coastal zones, in particular areas at the land–water interface, in supporting multiple economic activities and environmental functions is now well established. The concept of integrated coastal zone management stemmed from earlier developments with respect to 'integrated watershed management' and 'integrated river basin management'. It emerged from the realization that coastal zone components (land, water, forests) and the range of economic activities they supported could not be managed in isolation from one another. In the wake of the UNCED Rio Conference in 1992, which underlined the need for new, integrated and precautionary approaches to the management of coastal areas (Chapter 17 of Agenda 21), many international agencies embraced the concept of integrated coastal management and advocated its implementation. This contributed to growing awareness of the sensitivity of land–water fringes to human interventions, in particular in poverty-stricken areas, to which many guidelines for implementing integrated coastal management principles were oriented (e.g. Clark, 1994; Scialabba, 1998). Because of the importance of fisheries in these areas, from either capture or culture systems, the FAO took an active role in the promotion of sustainable coastal manage-

[3] *Externality* has been defined as an agent's unidirectional activity on the environment, which results in an uncompensated loss of welfare for other agents (Pearce and Turner, 1990).

ment (e.g. FAO, 1999 for aquaculture; Willmann and Insull, 1993 for fisheries). The Code of Conduct for Responsible Fisheries, adopted by member countries in 1995, specifically refers to the fragility of coastal ecosystems and the need for consultative management (Article 10.1).

Implementation of integrated coastal management: success and limitations

Although objectives of integrated coastal zone management differ from region to region based on local practices and uses, they usually address habitat restoration, resource allocation and development of various economic sectors (Clark, 1994). Consultation mechanisms must be activated to resolve two intertwined problems: one of conflict and one of sustainability (i.e. sustainable environmental resource use). Such consultation mechanisms, emanating from bottom-up, cross-sectoral participation and collective action at micro and meso levels, have been widely adopted. Community-based management, co-management initiatives (featuring prominently in coastal fisheries management) and the constitution of organizations promoting environmentally sustainable use of coastal resources have been hailed as suitable alternatives for the management of disputed resources (e.g. Arriaga *et al.*, 1999 for Ecuador; Masalu, 2000 for Tanzania; Rouf and Jensen, 2001 in the case of Bangladesh Sunderbans). Criteria for successful collective action in managing common resources are not new (Olson, 1971; Ostrom, 1990): reasonable group size, homogeneity of participants and agreed purpose are conditions for the successful organization of stakeholders in associations, cooperatives or resource-user groups for collective action at a local level.

However, unless group members had a financial stake in actions undertaken, cooperatively run associations were also reported to fail. They suffered from organizational problems, bureaucratic inefficiencies and lack of clearly defined property rights, which triggered opportunistic behaviour (e.g. free-riding, power influence affecting wealth distribution among members) and lack of investment incentives, in both time and financial terms, in the structure (Cook and Iliopoulos, 2000). Many integrated coastal management schemes in the tropics have failed at the implementation stage (Westmacott, 2002). Others have been bound with difficulties typically found in developing countries, such as information and communication gaps, restricted technical and financial capacity, strong sectoralism and limited democratic representation (Windevoxhel *et al.*, 1999).

On Conflicts at the Land–Water Interface: Perspective from New Institutional Economics on the Limitations of Integrated Coastal Management

Overall limitations of integrated coastal management initiatives can be traced to two interlinked factors: people and the *institutions* they create. It is not physical resources *per se* that need to be the focus of management, but rather human behaviour (FAO, 1998). The land–water interface, where resources 'overlap' and where access and use rules are complex, if not ill defined, is an ideal context in which to study how human interactions and institutions function and influence environmental sustainability.

Institutions and New Institutional Economics: defining the concepts

Institutions have been defined as a set of formal rules (laws, contracts, political systems, organizations, markets, etc.) and informal rules of conduct (norms, traditions, customs, value systems, religious beliefs, etc.) that facilitate coordination or govern relationships among individuals or groups (North, 1990). They influence human behaviour and therefore economic outcomes such as economic efficiency, economic growth and development, which reciprocally often result in changes in institutions (Kherallah and Kirsten, 2001).

New Institutional Economics rose from the

questioning of the fundamental assumptions of neo-classical economics,[4] and examines the role of institutions in furthering or hindering economic growth. According to North (1993), NIE 'adds institutions as a critical constraint and analyses the role of transaction costs as the connection between institutions and costs of production'.[5] In a world of incomplete information and limited cognitive capacity, humans impose constraints (rules, etc.) to structure their exchanges. The formation of institutions is thus determined by the costs of transacting. The critical role of institutions is to create stable structures for human interactions while minimizing costs and uncertainty in transactions.

How does New Institutional Economics relate to coastal conflicts?

Conflicts over natural resources exist because human beings compete for the same scarce resources to maximize their utility, in other words, because their individual interests and needs cannot be simultaneously satisfied. Because of interdependence,[6] one agent's decision to physically modify or use one resource affects other agents' options of use, thereby reducing other users' utility or satisfaction and resulting in conflicts. The ecological fragility of coastal areas can only exacerbate conflicts and losses incurred.

Although conflict occurrence is often linked to ethnic or social rivalries and their roots are understood from a cultural perspective, economic factors and interests often lurk in the background (Bardhan, 1997). These interests can manifest themselves in the form of conflicts: (i) *within sectors* (e.g. large- *versus* small-scale fishers or aquaculture operators); (ii) *between sectors* (e.g. between fisheries and other sectors); or (iii) *between objectives* (e.g. planning agencies with diverging objectives such as environmental protection, economic development, social equity) (Béné *et al.*, 2004). The origin of each type of conflict can be better comprehended in the light of NIE concepts.

Conflicts within sectors: interdependence and ill-defined property rights

Conflicts within sectors are often closely linked to the allocation, perceived as unfair, or ill definition of property rights, which are a form of institutional arrangement. Interdependence of agents with incompatible interests, that is, the fact that the choice of one agent influences that of another (Paavola and Adger, 2002), is a first cause of conflicts. This is illustrated in the case of the Indian shrimp industry below. Collective environmental choices are necessary to resolve conflicts and disagreements and these choices imply affirmation or redefinition of endowments, that is, property rights or environmental regulations (Coase, 1960). For Coase (1960), externalities can be internalized through negotiation and bargaining if property rights are well established and if transaction costs are absent. As these two criteria are not usually simultaneously met in developing countries, rational economic decision-making is hampered and conflicts linked to land appropriation and exploitation are exacerbated. From ill definition of formal land property regimes stem long-lasting conflicts and, at the macro level, slow economic advancement (De Soto, 2000; Zak, 2001).

Conflicts between sectors: imperfect information and high transaction costs

Conflicts between sectors are associated with the theory of imperfect information: all institutional arrangements, including both for-

[4] The fundamental assumptions of neo-classical economics are that: (i) people have rational preferences among outcomes; (ii) individuals maximize utility and firms maximize profits; and (iii) people act independently on the basis of full and relevant information.

[5] *Transaction costs* are made up of three types: (i) search costs (the costs of locating information about opportunities for exchange); (ii) negotiation costs (costs of negotiating the terms of the exchange); and (iii) enforcement costs (costs of enforcing the contract) (North and Thomas, 1973).

[6] Interdependence and externalities are closely related concepts, the former belonging to neo-classical economics, the latter to NIE. *Externality* was defined in footnote 3. *Interdependence* is the reciprocal nature of relationships that underlie what are conventionally regarded as externalities (Coase, 1960).

mal and informal contracts, are explained in terms of strategic behaviour under asymmetric information among the different parties involved (Bardhan, 1989). In coastal areas where sectoral 'special interests' (Dixit, 2003) are competing, many decisions regarding resource allocation and use are based on incomplete information (some ecological processes are still poorly understood). They are also influenced by high transaction costs arising from the unbalanced game of economic power forces because necessary institutional arrangements, such as adequately defined property rights, are not in place to minimize them. Transaction costs due to absent or underdeveloped institutions have been found to be higher in developing countries (North, 2000), and this may be a reason for the prevalence of unresolved conflicts in the developing world between fishing, farming, aquaculture and other activities.

Conflicts between objectives: problem of common agency

Diverging objectives within a central planning agency responsible for coastal management can be seen in the framework of common agency. The problem of common agency refers to what are called 'principal-agent' relationships. *Principals* are the actors within a hierarchical relationship in whom authority rests. *Agents* are those linked to the principals and conditionally designated to perform tasks in the name of principals (Lyne and Tierney, 2002). Dixit *et al.* (1997) define *common agency* as 'a multilateral relationship in which principals simultaneously try to influence the actions of one agent'. It is a problem because one agent who has to respond simultaneously to several principals whose interests are not necessarily aligned (Dixit, 2003). Lack of cooperation between stakeholders, multiple tasks and diverging objectives as encountered in coastal areas distort attempts towards integrated planning and management.

Although these issues have been investigated from a qualitative perspective, some researchers have attempted to explore in a quantitative manner some aspects of NIE, in particular the relationship between the existence – or lack – of established property rights and investment incentives (e.g. Besley, 1995; Deacon, 1999), economic growth and conflicts (e.g. Gonzales, 2004).

Weak institutions and institutional arrangements in developing countries

The next step in using NIE is to inquire why, where institutions exist, they have failed, and indeed did so predominantly in developing countries. The NIE literature helps in identifying a number of 'exogenous' factors that have contributed to the limited efficiency of institutions in developing countries. The first one relates to their historical evolution. The addition of developing countries' own 'improvements' to already complex administrative structures inherited from their colonial past increased their complexity and progressively took them out of the control of those operating them (Dixit, 2003). A second factor is the impact of global commodity and capital markets, as this impact reduces policy options for the state, disrupts the process of building institutions that govern the national economy and weakens the capacity of the state to mediate conflicts. Characteristics of developing countries, such as lack of human capital and suitable communication infrastructure to provide effective administrative, managerial and enforcement services, can also be considered responsible for the limited efficiency of institutions (Bardhan, 1997). Furthermore, the often limited human capacity within sectors can impede good sectoral management: the fisheries sector, one of the prime stakeholders in coastal zone management, typically lacks capacity to manage itself (Willmann *et al.*, 1999). Intertwined with these are other hindering factors such as opportunistic behaviour and plural motivations, which indirectly relate to the common agency problem.

The example of brackish-water shrimp aquaculture in India

Kurien's (1999) description of the 'State and Shrimp' situation in India is an eloquent account of the 'boom and bust' of coastal aquaculture activities, which had been long practised extensively in tidal-based systems 'surrounded by a regime of overlapping rights for the fish farmers and the communities surrounding farms' (Kurien, 1999, p. 241). The inadequate institutions contributed to the decline of the shrimp industry more than had viruses, cyclones or excessive pollution. Effects of these could not be prevented, or contained, because 'suitable rules and regulations did not exist or the authorities did not have the means or the political clout to enforce them' (Dehadrai, Deputy Director General of the Ministry of Agriculture in NABARD, 1994, quoted in Kurien, 1999). The chronology of events in the development of the shrimp industry in southern Indian states (Karnataka, Andhra Pradesh, Tamil Nadu, Kerala) is reported here as in Kurien's case, with emphasis on the characteristics of institutional failure.

Justification for developing shrimp culture

The official reason for promotion of shrimp farming was based on a simple and logical deduction: 'shrimp harvests from nature [have] remained more or less stagnant. Thus it is logically concluded that the culture [of shrimp] is the only alternative for augmenting export production of shrimp' (MPEDA, 1992: p. 1, cited in Kurien, 1999). *Links with NIE concepts*: Practised in tidal zones, property or access rights were initially ill-defined but did not constitute a problem given the extensive nature of the activity. However, the sectoral decision to increase yields was made based on inadequate information to assess market variables and future trends. Yet, under the pressure of global economic liberalization and the opening of India to foreign investors, this decision was not questioned because of the influence of personal/corporate interests, power forces and the impact of global commodity and capital markets on the state's capacity to mediate conflicts (Bardhan, 1997). Potential for the emergence of interdependencies arising from shrimp production in isolation from other resource uses and economic activities was not considered.

Government actions to promote the activity

The government enacted a lease policy of government lands. Although priority for a lease had been traditionally given to poor fishermen at a very low price, the price of land rents to private entrepreneurs also remained derisorily low. In addition to the quick issuance of licences for new aquaculture units, the government heavily subsidized start-up activities and provided investors with financial incentives. However, small-scale shrimp farmers were denied all forms of government support on the grounds that their land tenure for aquaculture was not secure. In 1991, a Coastal Zone Regulation Notification was passed as an attempt to regulate coastal developments. However, the rules of the Notification were 'complex and sometimes unclear, as if the government did not want to compromise all economic activities' (van Houtte-Sabbatucci, 1999). *Links with NIE concepts:* Through the leasing of lands under its ownership, the government contributed to the distortion of land market values at the expense of disadvantaged groups, without benefiting itself from revenues of leased lands. This allocation of land was, however, open only to entrepreneurs entering the shrimp business and subject to minimum requirements (4 ha for small entrepreneurs, 40 ha for progressive entrepreneurs), automatically excluding subsistence activities. Although it resulted in a form of property rights for commercial shrimp-farming activities, small-scale shrimp farming and traditional fishing remained unregulated by formal access rights to both land and water. The Coastal Zone Regulation Notification suffered from an information bias as its implementation was subject to interpretation. This left the credibility of the institution, as well as its commitment to applying the regulation, open to question.

Corporate power

Corporate shrimp farming grew rapidly. In comparison with small producers, large corporations had well-defined property rights (through government land leases). This position of strength allowed them to lease more land or encroach if necessary. They seized open-access lands such as beaches, grazing areas and mangroves at the expense of communities of fisherfolks who disproportionately bore the impacts of shrimp businesses (depletion of fishing grounds through mangrove destruction and trawling for larvae, direct interference with fishing activities because of increased water turbidity, loss of traditional access rights to fishing grounds, contamination of drinking water sources, ICSF, 1999). *Links with NIE concepts:* the dominant position of corporations increased their bargaining power to the point where they overtook the role of the institution in place: the government. They nevertheless did not replace it. Entrepreneurs' opportunistic behaviour and disregard of the impact of interdependencies resulting from shrimp activities increased tensions between the two interest groups.

The 1996 crisis

Virus attacks, combined with cyclones and excessive pollution generated by shrimp operations, resulted in a drop in production levels equal to those before the activity took off (1990–1991). The response to why the state did not regulate the industry was quoted at the outset of this section: its lack of capacity and political will to apply and enforce rules. This led the judiciary to come to the forefront of the crisis to enforce the laws that existed and recognize the voice of those whose livelihoods had been negatively affected. *Links with NIE concepts*: although enforcement was within the remit of the executive and legislative arms of the state, they could not be activated because of the progressive accumulation of institutional failures that led to the collapse of the industry.

The Supreme Court intervention

A case was put before the Supreme Court of India questioning the 'right of corporate entities to inflict both direct and indirect threats to life and livelihood of the coastal communities, by being unconcerned about the uni-directional externalities which they impose in the course of their business activities' (Kurien, 1999, p. 248). The Supreme Court Judgement (11 December 1996) ordered the demolition of all farms set up within the Coastal Zone Regulation and ordered the creation of the Aquaculture Authority of India in 1997, whose primary objective was to regulate shrimp farming. *Links with NIE concepts*: the establishment of the Aquaculture Authority of India was a positive step forward. However, the sustainability of an institution established as a response to a crisis has difficulties linked to the lingering of power forces and conflicting interest groups (Dixit, 2003). This may explain why, despite some improvements in the situation, the Aquaculture Authority of India is still struggling and requires a mix of policy measures such as strengthening of licensing rules, effective enforcement, judicious use of economic incentives and increased monitoring of environmental and social impacts (Hein, 2002). Bhat and Bhatta (2004), in their study of optimal land allocation for shrimp or crops following various government interventions (including the 1991 Coastal Zone Regulation and the 1996 Supreme Court order), showed that the current legal framework did not adequately address off-site effects of shrimp farms on coastal resources, on-site self-pollution and, more importantly, equity concerns of crop farmers and water users.

Neiland *et al.* (1999) pointed out that the lack of independent analysis of the factors affecting production strategies (i.e. inadequate information) reflected the polarization of viewpoints over the roles of different stakeholders in the shrimp industry. To overcome this polarization, these authors suggested the participation of all stakeholders in a policy formulation process geared towards the achievement of a common goal: the sustainability of the activity, now widely agreed upon and promoted through best management practices (BMPs). The following section takes a closer look at how institutions and their arrangements could be developed or strengthened, based on the NIE framework, before the Indian case study is re-explored in the light of these observations.

Decentralization, Devolution and Institutional Strengthening for Improved Integrated Coastal Planning and Development

Decentralization and devolution as institutional processes

Decentralization and devolution of management competencies and responsibilities stem from the increasing emphasis given to participation and representation in planning processes. Decentralization, the transfer of decision-making and financial responsibilities from a central authority (the state) to lower levels of government, has been motivated by two arguments. The first is increased efficiency as a central state authority usually lacks capacity to implement policies and programmes that reflect people's real needs and preferences. The second is improved governance through enhancement of accountability and monitoring of government officials and decision-makers (Jütting *et al.*, 2004). In the context of conflict, decentralization not only deflects tension away from the source of the conflict but also reduces the power of central bureaucracies (Bardhan, 1997). Devolution is a related reform but involves the transfer of rights and responsibilities to user groups at the local level. These organizations are accountable to their membership (the resource users) but do not represent others in the community, or the society at large (Ribot, 1999). The relation between decentralization and devolution in the context of coastal resources is schematically shown in Fig. 20.1.

However, from the perspective of NIE, the willingness to decentralize and devolve responsibilities is a political process influenced by the costs of negotiating and implementing agreements, coping with information asymmetries and making commitments credible. Reforms must 'alter or adapt institutions and organizations in the desired direction: to do this successfully, they must anticipate and make provision of the transaction costs that inevitably arise in the operation of the new or modified procedures' (Dixit, 2003).

Requirements for successful institutional strengthening in the coastal zone

Human resources development and *institutional strengthening* are two prerequisites for integrated planning, at all levels (Willmann *et al.*, 1999). However, to examine the reasons linked to institutional failure, a number of additional factors deserve attention. They are

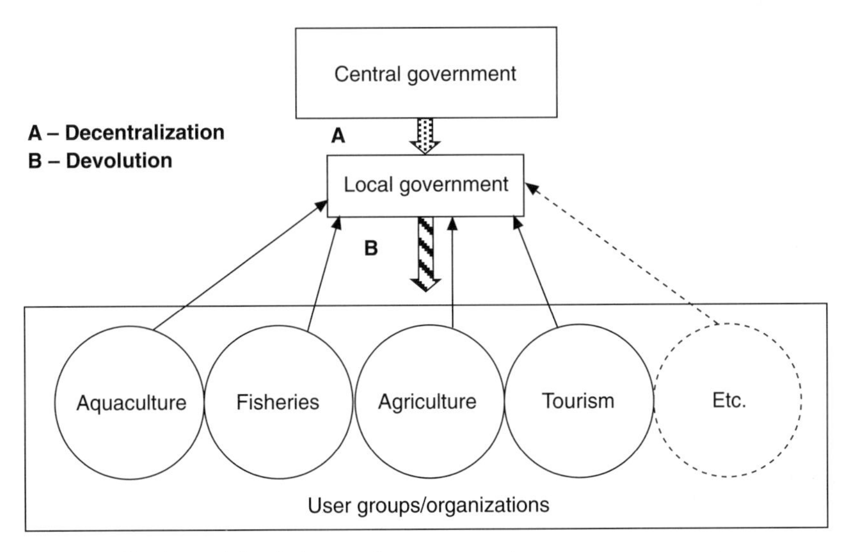

Fig. 20.1. Decentralization and devolution in relation to coastal resource management.

placed in the contexts of decentralization and devolution.

In the context of decentralization

TIMING AND TIME FACTOR Since socially beneficial reforms may not benefit from activation in a post-crisis situation, decentralization as a normative reform should be 'opportunistic', that is, initiated whenever openings and opportunities appear (Dixit, 2003). Implementation of coastal management initiatives takes time as it requires awareness raising and capacity building. The importance of this time factor was illustrated in a case study in the Philippines (Courtney *et al.*, 2002), in which more than a decade of efforts in building local government capacity was necessary to achieve the intended result of delivering coastal resource management as a basic service.[7]

ADAPTIVE AND COORDINATED NEW INSTITUTIONAL ARRANGEMENTS FOR RESOURCE MANAGEMENT To address constraints linked to weak incentives and the problem of common agency, a form of organizational reform achieved through reallocation and grouping of complementary tasks may be a promising approach (Dixit, 2003). In regional coastal development, a regional authority would be responsible for all projects and development initiatives falling within its geographic area of responsibility. Multisectoral integration can bring together and help coordination among agencies with a stake in coastal management to work together towards mutually agreed-upon goals. Structural integration, whereby an entirely new institutional structure with its own rules is created and is responsible for all development and policies linked to coastal use, can be an alternative (Scialabba, 1998). Although suggested in the context of coastal resource management in Tanzania (Masalu, 2000), the latter tends to be more difficult to put into practice given the protective attitude of ministries with regard to their power base and funding.

VISIBILITY, CREDIBILITY AND FLEXIBILITY OF NEW INSTITUTIONAL ARRANGEMENTS Visible mechanisms are necessary to ensure that the sacrifices required by large-scale adjustment programmes are equitably shared (Bardhan, 1997). The credibility of institutions is also paramount: institutional solutions will not work unless they are supported by reputational considerations, that is, they have been put in place through a widely recognized and supported, democratic and balanced process. The general principle of superiority of flexible rules over inflexible ones has to be balanced with arbitrary discretion. Although flexibility is required to respond to specific circumstances, the way in which it is applied has to be announced in advance and adhered to *ex post* (Dixit, 2003), dictating transparency and accountability as prerequisites.

BUILT-IN INSTITUTIONAL MECHANISMS FOR CONFLICT RESOLUTION Negative impacts of large-scale development projects could be lessened by the establishment of mechanisms through which grievances of those negatively affected or displaced could be voiced and given adequate weight and recognition in project evaluation and direction (Bardhan, 1997).

IMPROVED INFORMATION AND COMMUNICATION CHANNELS Because institutions often rely on incomplete and asymmetric information, improved negotiation outcomes could be achieved if more information, in both quantity and quality, were collected and equally shared (Dixit, 2003).

In the context of devolution

COLLECTIVE ACTION AND SOCIAL CAPITAL As relationships and networks, formal and informal, information flows, agreements, etc., have gained attention, in particular in relation to traditional management of common property resources, Paavola and Adger (2002) argue that social capital should be an

[7] This time requirement may also explain why we are only starting to find concrete results of the achievements of integrated management in coastal areas reported in the literature.

integral part of the NIE approach to the environment. Social capital and institutions are intimately linked: the latter will condition the capacity of social groups to act collectively (Woolcock and Narayan, 2000). Reciprocally, collective action takes place within an institutional framework where choices are made according to specific decision rules (Olson, 1971). This implies that institutions that respect and enhance all forms of social capital are necessary to foster successful collective action. Traditional management rules of environmental resources such as fisheries are a manifestation of social capital (Berkes *et al*., 2000). 'New generation cooperative' models that address the weaknesses of traditional cooperatives by strengthening the assignment of property rights to their individual members and by reducing the incentives for opportunistic behaviour have been advocated by Cook and Iliopoulos (2000).

CAPACITY BUILDING The assumption that sufficient local knowledge exists for managing resources sustainably, upon which many participatory management approaches are founded, is not always verified (Meinzen-Dick and Knox, 1999). For this reason, capacity building of the smallest units, that is, user groups such as fishers' cooperatives, aquaculture operators' associations, etc., is required so that their interests can be adequately represented in decision-making regarding resource allocation and management, thereby ensuring the likelihood of success of devolution programmes.

Policy reforms towards decentralization and devolution of resource management should occur concomitantly. Mechanisms to link user groups with local government authorities (thin feedback arrows in Fig. 20.1) should be established to avoid overlap of competencies and powers, and resurgence of conflicts. This implies that capacity needs to be strengthened on both sides: first building capacity in the user groups, that is, in a manner that can be seen as 'sectoral', to ensure that those groups base their activities on good-quality information and are adequately represented in decision-making processes at local government levels. The second step is to build capacity at the level of local government authorities to increase their capacity for dealing with the multiple interests generated by devolution initiatives and for managing coastal resources in a more integrated way. Thus, sectoral capacity building can have a role to play in making decentralization and devolution succeed when addressing coastal zone users' concerns.

Indian shrimp aquaculture case study revisited

Current institutions and their arrangements to deal with shrimp and coastal development in India, as schematically represented in Fig. 20.2, are now re-examined in the light of the above considerations.

The lack of an integrated approach on behalf of agencies and their respective arrangements dealing with shrimp development in isolation from coastal management, including protection, is evident (represented by the bold horizontal line dividing Fig. 20.2). Three main central agencies, often with overlapping mandates but no regulatory capacity, had been established by the government prior to the creation of the Aquaculture Authority of India: (i) the Marine Products Export Development Authority (MPEDA), an autonomous body under the Ministry of Commerce, in 1972; (ii) the Brackish-water Fish Farmers' Development Agencies (BFDAs) set up during the Seventh Plan Period (1985–1990) in coastal states and union territories to provide technical, financial and extension support to shrimp farmers; and (iii) the Central Institute of Brackish-water Aquaculture (CIBA), a fully fledged research institute, in 1987. Established in 1997, the Aquaculture Authority of India was indeed an attempt to create a cross-sectoral agency with regulatory powers. However, mandates and responsibilities of existing agencies were not modified accordingly and all pursued their work independently. This has been exemplified by the release of multiple guidelines for

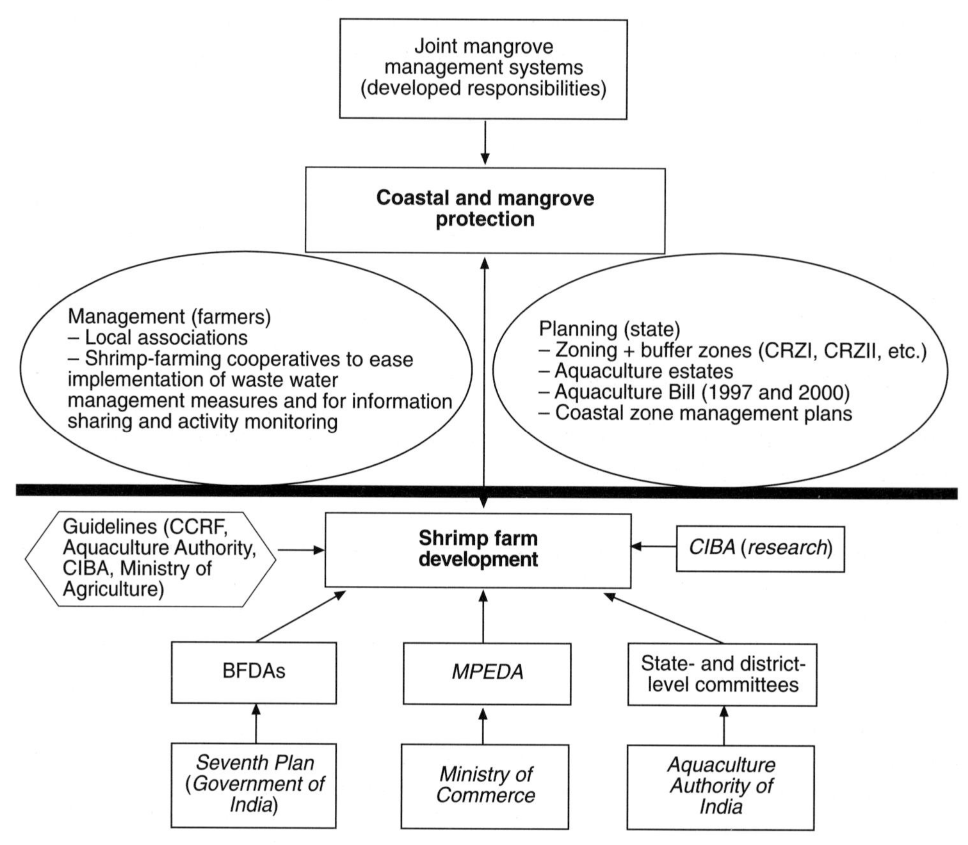

Fig. 20.2. Schematic representation of current institutional environment and arrangements influencing and regulating the planning and management of shrimp farms and coasts in India. In italics are centralized agencies. Rectangular boxes indicate institutional arrangements and agencies created to support the development of shrimp farming, whereas the hexagonal box denotes more specific guidelines (which can be assumed here to be the institutional 'environment') formulated to guide the development of the activity.

improved farm management.[8] In addition, to facilitate the implementation of its Rules and Procedures for issuing licences for shrimp farms, the Aquaculture Authority constituted state- and district-level committees to appraise licence applications (prior to a final decision for issuance by the Aquaculture Authority). The ovals in Fig. 20.2 show measures taken after the shrimp crisis and currently suggested to influence both shrimp management and coastal protection by the above-mentioned agencies. Zoning was a concrete mitigation measure proposed in the coastal states and union territories' Coastal Zone Management Plans submitted to the central government and Ministry of Environment and Forests. These resulted in the categorization of four Coastal

[8] 'Guidelines for adopting improved technology for increasing production and productivity in traditional and improved traditional systems of shrimp farming' and 'Guidelines for setting up of effluent treatment system in shrimp farms' by the Aquaculture Authority; guidelines to develop eco-friendly and economically viable culture technologies for greater productivity of fish, shellfish and other aquatic organisms in brackish-water areas by the CIBA; 'Guidelines for classification, use and lease of brackish-water lands' and 'Guidelines for sustainable development and management of brackish-water aquaculture' by the Ministry of Agriculture.

Regulation Zones of varying environmental sensitivity (CRZ I, CRZ II, etc.), in which restricted levels of development were allowed. Yet, this measure, although welcomed by environmental groups, has been only partially respected by state governments that fully supported commercial shrimp activities (Halim, 2004), and has been further threatened by the submission of the Aquaculture Authority Bill, first in 1997, then again in 2000, permitting the by-passing of farm-siting regulations.

In the Environmental Impact Assessment Report it submitted to the Supreme Court in support of the adoption of the Aquaculture Authority Bill, the Aquaculture Authority (2001) suggested several measures for promotion of the sustainable development of shrimp farming along Indian coastlines. These again included zoning for different activities and the establishment of buffer zones between shrimp farms to prevent salinization of soil and aquifers and to limit pollution. It was also suggested that states consider setting up 'aquaculture estates' in areas suitable for shrimp farming and where supporting infrastructure (roads and electricity) was already in place. Only small provisions were made, however, for farmers in the plan of action: they were encouraged to form local associations and organize themselves in shrimp cooperatives to address their production constraints. The formation of 'aqua-clubs' in Andhra Pradesh and Tamil Nadu was a significant step in this direction (Yadava, 2002). Although the provision of technical and extension support was recommended, it was nevertheless recognized that the capacity to do so by both the Aquaculture Authority and the BFDAs was limited. This illustrates two shortcomings: one linked to the fact that the decentralized agencies remained as fragmented as their central counterparts, and the other linked to the disassociation between local shrimp farmers' groups and local government structures, even if decentralized, as farmers' increased duties and responsibilities with regard to the implementation of sustainable farm management practices were not matched by capacity enhancement or by legal recognition of their rights.

Regarding mangrove protection, interesting initiatives have been made with the development of 'Joint Mangrove Management' systems (JMM), implemented by State Forest Departments and a research foundation, in Tamil Nadu, Andhra Pradesh and Orissa (M.S. Swaminathan Research Foundation, 2004). The success of these community-based mechanisms has relied on the capacity of a decentralized management scheme to make the necessary provisions for communities to assume full responsibility for the management and protection of their resources, and in particular to design and implement their own and shared system of rule enforcement (the *Thengapali*).

The overall system could be improved in the following manner. First, the Aquaculture Authority could be maintained as it has regulatory powers, but its responsibility could be expanded to encompass the mandates of other central agencies such as CIBA and MPEDA to reduce the problem of multiple agents.[9] Addressing environmental concerns such as mangrove protection would also be part of the new, re-defined responsibilities of the Aquaculture Authority. Articles 9 (aquaculture development) and 10 (integration of fisheries into coastal development) of the FAO Code of Conduct for Responsible Fisheries could be used as an all-encompassing guideline for the work of the Aquaculture Authority. Second, at a decentralized level, existing structures, such as the state- and district-level committees dealing with shrimp farm licences and BFDAs in their role of training and financial support providers, could be maintained but should be placed under the direct jurisdiction of the Aquaculture Authority, with the same legal power. In this process, environmental interests would also be represented at the local level. Third, the link between the decentralized form of Aquaculture Authority and shrimp farmers' associations (small-scale and/or commercial), Joint Mangrove Management committees and other local groups such as shrimp larvae collectors,

[9] This could imply the disappearance of these two agencies.

fisherfolks, etc., could be strengthened by involving *Gram panchayats*, which are village-level self-governance bodies. These could be made responsible for village-level implementation of the policies and measures developed by the decentralized authority (Bhatta and Bhat, 1998).

The next two key issues to examine are capacity building and enforcement of the rules. Sectoral capacity building of individual associations (or the 'smallest unit') was suggested earlier as a stepping stone towards improved participation and representation of common interests in planning processes. It should also be carried out at the level of the decentralized authority to increase its capacity for dealing with multiple interests and to ensure that the rights and duties of individual groups are formally recognized and enforced. Enforcement of rules should also take place at the level of individual associations, possibly along the lines of an arrangement similar to the *Thengapali* system developed under the umbrella of the Joint Mangrove Management scheme. Enforcement mechanisms are likely to be more easily accepted and respected if designed and implemented by beneficiaries themselves. Gibson *et al.* (2005) showed that it matters less which rules a community group or, at a higher level, a district, state or even a country adopts than how well it monitors or enforces the rules set, as long as those rules are clear and credible while remaining flexible enough.

Despite these suggestions, the limitations of decentralization should nevertheless not be forgotten. First, governments' willingness to partake in decentralization processes is a *sine qua non* condition for reforms to take place and power to be transferred to lower units. Secondly, decentralization and devolution will be pointless if they are not carried out in parallel with capacity building, as underlined above. Thirdly, the positive impacts of decentralization have yet to be demonstrated. Improper decentralization processes may potentially result in the 'explosion' of a central problem into a myriad of smaller, yet similar, problems. Finally, devolution is incomplete if people are given duties without being given rights. The allocation of these rights, as we have seen, is an ethical, economic and ultimately institutional issue.

Concluding Remarks

Although Westmacott (2002) concluded that conflicts in tropical coastal areas should be at the focus of management measures, this chapter has tried to demonstrate that conflicts were the mere outcome of institutional failure, or stemmed from the simple lack of suitable institutions (environment and arrangements) for dealing with coastal zone issues. Consequently, it is suggested that emphasis on integrated coastal management should shift from conflict resolution to the design and building of suitable institutions where they are absent, or strengthening of those that already exist. Some integrated, participatory coastal management initiatives have failed to examine both environmental and conflict issues because 'new' institutional arrangements put in place to deal with these issues mirrored, through power games, incompatible specific interests and high transaction costs – the sectoral management approaches that existed before.

The consideration of New Institutional Economics principles has allowed pinpointing of the precise causes for the conflicts that still prevail in environmentally sensitive areas such as coastal zones. If the root of conflicts is human competition for natural resources, failure of institutional arrangements is the reason why they remain unresolved.

Decentralization and devolution of management responsibilities are key reform processes, but can only be successfully achieved if the institutions created or remodelled from this process meet some basic requirements to ensure their good 'functioning'. An immediate step for future action is capacity building. In the light of today's experiences with coastal management and the fact that planning capacity is often limited in developing countries within disciplines (be it for the fisheries, aquaculture or agricultural sector), it is suggested that strengthening the capacity of individual sectors is a more efficient and practical move in the direction of the broader goal of integrated management. Strengthening the capacity of local governments to deal with and coordinate those multiple interests in the elaboration of integrated management plans should be the next step. Of course, these efforts should not preclude ongoing work related to improvements in information quality and communication channels, governance and accountability, along with current efforts to better represent the interests of direct and indirect users in decision-making processes regarding coastal resource allocation.

The issuance of property rights on common resources and their acceptance remains a sticky issue, in particular when the problem is disaggregated between *use* rights (access and withdrawal) and *control* rights (management, exclusion, alienation) (Schlager and Ostrom, 1992). Although they are being advocated in the context of capture fisheries where they have potential to revert overexploitation of fish stocks (Hannesson, 2004), they may be more difficult to implement in the case of coastal resources because of the multiplicity, divergence and competition of all interests at stake. In addition, such rights are worthless if they are not, or cannot be, enforced.

Finally, outcomes of decentralization and devolution processes, in terms of efficiency of resource use and conflict mitigation and the long-term financial, as well as environmental, sustainability of coastal resource management systems, need further documenting. The issue of time, however, is not to be dismissed: it takes time to make decisions, time to accept them, and even more time to see and measure their impacts.

References

Aquaculture Authority (2001) *Shrimp Aquaculture and the Environment – An Environment Impact Assessment Report Submitted to the Supreme Court of India.* Aquaculture Authority of India, Chennai. <http://www.aquaculture.tn.nic.in/pdf/FAO%20Aqua80–114.pdf>

Arriaga, L., Montaño, M. and Vásconez, J. (1999) Integrated management perspectives of the Bahía de Caráquez zone and Chone River estuary, Ecuador. *Ocean and Coastal Management* 42, 229–241.

Bardhan, P. (1989) The New Institutional Economics and development theory: a brief critical assessment. *World Development* 17(9), 1389–1395.

Bardhan, P. (1997) Method in the madness? A political-economy analysis of the ethnic conflicts in less developed countries. *World Development* 25(9), 1381–1398.

Béné, C., Macfayden, G. and Allison, E. (2004) Increasing the contribution of small-scale fisheries to poverty alleviation and food security. *FAO Technical Guidelines for Responsible Fisheries.* Draft Report. Food and Agriculture Organization, Rome.

Berkes, F., Colding, J. and Folke, C. (2000) Rediscovering of traditional ecological knowledge as adaptive management. *Ecological Applications* 10, 1251–1262.

Besley, T. (1995) Property rights and investment incentives: theory and evidence from Ghana. *Journal of Political Economy* 103(5), 903–937.

Bhat, M.G. and Bhatta, R. (2004) Considering aquacultural externality in coastal land allocation decisions in India. *Environmental and Resource Economics* 29, 1–20.

Bhatta, R. and Bhat, M.G. (1998) Impacts of aquaculture on the management of estuaries in India. *Environmental Conservation* 25(2), 109–121.

Clark, J.R. (1994) Integrated management of coastal zones. *FAO Fisheries Technical Paper*, No. 327. Food and Agriculture Organization, Rome.

Coase, R.H. (1960) The problem of social cost. *Journal of Law and Economics* 3, 1–44.

Cook, M.L. and Iliopoulos, C. (2000) Ill-defined property rights in collective action: the case of US agricultural cooperatives. In: Menard, C. (ed.) *Institutions, Contracts and Organizations: A New Institutional Economics Perspective.* Edward Elgar Publishing Ltd., Cheltenham, UK, pp. 335–348.

Courtney, C., White, A.T. and Deguit, E. (2002) Building Philippine local government capacity for coastal resource management. *Ocean and Coastal Management* 30, 27–45.

Deacon, R.T. (1999) Deforestation and ownership: evidence from historical accounts and contemporary data. *Land Economics* 75(3), 341–359.

De Soto, H. (2000) *The Mystery of Capital: Why Capitalism Triumphs in the West and Fails Everywhere Else.* Basic Books, New York.

Dixit, A. (2003) Some lessons from transaction-cost politics for less-developed countries. *Economics and Politics* 15(2), 107–133.

Dixit, A., Grossman, G. and Helpman, E. (1997) Common agency and coordination: general theory and application to government policymaking. *Journal of Political Economy* 105(4), 752–769.

FAO (1998) *The State of Food and Agriculture*. Food and Agriculture Organization, Rome.

FAO (1999) Papers presented at the Bangkok FAO Technical Consultation on policies for sustainable shrimp culture. Bangkok, Thailand, 8–11 December 1997. *FAO Fisheries Report*, No. 572, Suppl. Food and Agriculture Organization, Rome.

Gibson, C.C., Williams, J.T. and Ostrom, E. (2005) Local enforcement and better forests. *World Development* 33(2), 273–284.

Gonzales, F.M. (2004) *Effective Property Rights, Conflicts and Growth*. University of British Columbia. <http://www.econ.ubc.ca/gonzalez/paper3.pdf>

Halim, U. (2004) *Shrimp Monoculture in India: Impact on the Livelihood of Coastal Poor*. Institute for Motivating Self-Employment (IMSE). <http://www.landcoalition.org/pdf/ps04imse.pdf>

Hannesson, R. (2004) *The Privatization of the Oceans*. The MIT Press, Cambridge, Massachusetts.

Hein, L. (2002) Toward improved environmental and social management of Indian shrimp farming. *Environmental Management* 29(3), 349–359.

ICSF (1999) Submission of the International Collective in Support of Fishworkers (ICSF). In: FAO (ed.) Papers presented at the Bangkok FAO Technical Consultation on Policies for Sustainable Shrimp Culture, Bangkok, Thailand, 8–11 December 1997. *FAO Fisheries Report*, No. 572, Suppl. Food and Agriculture Organization, Rome, pp. 197–200.

Jütting, J., Kauffmann, C., McDonnell, I., Osterrieder, H., Pinaud, N. and Wegner, L. (2004) Decentralisation and poverty in developing countries: exploring the impact. Working Paper No. 236, OECD Development Centre, Organisation for Economic Cooperation and Development, Paris.

Kherallah, M. and Kirsten, J. (2001) *The New Institutional Economics: Applications for Agricultural Policy Research in Developing Countries*. Markets and Structural Studies Division. MSSD Discussion Paper No. 41. International Food Policy Research Institute, Washington, DC.

Kurien, J. (1999) State and shrimp: a preliminary analysis of the economic and ecological consequences of India's fisheries policies. In: FAO (ed.) Papers presented at the Bangkok FAO Technical Consultation on Policies for Sustainable Shrimp Culture, Bangkok, Thailand, 8–11 December 1997. *FAO Fisheries Report*, No. 572, Suppl. Food and Agriculture Organization, Rome, pp. 235–255.

Lyne, M. and Tierney, M. (2002) Variation in the structure of principals: conceptual clarification for research on delegation and agency control. Memo presented at the Conference on Delegation and International Organizations, 3–4 May 2002, Park City, Utah.

M.S. Swaminathan Research Foundation (2004) *Atlas of Mangrove Wetlands of India*. M.S. Swaminathan Research Foundation, Chennai, India.

Masalu, D.C.P. (2000) Coastal and marine resource use conflicts and sustainable development in Tanzania. *Ocean and Coastal Management* 43, 475–494.

Meinzen-Dick, R.S. and Knox, A. (1999) Collective action, property rights, and devolution of natural resource management: a conceptual framework. Paper presented at the International Workshop on Collective Action, Property Rights and Devolution of Natural Resource Management: Exchange of Knowledge and Implications for Policy, 21–25 June 1999, Puerto Azul, Philippines.

Neiland, A., Soley, N. and Baron, J. (1999) A review of the literature on shrimp culture. In: FAO (ed.) Papers presented at the Bangkok FAO Technical Consultation on Policies for Sustainable Shrimp Culture, Bangkok, Thailand, 8–11 December 1997. *FAO Fisheries Report*, No. 572, Suppl. Food and Agriculture Organization, Rome, pp. 209–229.

North, D.C. (1990) *Institutions, Institutional Change and Economic Performance*. Cambridge University Press, New York.

North, D.C. (1993) The New Institutional Economics and Development. Economic History No. 9309002. Economics Working Paper Archive at WUSTL. Washington University at St Louis, Missouri. <http://www.econwpa.wustl.edu/eps/eh/papers/ 9309/9309002.pdf>

North, D.C. (2000) A revolution in economics. In: Menard, C. (ed.) *Institutions, Contracts and Organizations: A New Institutional Economics Perspective*. Edward Elgar Publishing Ltd., Cheltenham, UK, pp. 37–41.

North, D.C. and Thomas, R.P. (1973) *The Rise of the Western World: A New Economic History*. Cambridge University Press, Cambridge.

Olson, M. (1971) *The Logic of Collective Action: Public Goods and the Theory of Groups*. Shocken Books, New York.

Ostrom, E. (1990) *Governing the Commons: The Evolution of Institutions for Collective Action*. Cambridge University Press, New York.

Paavola, J. and Adger, N.W. (2002) New Institutional Economics and the environment: conceptual foundations and policy implications. CSERGE Working Paper EDM 02–06. University of East Anglia, Norwich, UK.

Pearce, D.W. and Turner, R.K. (1990) *Economics of Natural Resources and the Environment*. Harvester Wheatsheaf, Hemel Hempstead, UK.

Ribot, J.C. (1999) Integrated local development: authority, accountability and entrustment in natural resource management. Working Paper prepared for the Regional Program for the Traditional Energy Sector (RPTES) in the Africa Technical Group (AFTGI – Energy) of the World Bank. The World Bank, Washington, DC.

Rouf, M.A. and Jensen, K.R. (2001) Coastal fisheries management and community livelihood: possible strategy for the Sunderbans, Bangladesh. *ITCZM Monograph Series*, No. 4. Asian Institute of Technology, Pathum Thani, Thailand.

Schlager, E. and Ostrom, E. (1992) Property-rights regimes and natural resources: a conceptual analysis. *Land Economics* 63(3), 249–262.

Scialabba, N. (ed.) (1998) Integrated coastal area management and agriculture, forestry and fisheries. *FAO Guidelines*. Environment and Natural Resources Service, Food and Agriculture Organization, Rome.

van Houtte-Sabbatucci, A. (1999) Salient legal and institutional features with regards to the development of shrimp culture in a few countries. In: FAO (ed.) Papers presented at the Bangkok FAO Technical Consultation on Policies for Sustainable Shrimp Culture, Bangkok, Thailand, 8–11 December 1997. *FAO Fisheries Report*, No. 572, Suppl. Food and Agriculture Organization, Rome, pp. 256–266.

Westmacott, S. (2002) Where should the focus be in tropical integrated coastal management? *Coastal Management* 30, 67–84.

Willmann, R. and Insull, D. (1993) Integrated coastal fisheries management. *Ocean and Coastal Management* 21, 285–392.

Willmann, R., Halwart, M. and Barg, U. (1999) *Increasing food security: integrating fisheries and agriculture to enhance food production*. GATE, 1/99. Deutsche Gesellschaft für Technische Zusammenarbeit (GTZ), Eschborn, Germany.

Windevoxhel, N.J., Rodríguez, J.J. and Lahmann, E.J. (1999) Situation of integrated coastal zone management in Central America: experiences of the IUCN wetlands and coastal zone conservation program. *Ocean and Coastal Management* 42, 257–282.

Woolcock, M. and Narayan, D. (2000) Social capital: implications for development theory, research and policy. *World Bank Research Observer*, 15, 225–249.

Yadava, Y.S. (2002) Shrimp farming in India: lessons and challenges for sustainable development. *Aquaculture Authority News* 1(1), 1–4.

Zak, P.J. (2001) Institutions, property rights, and growth. *The Gruter Institute Working Papers on Law, Economics and Evolutionary Biology* 2(1): Article 2.

21 Responding to Coastal Poverty: Should we be Doing Things Differently or Doing Different Things?

J. Campbell, E. Whittingham and P. Townsley
IMM Ltd., Innovation Centre, University of Exeter, Exeter, United Kingdom, e-mail: j.campbell-IMM@ex.ac.uk

---

Abstract

At the interface between land and sea, the coast is arguably one of the most complex and dynamic environments on this planet. Composed of a diversity of interacting natural, socio-cultural, economic and political systems, the coast is in a constant state of change, not only as a result of the constant biophysical forces operating at the coast but also as a result of the significant longer-term changes – population growth, industrial and tourist development, pollution, habitat and biodiversity loss, changes in access rights, markets and technology and the growing reality of climate change – that are increasingly threatening the future sustainability of coastal environments. Although many of these changes occur in other ecosystems, they are particularly concentrated on the coast.

In the past, coastal people, and particularly the coastal poor, have adapted to the intrinsically dynamic nature of the coast, but they now find themselves having to respond and cope within an increasingly competitive environment, in which access to the resources they depend on is becoming more and more restricted and opportunities based on the use of natural resources in general are becoming increasingly limited. For many coastal people, particularly those dependent on natural resources, current changes mean that they must adapt or face increased marginalization and displacement from the coastal resources on which they depend.

This chapter reviews the impact of current changes on the poor in coastal fishing communities, with examples from around the world, and examines existing responses to assist the poor in coping with change on the coast and finding 'alternative livelihoods'. It asserts that current responses supporting the poor to develop their livelihoods have had limited success because of the lack of understanding of who the poor are, the nature of their existing livelihoods and the wider economic, institutional, political and social influences. The use of the sustainable livelihoods approach is discussed as a means of improving our understanding of coastal poverty and linking support for livelihood diversification and enhancement with the livelihoods of the poor, their needs and aspirations, and within the context of local and wider development. Some broad principles to guide more systematic and participatory approaches to interventions are proposed.

Background to Coastal Poverty and Coastal Complexity

In order to respond to poverty on the coast, we first need to understand how coastal complexity interacts with poverty and how that poverty might be defined. Coastal ecosystems are often characterized by their high degree of complexity associated with a number of distinctive features, including their diversity and dynamism, the fugitive nature of some of the resources available, their open-access nature, the concentration of externalities and people on the coast, and the often hostile nature associated with these features.

Coastal diversity and dynamism

Coastal diversity reflects the environmental, social, economic and political processes and systems that operate there and the interactions among them, which lead to a high degree of complexity and the potential to generate both opportunities and problems. One of the key aspects of the coast for the poor is the species and ecosystem diversity that provides a wide range of productive opportunities that do not lend themselves to scale economies. The coast is also arguably one of the most dynamic environments on the planet. The daily tidal changes and the seasonal weather patterns are regular factors affecting this dynamism; in addition, the coast is the focus of many sudden weather hazards such as tsunamis, cyclones and tidal floods. It is also subject to the changing patterns of river run-off from the land, which, in places such as Bangladesh, can have a very profound effect on the coast. These effects are worsened when externalities (such as pollution, sedimentation, water abstraction for irrigation and irregular flooding from dams) of upstream industries or countries are concentrated in coastal waters.

Access complexity and dynamism

Many of the resources that sustain opportunities on the coast for the poor are common-pool resources that the poor are able to exploit because they are open-access or subject to sets of use rights that are poorly defined or largely unenforced. From one point of view, this is an advantage for the poor as it leaves open a 'window of opportunity' to use resources (such as fish, forests, land and wildlife) that are often not available for them in other, non-coastal areas. However, this advantage has a limited lifespan and the poor often find their access to these common-pool resources increasingly restricted for a number of reasons.

Although people on the coast depend on a wide range of natural resources, fish make up an important part of this. Two of the key features that affect fish-dependent livelihoods are the fugitive nature and perishability of fish. Both these factors introduce high levels of risk and uncertainty into the livelihoods of the people who depend upon them, and these risks are compounded by the open-access (and thus potentially competitive) systems that prevail in such areas.

Livelihood diversity

The dynamic and complex features of the coastal environment represent areas of opportunity for poorer sections of society, as they have no choice but to accept the risks associated with living in marginal coastal areas and the difficult work often involved in exploiting coastal resources. At the same time, because other, better-off groups are unwilling to accept these risks, the coast has often attracted the poor as they may (at least initially) experience less competition for resources there compared with other areas in the hinterland. The wild shrimp seed collection, to satisfy the demand of the growing shrimp aquaculture industry in coastal India and Bangladesh, is a good example of where poorer sections of society have entered difficult and dangerous work in areas prone to cyclones, which are unattractive to the better-off.

The dynamic nature of the coast, the threats it is subject to and the constant changes in the environment faced by the people who live there mean that adaptability has always been an essential feature of the livelihoods of coastal people. This is particularly true for the poor and adaptability provides a variety of benefits. The most obvious benefit flows are the food, income and employment that fisheries resources provide. However, benefit flows from coastal resources to the poor are much more complex than simply providing a source of food, income and jobs (see Box 21.1).

Stakeholder diversity and poverty

In addition to the complexity that surrounds them, the poor make up a diverse group of people with different skills, knowledge, attitudes, traditions, beliefs and histories. Their interactions with the world they inhabit vary between different groups, and often between different households and individuals. Of particular importance in determining how poverty manifests itself amongst different groups are the gender, age and class or caste of those people.

Poverty can be defined in many ways and it is not the role of this chapter to explore this issue in depth. From our research, coastal poor people, like the poor in most other locations, describe their poverty through a diversity of measures rather than solely an absence of money. In a paper titled *Exploring the Links*, UNEP-IISD (2004) link poverty and well-being to the presence or absence of a range of key determinants:

- adequate nourishment;
- freedom from avoidable disease;
- an environmentally clean and safe shelter;
- adequate and clean drinking water;
- clean air;
- energy for cooking and warmth;
- availability of traditional medicine;
- continuing use of natural elements found in ecosystems for traditional cultural and spiritual practices;
- ability to cope with extreme natural events, including floods, tropical storms and landslides; and
- making sustainable management decisions that respect natural resources and enable the achievement of a sustainable income stream.

It is useful to think of poverty and well-being in these complex terms to avoid seeking simple solutions to poverty; this is particularly important on the coast, where the complexity of the ecosystem adds to the complexity of poverty. Even though frameworks for understanding poverty can be developed to high levels of complexity, they do not detract from the fact that the only people who can really define what poverty means are the poor themselves.

Although poverty in some coastal areas is clearly visible (e.g. in coastal Bangladesh), much coastal poverty is hidden from view by the development that is a growing part of much of the coast (especially in urban areas). While some of the poor are able to benefit from these developments, finding paid employment in mechanized fisheries, aquaculture, salt pans or industrial developments along the coast, many are less able to adapt and find themselves excluded from the development processes going on around them. Often they are hidden by the wider coastal development process; they, in effect, become the interstitial poor (Jazairy *et al.*,

Box 21.1. Benefit flows to reef users

For reef-users, coral reefs provide seasonally stable sources of food, building materials, a medium of exchange, medicines and a source of income and status. It is the coral reef that often gives rise to islands that provide habitats for people and lenses of fresh water for drinking and agriculture. The reef also protects coastal villages from storms and wave action and provides shelter to lagoons and other productive areas, such as sea grasses and mangroves, which in turn provide a reserve of food in all weather conditions. The physical structure of the coral reefs dictates that many activities are performed communally and the traditional linkages between reef resources and the spirit world mean that reefs can be socially and spiritually unifying (Whittingham *et al.*, 2003).

1992), who live in 'pockets' side-by-side with the better-off and within areas where there may be relatively high levels of development (IMM, 2003a). This characteristic of poverty in coastal areas needs to be distinguished from that found in some inland areas where natural conditions, such as chronic drought, may create a far more generalized condition of relative poverty.

Responding to poverty on the coast requires us to understand not only the nature of poverty found there but also the wider influences that affect the poor. Coasts around the world are characterized by significant changes. The following sections of this chapter review the changes taking place on the coast, the different impacts these changes are having on the coastal poor and current responses to coastal poverty, focusing finally on livelihood enhancement and diversification as a possible way forward.

Evidence is presented from around the world, but focuses mainly on research on the livelihoods of coastal communities in India, Bangladesh, Sri Lanka, Ghana and Cambodia, which was implemented by IMM Ltd. of the UK and its local partners from 2000 to 2004. This work, funded by the UK's Department for International Development (DfID), focused on the livelihoods of poor people on the coast of those countries, how those livelihoods were changing and what might be done to improve them. Although the work covers a broad spectrum of people involved in different coastal livelihood activities, fisheries inevitably form a major part of this work.

Changes on the coast

In addition to the regular changes that constitute the dynamic nature of the coast, coasts around the world are also undergoing major longer-term changes:

- population growth,
- coastal urbanization and industrialization,
- increasing habitat destruction,
- increasing pollution and sedimentation,
- development of coastal aquaculture,
- increasing access to global markets,
- increasing use of new technologies in resource exploitation,
- declining resource productivity,
- overfishing,
- climate change.

Changing population

Perhaps one of the most pressing changes on the coast is the rising population. In 1995, it was estimated that almost 40% of the world's population lived within 100 km of the coast (Burke *et al.*, 2001), and this number is increasing from both natural population growth and inward migration. In fishing alone, the numbers are impressive: from 1990 to 2000, the number of people directly employed in fisheries and fish farming globally rose from 28 million to 35 million, and 85% of those people live in Asia (FAO, 2002).

Increasing habitat destruction and pollution

Increasing coastal populations have led to increasing coastal urbanization with the associated habitat destruction, sewage pollution and increasing freshwater use. Coastal industrialization has added to these pressures, as has port development and coastal tourism. An increasing cause of pollution on the coast is agriculture (Burke *et al.*, 2001), which introduces ever-increasing quantities of agricultural chemicals into coastal waters through run-off. Coastal aquaculture has also impinged upon important coastal habitats, especially through the development of shrimp farms (Pillay, 1992), and in some cases has promoted unsustainable harvesting practices such as that for wild shrimp seed in India and Bangladesh. Deforestation in inland areas has led to increased flooding in coastal areas and to increased sediment loads, resulting in rising siltation of coastal areas affecting inshore habitats and fisheries. As a result of these different forces, critical coastal habitats are under threat, especially mangroves, seagrass areas, coastal wetlands and coral reefs (Burke *et al.*, 2002). In addition, biomass and biodiversity are in decline.

Increasing competition and unsustainable exploitation

As more people seek employment in coastal areas and as land areas held by individual households decline, the competition in many parts of the world for traditional employment opportunities on the coast, such as agricultural labour, is increasing. Falling farm employment options push more people to depend on the very common-pool resources (e.g. forests, fisheries, shoreline areas, swamps, mangroves and coral reefs) that the poor have traditionally used to eke out a subsistence living. But many of these resources are declining in productivity as the environmental carrying capacity declines with increasing coastal pollution, habitat destruction, resource depletion and changing water quality and movement patterns. Common-pool resources that historically maintained sustainable levels of communal use are now becoming open-access resources, whose use has become uncontrolled and often unsustainable.

Not only do increasing numbers of people bring with them increasing industrialization and its associated problems, they also increase the harvesting pressure on the available resources. Population pressure, improved access to markets, rising global demand for fish and changes in technology have resulted in many coastal fish resources being at or beyond the point of maximum sustainable yield. Globally, only 25% of the available major fish stocks are underexploited or moderately exploited, 47% are fully exploited and 18% of stocks are overexploited (FAO, 2002).

Changing markets and technology

New global markets are providing opportunities for selling goods and services far and wide (e.g. Asian vegetables and fish for the European market). However, these often require the adoption of new technologies to ensure uniformity of product or product quality (e.g. woven baskets produced with traditional skills in the village are now being displaced by plastic containers made in factories in cities). Often these technologies are efficient only if they operate with certain economies of scale (e.g. mechanized fishing boats) and this often requires a concentration of capital ownership in the hands of fewer people and, more often than not, in fewer locations. These technologies also often require skills different from those needed in traditional village industries and so new skills must be learned.

Climate change

Further change is likely in the future as a result of climate change. The Working Group of the Intergovernmental Panel on Climate Change (IPCC, 2001) has identified a series of probable impacts and vulnerabilities of climate change in relation to coastal areas. These areas are expected to become progressively inundated and many small islands are predicted to become partially or wholly submerged. Coastal areas will also be subjected to increased cyclonic weather patterns and increased variability and unpredictability of general weather patterns. Linked to the direct effects of climate change are the likely changes in coastal agricultural activities adjacent to tropical coastal areas. It is projected that these will exhibit a general reduction in crop yields, compounded by declining water availability and a widespread risk of flooding (from both changing precipitation and sea-level rise). In the short term, these climate-induced changes may lead to ever greater dependence on coastal resources (IPCC, 2001).

The ability of human systems to adapt to these changes is highly variable and those with the fewest resources have the least capacity to adapt and are the most vulnerable. Thus, impacts are expected to fall disproportionately on the poor.

Impacts of Change on the Poor

Changes in coastal areas have had a range of impacts on poor people living on the coast: financial insecurity, increasing employment insecurity and underemployment, loss of rights, exclusion and criminalization of their livelihoods, increased use of child labour and increasing gender imbalances. These impacts have not all occurred in all locations but examples of each are widespread. These are outlined below.

Financial insecurity

Many poor people on the coast are suffering from declining incomes and increasing financial insecurity, mainly as a result of increased competition. Increasing numbers of people are interacting with each other to catch the declining fish stocks: more people are chasing fewer fish and conflicts among people are on the increase. In some cases, declining catches per fisherman is a key factor contributing to declining incomes, such as in the coastal fisheries of Ghana (Ward *et al.*, 2004). People are also competing for available land for agriculture and trees for fuelwood and building materials. The profitability in fishing in many parts of the world is now decreasing and engine failure, gear loss, illness or indebtedness can quickly lead to a loss of fishing assets to moneylenders and middlemen/-women. This is now a major concern of coastal fishermen in Cambodia, where marine catches are under increasing pressure and indebtedness is increasing. In some cases, such as on the east coast of India, the decline in catch rates has been offset by rising market demand, resulting in stable incomes for fishermen, but also in increasing vulnerability as fish stocks become threatened (IMM, 2003a).

For poor processors and traders in India, the declining local supplies of fish and increasing competition have meant that transaction costs have increased, thus reducing their already marginal incomes. This is further worsened where larger-scale outside operators have moved into coastal areas and taken over market access for locally produced products and for credit and microfinance, thus increasing transaction costs and reducing market access for the poor (IMM, 2003a). Likewise, in Cambodia, the market linkages between small-scale producers and processors and urban and export markets are becoming dominated by middlemen, who control prices and ensure their access to fish supplies through credit provision (CFDO-IMM, 2005). Although such intermediary activities can be beneficial for the poor in the absence of alternatives, they can also become exploitative where competition is limited, as it is in many parts of Cambodia, thus leading to increased transaction costs (Yim Chea and McKenney, 2003a), increased costs of credit and lowered prices paid to the poor fish suppliers. Additional transaction costs are incurred through the imposition of informal taxes, by corrupt officials, on the movement of products (Yim Chea and McKenney, 2003b) – a highly effective means of extortion in the case of fish, given their high perishability.

Employment insecurity and underemployment

Globally, the growing demand and competition for fish have increased investment in catching technology, resulting in bigger nets and boats. These in turn require better shore facilities and safer harbours, and so more fish are landed at fewer landing sites. These changes at sea have started to affect people on land. With fewer fish being landed at the smaller landing sites, traditional traders and processors are finding fish harder to come by and employment in the harvesting sector is becoming less secure. In India, increased pressure of people wanting to join fisheries and declining fishing opportunities have led to increased underemployment in the sector (IMM, 2003b). In Cambodia, agricultural labour is also increasing in supply and, with little expansion in demand for this labour, levels of underemployment are high (Sarthi Acharya, 2003). In India, conflicts between small-scale and larger-scale processors and traders in response to changes in fish supply, to access to bigger and fewer landing sites, to changing investment patterns and to changing market access and technology have been identified. The poor are generally least able to adapt and cope with these changes, resulting in increased employment insecurity (IMM, 2003a). Likewise, in Ghana, employment insecurity and underemployment are on the increase among fish processors as catches decline or become uncertain and competition between buyers increases (Ward *et al.*, 2004).

Loss of rights, exclusion and criminalized livelihoods

For many coastal resources, tenure rights are unclear or not defined (Luttrell, 2001). Where conflicts over access rights to common-pool resources have increased with the growth of competition in coastal areas, the poor are often the least able to defend their livelihoods or to establish tenure rights that are supported by the legal system and effectively documented. At times, this is because the poor are simply unaware of the rights that they may have and are unfamiliar with the mechanisms through which they can assert their rights. On other occasions, the rights of the poor are actively suppressed in favour of those who have better access to the prevailing legal and political system, are able to understand the language used and can influence the attribution of exclusive rights in their favour.

For migrants, who are commonly among the poorest groups to be found in coastal areas, this situation is particularly marked as they often lack local language skills, are politically weak and have no linkages with local systems of patronage or support, and may often be regarded with suspicion by a large proportion of local people (Box 21.2). This often translates into a systematic trampling of even the most basic rights of some of these marginal migrant groups. Following the devastating cyclone in Orissa, India, where the legal status of people affected their access to rehabilitation support (IMM, 2001), this situation became particularly marked.

The fact that many coastal resources are open-access – the very feature that has enabled the poor to use them – may also lead to their alienation from the poor. While the poor may, initially, be the only group willing to engage in the exploitation of 'difficult' resources such as mangrove swamps, the open sea or coastal areas, once the economic incentives and technology make it viable for others to exploit those resources as well, the poor often find themselves in competition with a wider group of people who have a comparative advantage over them. So, fish resources, previously the exclusive preserve of artisanal fishers, come to be increasingly monopolized by those using more efficient mechanized trawlers, leading to the progressive exclusion of artisanal fishers from the use of the resource. The resulting conflicts between technology levels, such as those off the coast of Kerala in India, are well documented (Kurien, 1992).

In addition, coastal swamps and mangroves used by the poor for a variety of livelihood activities have been progressively converted for aquaculture, agriculture or industrial development, again removing these resources from the common pool and from the uses of the poorer sections of coastal society. With the growth of aquaculture in Andhra Pradesh in India, for instance, traditional occupations such as fish drying and fishing with beach-seines, which required large open areas, are confined to increasingly congested areas. Even fish landings that took place all along the coast are now concentrated in smaller areas (IMM, 2003a; IMM-ICM, 2003). The poor often find that coastal development operates faster than they can cope with. In Vietnam, for example, some poorer farmers have been forced to sell their rice fields to people who are able to make the needed investments for shrimp culture (Luttrell, 2001).

The expansion of coastal tourism in some areas has also led to conflicts with local people; efforts to exclude fishermen from conservation areas have also created tension in some areas. For many of the poor, changes in policies or management plans favouring conservation priorities have left parts of their livelihoods on the wrong side of the law. With few options available to many coastal poor people, they tend to remain in those criminalized livelihoods and absorb the additional risk that this incurs.

For example, some marine protected areas in India, focused on biodiversity conservation and tourism, are off-limits for local fishery activities. In the Gulf of Mannar Marine Biosphere Reserve, access to and

Box 21.2. The lost rights of coastal migrants

In India, the shoe-dhoni fishermen of Andhra Pradesh have traditionally migrated to areas along the coast from their base in the Godavari delta. Although many of the shoe-dhoni people have been migrating for decades, they have few rights in the areas where they spend much of the year and they are losing them in their home area. They have never been allowed by local people to build even temporary homes on the land as there is a general fear that this will lead to attempts to acquire other rights to local resources that are already fiercely contested among local inhabitants. So, the shoe-dhoni people simply live on their boats when they migrate (IMM, 2003b).

exploitation of shallow reef and sea-grass areas surrounding the 21 coralline islands in the Gulf are prohibited and the Wildlife Protection Act (1972) prohibits the collection of many reef species. For the majority of poor reef stakeholders living along the coast of the Gulf of Mannar, these restrictions place severe restraints on their livelihoods. With no viable alternatives, poor reef stakeholders continue to access prohibited reef resources at great risk and increasing transaction costs (Rengasamy *et al.*, 2003). In Cambodia, the decline in the availability of coastal fuelwood relative to the demand is causing more people to harvest forest resources from protected areas (CFDO-IMM, 2005). Many people have few alternatives to continuing with the now illegal harvests and to facing the consequences: fines or bribes.

Criminalized livelihoods are also associated with the declining returns from fishing globally, which are encouraging an increase in fishing effort to compensate and in many places the use of illegal fishing gear. In India and Bangladesh, the harvesting of wild shrimp seed has become illegal, either directly or through changes in gear restrictions (IMM, 2003a). In Ghana, declining catches are forcing many fishermen to use light fishing, which is illegal (Ward *et al.*, 2004). This has, in effect, criminalized the livelihoods of many thousands of coastal poor people, but they have few alternatives but to continue with these fishing practices.

Using child labour

The use of child labour in coastal communities is a common strategy for many of the poorer households coping with declining income and increasing vulnerability. Children on the coast of Ghana, for example, were found to play an important role in different postharvest fishery activities, providing labour in fishing, processing and trading (Ward *et al.*, 2004). In many cases, child labour takes place in resident communities; however, in some cases it was also reported that children from coastal households migrate to work. In Ghana, coastal children are found working in the inland Lake Volta fishery. Although illegal and discouraged, child labour is often a key strategy for the poorest households.

In Cambodia, in a survey implemented in Krong Preah Sihanouk, 1678 children were found working in three fishing areas, involved in a diversity of work on fishing boats, repairing nets, peeling shrimp, removing crab meat from the shell, freezing fish and transporting fish (NIS, 2001). Although often a key livelihood strategy among the poor, child labour is often seen as a problem as it can have the double effect of depriving the children of their education and limiting their future chances of getting out of poverty through better employment. However, for many poor people, the pressures of today far outweigh the benefits of tomorrow.

Gender imbalances

Change on the coast is also affecting the balance in the roles played by men and women both within the household and in wider society.

Men and women are able to respond to development opportunities in different ways in different cultural and religious situations. Opportunities that benefit men may disadvantage women (and vice versa) and the benefit balance of the household may not be improved overall. For example, in India, the increased landings of fish at larger landing sites and the reduction in landings at local village markets have been easier for men to access as they are more able to travel longer distances from their homes (IMM 2003b). By contrast, women have often found it more difficult to access landings that are distant from their home villages and face greater problems in dealing with the more competitive environment often found at larger fishing harbours. In addition, increased mechanization in fisheries has displaced men from fishing activities in some areas and more are now entering the fish trade, and thus increasing competition with traditional women traders and processors (IMM, 2003a). Likewise, in Cambodia, both occupational and geographic mobility of women are less than for men, thus affecting their ability to take up new employment opportunities (Sarthi Acharya *et al.*, 2003).

Closely linked to this are the different ways in which migration affects men and women. Although seasonal migration has been a traditional response to fugitive fish stocks in some coastal communities, fishermen and labourers, in as diverse a range of places such as Ghana, India and Cambodia, are changing their migration patterns in search of a better life. Migration is now taking on new dimensions: people (especially men) are moving for longer periods, they often set up semi-permanent second homes and they are exploring new areas to migrate to where they do not have traditional social safety nets. In many cases, women are left behind to become, de facto, heads of households. On the east coast of India, part or all of some fishing communities migrated along the coast to either avoid bad weather or to follow mobile fish resources to areas where they were more abundant. More recently, fishers have been migrating to the west coast for part of the year, but they return home during the period of the southwest monsoon when fishing in the Arabian Sea is much more difficult (IMM, 2003a). In Sri Lanka, seasonal migrations from the western to the eastern and northern coasts were severely affected by the security situation, which effectively removed the associated fishing grounds from access (IMM, 2003a). In some cases in the region, such seasonal migrations are met more and more by hostility from the local population as competition for resources increases (IMM, 2003a).

The rapid changes that characterize many coastal areas create opportunities, but people's ability to respond to those opportunities is variable. The coastal poor are often the least able to adapt to, or to cope with, these changes. They have limited access to new technologies, to the skills, knowledge, confidence or education to use them, or to the finances to purchase them. Nor do they have the time to invest in their development or the financial reserves to take the risks associated with them. They also characteristically lack the networks to access new knowledge, skills, technologies, finances and markets. In many instances, the poor, who are least able to adapt, face increased marginalization and displacement. Many development agencies on the coast have the remit to respond to these changes and their responses, along with those of the poor themselves, are discussed in the next section.

Responses to Coastal Poverty

It is recognized that, barring major philosophical and policy shifts, the situation of degradation of ecosystems and poverty within and between countries is likely to get worse (UNEP-IISD, 2004). This situation requires urgent action on the part of all stakeholders, from the poor to the international community. The poor have actively been trying to escape poverty, or at least cope with an increasingly bad situation, by adapting their livelihoods. Governments and aid agencies have also attempted to address these issues through development programmes and the targeting of poverty reduction strategies. Conservation efforts have tried to limit the damage to key ecosystems, fisheries management has attempted to move towards more sustainable use of resources and integrated coastal management initiatives have tried to link development and environmental concerns specific to the coast. More recent interventions have tried to balance poverty and conservation through pro-poor conservation initiatives. There have been successes and failures in all of these.

The responses of the poor

The poor do not sit idly by and passively accept the changes that are occurring on the coast. Often they cope with the situation and adapt to the changes; sometimes they diversify their income-earning activities to both supplement their income and reduce risk, or even move into new alternatives (Barrett *et al.*, 2001; Luttrell, 2001). Aquaculture has sometimes been adopted as an alternative to fishing or to agriculture (Luttrell, 2001), but turning fishers into fish farmers is not always easy (World Bank, 2004).

A frequent response to the pressures on the coast, particularly for men, is to migrate to new areas where their current skills will be needed. As discussed above, they often change their income/employment sources (e.g. farmers becoming fishermen, fishermen becoming traders). These may seem to be minor changes but they involve many adaptations and adjustments that relate to household roles, transportation, social networking, credit arrangements, market access and knowledge. Coping often involves accepting, at least in the short term, reduced livelihood expectations and probably increased risks, for example, through illegal fishing. It may also involve dipping into any savings they may have or selling capital, such as jewellery, cattle or boats. In India and Ghana, in some cases reduced returns from fishing have resulted in poor fishers giving up high-input technology and reverting to low-cost traditional technologies, such as sail power (IMM, 2003b; Ward *et al.*, 2004). This may mean that they move from selling the produce from their own small area of land or from their fishing boat to selling their labour on other people's land or boats. If children are at school, they may have to be moved into the workforce to contribute to household income (IMM, 2003a).

General development programmes

Although the poor are often unable to take up all the opportunities or respond to or cope with the threats that the changes on the coast generate, many have benefited from a diversity of general development programmes that have been aimed at poverty reduction and wider rural community development. Improved access to low-cost housing, improved local education, better water supplies, more accessible health services, group organization and an increased say in decision-making through local government reforms (such as the *Panchyati Raj* institutions in India) have had a marked effect on the lives of at least some of the poor. In other areas, the poor may be too poor to benefit from such wider development initiatives when these developments still involve costs (such as fees to access education and health services) that are unacceptable to the very poor.

Targeted responses

There have also been targeted responses to poverty on the coast. NGOs in particular have been very effective at working in villages to provide credit and training to allow the poor to invest in new opportunities or to expand existing ones. In Bangladesh, the NGO sector is very active at the community level in building support networks that provide opportunities and create safety nets for the poor; for example, the Grameen Motsho Foundation has rented land from the government to help a large number of poor people participate in aquaculture. However, the extent to which many of the NGO efforts have allowed the poor to leave poverty, rather than to be more secure within poverty, has yet to be adequately demonstrated on a large scale.

The creation of targeted delivery mechanisms alone has frequently failed to address the underlying causes and features of poverty that often make it extremely difficult for the poor to take advantage of services that are available. Obstacles for the poor include a lack of awareness of services and programmes available to help them, lack of confidence in dealing with bureaucracies and officials, and the assumption that they are 'too poor' to be eligible for assistance (IMM, 2003a). More recently, efforts by both government and NGOs have attempted to pay more attention to the capacity of the poor to take up services and programmes aimed at benefiting them, especially through self-help groups for micro-finance, for example, in coastal communities in coastal Bangladesh (D. Kumar, Bangladesh, 2003, personal communication).

Fisheries management

Fisheries management has been an important strategy for ensuring the sustainable use of resources in recent years, but the cases where this has been really successful are notable for their scarcity. Centrally planned and implemented fisheries management systems have failed for a variety of reasons, not least of which relate to the uncontrollable access arrangements of most fisheries, the cost of enforcement over many fishing units over many small landing sites, the lack of involvement of fishers in decision-making processes, conflicts between policy objectives and the gap between policies and action (World Bank, 2004).

Greater emphasis is now being placed on the greater involvement of fishing communities and fishers themselves in the management of fishing efforts through co-management systems.[1] The evolution of such measures is still in its early stages. Much work has been done on experimenting with different management arrangements for co-management, for example, for inland fisheries in Bangladesh (Department of Fisheries, 1999a,b; Middendorp *et al.*, 1999), from which lessons are being learned.

Co-management has the potential to play a significant role in future coastal development strategies, particularly if it embraces fisheries as a social policy instrument, rather than just as a contributor to national productivity. As a social policy instrument, it has the potential, through targeting, to address the socio-economic needs of communities that are highly dependent on fisheries for their survival, to address the needs of communities that are regarded as being very poor and to start to address wider environmental concerns.

Coastal Management

Attempts have been made to address the complex and conflicting demands on the coast through a range of integrated coastal area management (ICAM) approaches. These have a diversity of names such as integrated coastal management (ICM), integrated coastal zone management (ICZM) and coast conservation (CC), but they are broadly similar in approach. Under such approaches, the actual management of resource use tends to remain within relevant sector ministries such as fisheries, forestry, agriculture, etc., whereas ICAM implementation has tended to focus on the horizontal coordination function that links different ministries together and integrates policies and policy implementation.

These ICAM initiatives have generally started from a resource management perspective, working on the basis that, if resources are well managed and conflicts over resource use controlled, development is more likely to be sustainable. This management focus was not always intended to be the outcome of the ICAM process but was rather to be the strategy by which coastal development was to be enabled. The keynote address to the ASEAN/US CRM (coastal resource management) conference on managing ASEAN's coastal resources for sustainable development in 1990 mentioned that 'Government leaders ... should not view CRM as 'anti-development' or purely a 'conservationist' stance, but rather as a viable strategy for improving the quality of life of

[1] The term 'co-management' here is used to describe all fishery management arrangements where the community/fishers and the government share aspects of the management regime, including so-called community management measures.

the people whom they have vowed to serve, through increased food supply, alternative livelihoods, employment and investment opportunities and long-term solutions to multiple-use resource conflicts' (Macaraig, 1991). However, globally, the application of ICAM to address the needs of wider coastal development has proved less successful than was hoped. Olsen and Toby (1997) said, '... that despite a flowering of initiative and support for the idea of ICM, both investments and successes are puny compared to the forces worldwide causing coastal transformation. Worse yet, ICM projects, particularly in developing nations, are proceeding as isolated efforts with little or no communication between one project and another'. In an analysis of a number of ICAM projects in East Africa, Moffat *et al.* (1998) said, 'Projects focusing on biodiversity protection but neglecting local development are often only successful in the short term'.

Part of the problem is that the management strategies to achieve wider development policies on the coast have often become more dominant than the policies themselves and the development goal has sometimes been left behind. As Clark (1996) says of ICM, 'In its *management* mode, [it] assesses the environment and socio-economic impacts of specific development projects and recommends changes necessary to conserve resources and protect biodiversity'. In measuring the success of ICAM projects, Matuszeski (2001) said, 'The ultimate measures of success must be the recovery and sustained health of coastal resources themselves'. These statements reflect a shift in policy away from wider sustainable coastal development inclusive of people towards resource management as an objective in itself.

This is not to say that ICAM in its various forms is not the way forward. ICAM has an important role to play in coastal development, but we must not assume that the sound management of resources is synonymous with creating opportunities for the poor. Management, unless it is specifically targeted at helping the poor, is likely to remove opportunities from them.

Conservation of coastal resources

Much emphasis has been placed on conserving coastal resources to ensure their availability to future generations. However, in attempts to conserve the resources that the poor depend on, the good intentions of government and conservation NGOs have sometimes resulted in the poor being inadvertently excluded from the benefits of such conservation. The poor often depend on coastal resources in complex ways that are difficult to see or understand. Some rely on aquatic resources continuously, others only periodically when land-based opportunities are few (e.g. in the agricultural low season); some have alternatives that they can fall back on; others depend on aquatic resources for absolute survival; some depend on a wide range of benefit flows from the resources while others depend on only a few (Whittingham *et al.*, 2003). It is difficult to understand and accommodate these different levels and forms of dependence in conservation measures (Box 21.3).

Box 21.3. Marine protected areas

Marine protected areas (MPAs) are one of the key tools in the coastal management toolbox and they can be extremely useful for conserving coastal resources. However, unless they are applied with great care, they can also become major obstacles to the livelihoods of the poor by excluding them from key benefit flows from the protected resources, especially when dependence is critical and alternatives are few. For example, in the Lakshadweep Islands of India, mollusc collection from the coral reef is now banned. Mollusc collecting has traditionally been a major livelihood occupation of some of the poorest elderly women on the islands who have few options for other forms of income (Hoon, 2003).

Pro-poor conservation

The intrinsic links between poverty and the environment have been well known, discussed and debated for many years, and the need to respond to both simultaneously is stated in a diversity of global policy initiatives such as Agenda 21, the Convention on Biological Diversity (CBD), the International Coral Reef Initiative's Call for Action, the World Summit on Sustainable Development (WSSD), the Durban Accord and the Millennium Development Goals (MDGs). The practice, however, has slipped considerably behind the policy and very little attention, until recently, has been paid to implementing pro-poor conservation/management measures in meaningful ways. Reed (2001) wrote, '... during the 10 years since the World Conference on Environment and Development was held in Rio de Janeiro, very little attention has been given to the intimate relationship between rural poverty and the environment'.

More recently, pro-poor conservation/management has begun to emerge as a practical strategy. Part of this process has been the recognition that poverty is not just a lack of income and a move to a much more multidimensional view of poverty (UNEP-IISD, 2004). Linked to this is a growing recognition that responses to the poverty–environment nexus need to be multidimensional, and to step outside of the traditional conservation approach. It is also increasingly recognized that the relationship between poverty and the environment is not a simple downward spiral and that population increase does not necessarily lead to environmental degradation (UNDP-EC, 1999). There is also a growing recognition of the value of poor people's perceptions of their own development, resulting in a greater inclusion of participatory approaches in management measures, although these can range greatly from information extraction to being fully empowering (Campbell and Salagrama, 2001).

Alternative livelihoods

For many concerned with coastal management, alternative livelihoods are seen as the way to help and encourage people dependent on coastal resources to move away from unsustainable harvesting practices (e.g. the World Bank-funded COREMAP Project in Indonesia). The International Coral Reef Initiative, in its 'Renewed Call to Action', highlighted the importance of alternative sources of income as essential to the management of the coral reef resource (ICRI, 1999). As the poor are often the most dependent on the common-pool resources of the coast, it is often they who fall within the framework of the push for alternative livelihoods. In fact, 'alternative livelihoods' has become the new mantra for many in the coastal management world. This is often seen as the panacea for coastal problems, especially when linked to the restricted access rights associated with marine protected areas.

However, the complexity of initiating and sustaining diversified livelihoods is often underestimated. Those involved in coastal conservation and management are often from technical ministries, such as environment, fisheries, coastal engineering or forestry, and they do not always have either the remit or the skills to engage in developing these alternatives. In many cases, the suggestion alone that alternative livelihoods need to be found for those considered to be degrading the resources is as far as the response goes. In part, this is because alternative livelihoods require interventions across many overlapping and adjacent administrative jurisdictions, and as such they remain outside the remit of any particular one. In effect, they become the stepchild of government, donor and NGO efforts (Haggblade *et al.*, 2002).

Initiatives aiming at the promotion of alternatives have suffered from several common failings. Particularly where the search for alternative livelihoods has been driven by the desire to reduce the exploitation of natural resources, the needs and priorities of the poor themselves are often given less importance than the desire of the concerned agencies to protect particular resources or areas. Efforts to propose new or alternative activities have often concentrated on menus of 'new ideas' generated by outsiders rather than on encouraging the poor themselves to properly analyse their strengths, weaknesses, opportunities and threats and come to decisions regarding the options available to them based on that analysis, and on positive perceptions of where they wish to see their lives going. Many of these have been short-term interventions that have underestimated the time required to create capacity among the poor to take up new activities. Often efforts have just increased the diversity in the livelihood portfolio with none of the intended reduction in resource exploitation. In other cases, the alternative options offered are taken up more easily and more quickly by the wealthier members of the community, or taken up and abandoned, for example, in the case of agricultural opportunities for coral reef miners in Sri Lanka (Perera, 2003).

Livelihood Enhancement and Diversification – the Way Forward?

As can be seen from the discussion above, diverse problems face the poor on the coast but there is little agreement about how these problems should be addressed. Various initiatives, ranging from conservation of coastal resources to their sustainable and equitable use, are being tried, with varying degrees of success. The development of alternative income-generating opportunities (AIGOs) is now emerging as an important strategy, but success has been limited to date.

Where are we now?

From this discussion, two key points start to emerge. First, that the way we currently manage the coastal resource base is neither sustainable nor, in most cases, very equitable. Second, that harvesting from the existing natural resource base cannot provide employment and income for the ever-increasing coastal population at levels that will raise or keep those people out of poverty. The implication of this is that displacement from coastal resource dependency will increasingly have to become a key strategy in addressing sustainability, poverty and conflicts on the coast.

To put these concerns more positively, we need to manage current exploitation more effectively so that it is both sustainable and more equitable, and increase non-primary industry opportunities for people, either to reduce their dependency on the coastal resource base or to leave (or, for the young, not enter) primary industries and move into secondary and tertiary industries. In other words, *coastal people need to do things differently and do different things*. This can be represented on a spectrum, as shown in Fig. 21.1.

However, for those who do have the remit, resources and skills, the process of generating viable and sustainable alternatives is still not easy or straightforward. Understanding rural and coastal livelihoods and how and why rural people generally, and coastal people specifically, diversify their income-generating activities themselves is key to developing effective strategies to support this process. However, this is poorly understood and attempts to assist this process have tended to be based on only limited understanding of the factors and forces that are needed to ensure success. The few studies that exist suggest that the factors influencing successful livelihood enhancement and diversification are complex and that these include both endogenous and exogenous factors. Endogenous factors include the sense of identity that people have, their perceptions of risk, their ethnic origins, their culture and religion, their gender (see Knudsen and Halvorsen, 1997, for a discussion on the role of gender), their age and their education and wealth. Exogenous factors include the market for diversified

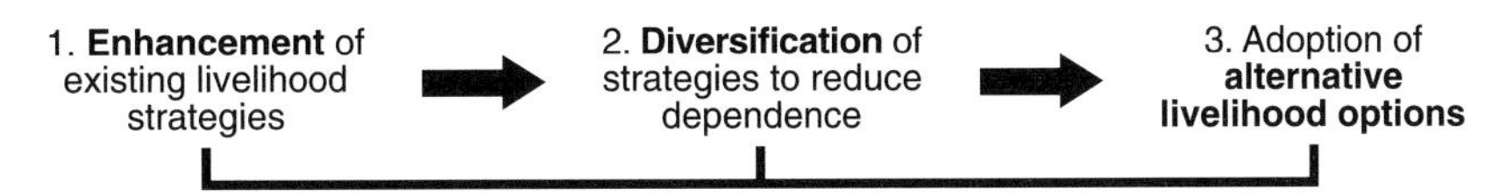

Fig. 21.1. Three broad approaches to improving coastal livelihoods.

goods and services (especially its size and seasonality), infrastructure, communications, finance availability, policies, legislation and patronage. Neither of these lists is exhaustive.

To develop sustainable livelihoods, the influence of these two groups of factors needs to be balanced; however, past development efforts have tended to focus on improving the environment for economic growth through improved exogenous factors, that is, creating opportunities for growth. As discussed earlier, all too often the poor are the least able to take up these opportunities, which are instead taken up by the more educated, wealthier and better connected members of rural communities. This can lead to a *development exclusion paradox,* in which coastal development and livelihood diversification are generally taking place, but the poor fail to benefit from these opportunities, falling in the gaps to become, as discussed above, the interstitial poor. Providing safety nets to compensate for those people left behind in the development process is one approach to addressing this. However, a more sustainable and effective route would be to better understand and address those endogenous factors that can enable the poor to compete more effectively in the marketplace for viable livelihood change.

Developing a way forward

As discussed above, the promotion of alternative livelihoods within coastal management efforts has often focused on external and short-term initiatives, which frequently fail to reach the poor. A key element affecting the likelihood of success seems to be the degree to which the complexity of the livelihoods of the poor is well understood and incorporated into the design of the intervention. Addressing livelihood diversification and enhancement must also address the changing nature of the coast, that is not so much a periodic phenomenon as *an ongoing process*. Today's viable alternative is unlikely to be sustainable unless it also incorporates the capacity to evolve with the changes around it.

The Sustainable Livelihoods Approach (SLA) offers a way of addressing coastal livelihood complexity and may provide an entry point to supporting coastal poverty and livelihood change. The SLA is a way of understanding the capacities and strengths of the poor and setting the objectives of development so that those objectives are based on what the poor already have and can do, and are responsive to the external policy, institutional and vulnerability context (see Carney, 1998, for an outline). The SLA has been in existence, and has been evolving, since the 1980s. It is used by several development agencies such as UNDP, IFAD, FAO, OXFAM, CARE and DfID (Ashley and Carney, 1999). It evolved from a wide array of participatory and other grass-roots approaches to working with the rural poor and in many ways it brings together past methods and best practice into a consolidated approach that is both comprehensive and fairly easy to understand and use. What is new about the SLA is that it embraces the complexity of rural livelihoods from the perspective of the poor. It takes the poor as the centre of the development process because they are the ones most in need of support and it provides a mechanism for enhancing empowerment of the poor (see Fig. 21.2).

Focusing on the approach rather than on the alternative income activities

Livelihood-based approaches have already begun to be adopted in systematically identifying and promoting alternative incomes. The details of these alternatives are specific to the location and groups concerned and are less relevant to this wider discussion than the broad approaches used. For example, an intervention in southern Africa, having recognized the complexity of the situation, carried out a detailed analysis of the livelihoods of the women concerned, then presented the results to the community and discussed ideas for an intervention. The strategies were then discussed with extension agents and, when interventions were agreed upon, the women themselves worked with the extension agents to prepare a feasibility study of

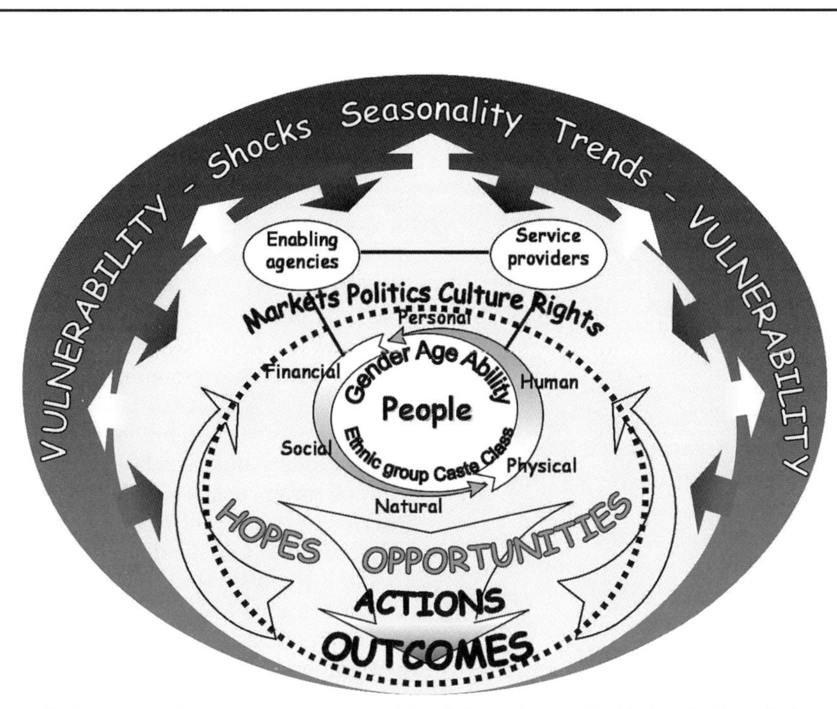

Endogenous factors are represented by all those factors *inside* the dashed circle.
Exogenous factors are represented by all those factors *outside* the dashed circle.

Fig. 21.2. A framework for understanding and responding to livelihoods of the poor. (From Julian Hamilton-Peach and Philip Townsley at www.itad.org/s/a/framework/index.htm)

the proposal (Due, 1991). Ellis (2000) uses the Sustainable Livelihoods Approach as a framework for understanding livelihoods, livelihood diversification and the factors affecting it; Allison and Ellis (2001) apply SLA to understanding the diverse livelihood strategies of fishers confronted with changing fishery resources. Dasgupta *et al.* (2004) have also used the SLA to systematically analyse the diversification of rural livelihoods in India. The research undertaken by IMM and referred to in this chapter has been implemented largely through livelihood approaches, providing new insights into the livelihoods of the coastal poor and the relationships between the poor, the surrounding coastal resources and the policy context in which they operate.

The application of livelihood approaches to coastal poverty and livelihood enhancement and diversification is based on several key principles:

- They should build on the strengths of the poor.
- They should relate to the needs and aspirations of the poor.
- They should be developed in participatory ways.
- They should make use of multidisciplinary approaches.
- They must be viable and sustainable (from economic, institutional, social and cultural perspectives).
- They should be appropriate for the number of people concerned.
- They should have acceptable (to the poor) levels of risk.
- They should not increase vulnerability.
- They should be in harmony with existing household livelihood strategies.
- They should complement the strategies of other people in the community.
- They should conform with national policies and legislation.
- They should enhance the independence of the poor.
- They should ensure the rights of the poor.

- They should ideally enhance the innovative capacity, vision and adaptability of the poor.

Approaches should build on the strengths of the poor and be implemented in participatory ways to accommodate the aspirations and needs of different stakeholder groups. The focus should be on a range of strategies, which include: (i) enhancing existing uses of aquatic resources to make them more sustainable and more equitable; (ii) diversifying people's livelihoods to reduce their dependence on natural resources; and (iii) developing alternatives, especially for the next generation about to join the coastal work force. Such livelihood change also needs to be accompanied by greater emphasis on building the capacity of people to respond to future change so that they will continue to innovate and maintain the viability of their livelihoods.

Such efforts will require multidisciplinary approaches that require networks that bridge the gaps between administrations and NGOs and allow them to work together to bring their different skills, knowledge and experience to the issues. Above all, coastal livelihood change needs to be mainstreamed over a long period of time rather than being a short-term activity or an afterthought to conservation or management initiatives. And, livelihood change for the poor needs to be at the centre of the mainstreaming process if the poor are to stand a chance of benefiting.

Conclusions

The situation on the coast is complex and that complexity has created opportunities for the poor. However, those opportunities are decreasing as coastal populations rise, access to global markets increases, technology becomes more advanced and greater investment is made in the coast. Efforts to conserve coastal resources or to improve the effectiveness of management regimes have often led to further marginalization of the poor and increasing conflicts. Even where AIGOs have been generated, the poor have not always benefited from them.

There is clearly a need to improve the way resource use is managed to overcome problems of both sustainability and equity. Co-management systems, better integration across sectors and agencies on the coast to achieve wider coastal development and propoor conservation all have roles to play in enhancing the way people interact with the coast. However, in the medium to long term, more people will have to be encouraged to move out of livelihoods that depend on primary production on the coast and be encouraged and supported to engage in secondary and tertiary industries. Efforts to address this issue have already started but success has been limited. The 'menu approach' to livelihood change (where specific livelihood options that are supposed to work in all situations are selected from a global list of coastal alternatives) has not proved very successful and addressing single issues as obstacles to diversification is an oversimplification.

There are no simple solutions – we need to incorporate all the different factors that affect the ability of people to identify, take up and sustain livelihood changes. These factors need to be understood and responded to systematically in ways that recognize and respond to the complexities of poor people's lives rather than using predetermined strategies. Sustainable Livelihoods Approaches are one way of doing this.

References

Allison, E.H. and Ellis, F. (2001) The livelihoods approach and management of small-scale fisheries. *Marine Policy* 25, 377–388.

Ashley, C. and Carney, D. (1999) *Sustainable Livelihoods: Lessons from Early Experience.* DFID, London.

Barrett, C.B., Rearden, T. and Web, P. (2001) Nonfarm income diversification and household livelihood strategies in rural Africa: concepts, dynamics, and policy implications. *Food Policy* 26, 315–331.

Burke, L., Kura, Y., Kassem, K., Revenga, C., Spalding, M.D. and McAllister, D. (2001) *Pilot Analysis of Global Ecosystems: Coastal Ecosystems.* World Resources Institute, Washington, DC.

Burke, L., Selig, E. and Spalding, M.D. (2002) *Reefs at Risk in South-east Asia*. World Resources Institute, Washington, DC, pp. 1–72.

Campbell, J. and Salagrama, V. (2001) New approaches to participation in fisheries research. *FAO Fisheries Circular* No. 965, 56.

Carney, D. (1998) Sustainable rural livelihoods: what contribution can we make? Papers presented at the Department for International Development's Natural Resources Adviser's Conference, 1998. DFID, London.

CFDO-IMM (2005) *The Cambodia Post-harvest Fisheries Overview*. Department of Fisheries, Phnom Penh, Cambodia.

Clark, J.R. (1996) *Coastal Zone Management Handbook*. CRC Press, Inc., Boca Raton, Florida.

Dasgupta, N., Klei, U., Marter, A. and Wandschneider, T. (2004) India: policy initiatives for strengthening rural economic development: case studies from Madhya Pradesh and Orissa, India. *NRI report* No. 2770, Natural Resources Institute, Kent, UK.

Department of Fisheries, Bangladesh (1999a) Papers presented at the National Workshop on Community Based Fisheries Management and Future Strategies for Inland Fisheries in Bangladesh. DOF, Dhaka, Bangladesh.

Department of Fisheries, Bangladesh (1999b) National Workshop on Community Based Fisheries Management and Future Strategies for Inland Fisheries in Bangladesh: Recommendations for Future Strategies for Inland Fisheries and Workshop Summary. DOF, Dhaka, Bangladesh.

Due, J.M. (1991) Experience with income generating activities for Southern African women. *Savings and Development*, Vol. 1, XV.

Ellis, F. (2000) *Rural Livelihoods and Diversity in Developing Countries*. Oxford University Press, UK.

FAO (2002) *The State of World Fisheries and Aquaculture*. FAO Information Division, FAO, Rome.

Haggblade, S., Hazell, P. and Reardon, T. (2002) Strategies for stimulating poverty-alleviating growth in the rural nonfarm economy in developing countries. International Food Policy Research Institute and the World Bank, Washington, DC.

Hoon, V. (2003) A case study from Lakshadweep. In: Whittingham, E., Campbell, J. and Townsley, P. (eds) *Poverty and Reefs: Volume 2 Case Studies*. For DFID, IMM and IOC-UNESCO by UNESCO, Paris, pp. 187–226.

ICRI (1999) *Renewed Call to Action: International Coral Reef Initiative 1998*. Great Barrier Reef Marine Park Authority, Townsville, Australia.

IMM (2001) *Learning Lessons from the Cyclone: A Study of DFID's Support for Post-Cyclone Livelihoods Rehabilitation in Orissa, India*. IMM Ltd., Exeter, UK.

IMM (2003a) *Sustainable Coastal Livelihoods: Policy and Coastal Poverty in the Western Bay of Bengal: Main Report*. An output from the Sustainable Coastal Livelihoods project funded by DFID's Policy Research Programme. IMM Ltd., Exeter, UK.

IMM (2003b) *Major Trends in the Utilisation of Fish in India: Poverty-Policy Considerations*. Output of Project R7799: Changing Fish Utilisation and Its Impact on Poverty in India, funded by DFID's Post-harvest Fisheries Research Programme. IMM Ltd., Exeter, UK.

IMM-ICM (2003) *Changing Fish Utilisation and Its Impact on Poverty in Andhra Pradesh: A Scoping Study*. Output of Project R7799: Changing Fish Utilisation and Its Impact on Poverty in India, funded by DFID's Post-harvest Fisheries Research Programme. IMM Ltd., Exeter, UK.

IPCC (2001) *Climate Change 2001: Impacts, Adaptation and Vulnerability – Summary for Policymakers*. A report of Working Group II of the Intergovernmental Panel on Climate Change (17).

Jazairy, I., Alamgir, M. and Panuccio, T. (1992) *The State of Rural Poverty: an Inquiry into Its Causes and Consequences*. Published for the International Fund for Agricultural Development, IT Publications, London.

Knudsen, A.J. and Halvorsen, K. (1997) Income-generating programmes in Pakistan and Malawi: a comparative review. *Journal of Refugee Studies* 10(4), 462–475.

Kurien, J. (1992) Ruining the commons and responses of the commoners: coastal overfishing and fishworkers' actions in Kerala State, India. In: Gai, D. and Vivian, J. (eds) *Grassroots Environmental Action: People's Participation in Sustainable Development*. Routledge, London.

Luttrell, C. (2001) Institutional change and natural resource use in coastal Vietnam. *GeoJournal* 54, 529–540.

Macaraig, C. (1991) Sustainable development of coastal resources: a challenge and a choice. In: Thia-Eng, C. and Scura, L.F. (eds) *Managing ASEAN's Coastal Resources for Sustainable Development: Roles of Policymakers, Scientists, Donors, Media and Communities*. Association of South-east Asian

Nations/United States Coastal Resources Management Project Conference Proceedings 6, Manila, Philippines.

Matuszeski, W. (2001) Coastal management: in search of success. *Intercoast Network, International Newsletter of Coastal Management* No. 40, 2001. Coastal Resources Center, University of Rhode Island, Kingston, Rhode Island.

Middendorp, H.A.J., Thompson, P.M. and Pomeroy, R.S. (1999) *Sustainable Inland Fisheries Management in Bangladesh.* ICLARM, Danida and the Ford Foundation. ICLARM, Manila, Philippines.

Moffat, D.M., Ngoile, M.N., Linden, O. and Francis, J. (1998) The reality of the stomach: coastal management at the local level in Eastern Africa. *Ambio* 27, 590–598.

NIS (2001) *Cambodia Child Labour Survey 2001.* National Institute of Statistics, Ministry of Planning, Phnom Penh, Cambodia.

Olsen, S. and Toby, J. (1997) *Avoiding an ICM Nightmare.* Intercoast Network No. 29, Fall 1997, Rhode Island.

Perera, N. (2003) *Alternative Livelihoods Through Income Diversification: An Option for Sustainable Coral Reef and Associated Ecosystem Management in Sri Lanka.* South Asia Cooperative Environment Programme in collaboration with Coral Reef Degradation in the Indian Ocean (CORDIO) Programme, Colombo, Sri Lanka.

Pillay, T. (1992) *Aquaculture and the Environment.* Fishing News Books, Oxford, UK.

Reed, D. (2001) Poverty is not a number, the environment is not a butterfly. *WWF 'Viewpoints on Poverty and the Environment'.* WWF, Washington, DC.

Rengasamy, S., Devavaram, J., Prasad, R. and Arunodaya, E. (2003) A case study from the Gulf of Mannar. In: Whittingham, E., Campbell, J. and Townsley, P. (eds) *Poverty and Reefs: Volume 2 Case Studies.* For DFID, IMM and IOC-UNESCO by UNESCO, Paris, pp. 113–146.

Sarthi Acharya (2003) Off-farm and non-farm jobs in SEATEs and Thailand: rationale and synthesis of country studies. In: Development Analysis Network. *Off-Farm and Non-Farm Employment in South-east Asian Transitional Economies and Thailand.* Cambodia Development Resource Institute, Phnom Penh, Cambodia.

Sarthi Acharya, Kim Sedara, Chap Sotharith and Meach Yady (2003) Off-farm and non-farm activities and employment in Cambodia. In: Development Analysis Network. *Off-Farm and Non-Farm Employment in South-east Asian Transitional Economies and Thailand.* Cambodia Development Resource Institute, Phnom Penh, Cambodia.

UNDP-EC (1999) *Attacking Poverty While Improving the Environment: Practical Recommendations.* UNDP, New York.

UNEP-IISD (2004) *Exploring the Links: Human Well-Being, Poverty and Ecosystem Services.* IISD, Winnipeg, Canada.

Ward, A.R., Bortey, A., Whittingham, E., Braimah, L.I., Ashong, K. and Wadzah, N. (2004) *Poverty and post-harvest fishery livelihoods in Ghana.* Output from the Post-Harvest Fisheries Research Programme Project R8111. IMM Ltd., Exeter, UK.

Whittingham, E., Campbell, J. and Townsley, P. (2003) *Poverty and Reefs.* Vols 1 and 2. DFID-IMM-IOC/UNESCO, Paris.

World Bank (2004) *Saving Fish and Fishers: towards Sustainable and Equitable Governance of the Global Fishing Sector.* Agriculture and Rural Development Department, Washington, DC.

Yim, Chea and McKenney, B. (2003a) *Domestic fish trade: a case study of fish marketing from the Great Lake to Phnom Penh.* Cambodia Development Resource Institute Working Paper No. 29. Phnom Penh, Cambodia.

Yim, Chea and McKenney, B. (2003b) *Fish exports from the great lake to Thailand: an analysis of trade constraints, governance, and the climate for growth.* Cambodia Development Resource Institute Working Paper No. 27. Phnom Penh, Cambodia.

22 Achieving Food and Environmental Security: Better River Basin Management for Healthy Coastal Zones

S. Atapattu and D. Molden
International Water Management Institute, Colombo, Sri Lanka, e-mail: s.atapattu@cgiar.org

Abstract

The Millennium Development Goals call for the eradication of extreme poverty and hunger while ensuring environmental sustainability. Water is essential for food production and sanitation – key factors in poverty reduction. At the same time, it is increasingly difficult to ensure clean water to maintain ecosystem services. The challenge is to find ways of managing water in order to meet both these goals in a balanced way. This water–food–environment dilemma is exceedingly clear in the coastal zone because of increasing urban demands for water, with agriculture remaining an important component of rural livelihoods, and the dependence on water by some of the most valuable ecosystems of the world that are essential for the well-being of society.

Driving forces affecting this water–food–environment nexus are increases in population and income and changing dietary patterns, which highly influence food production and water-use patterns. In response, there have been many positive trends such as overall better nutrition, an increase in overall food production and water productivity, and increased global trade in food products, thus helping in food distribution. On the other hand, there are some very disturbing trends. South Asia and sub-Saharan Africa still fall below the level required to 'lead a healthy and reproductive life' and water could be much better used for providing food security. An increased number of water bodies are becoming depleted, polluted and non-productive, negatively affecting both aquatic and terrestrial ecosystems.

Coastal zones are particularly vulnerable as they are susceptible to rapid changes both from within the area and from outside – shock events like tsunamis plus consistent pressure from upstream water development and use. Coastal zones bear the brunt of upstream water development, reduced flows and pollution loads, yet river basin water management has been concentrated in upstream areas and very often the coastal zone is neglected or considered separately. Thus, in order to achieve and maintain healthy coastal zones, agricultural water management has to be able to respond to rapid changes from within coastal zones to maintain a balance of food, livelihood and environmental interests, and coastal zone interests must be firmly represented in overall river basin management.

Introduction

Achieving the Millennium Development Goals (MDGs) on poverty, hunger[1] and environment[2] simultaneously is a challenge, as the goals seem to be in direct conflict with each other, particularly when reflecting on the role of water management. This is because a reduction in poverty and hunger requires water for agriculture while environmental sustainability requires sufficient water for ecosystems to prosper. The use of water for agriculture has imposed stress upon many important ecosystems – especially rivers and coastal systems. There is therefore a need for more water to feed people, more water to reduce poverty and more water to sustain natural ecosystems. And because of increasing water demand for cities and industries, there will be less water to go around, especially with the predicted 2 billion more people to feed in the next 25 years. The challenge is to find ways to manage water to meet the MDG targets on poverty, hunger and the environment.

This dilemma is exceedingly clear in coastal zones. These areas bear the brunt of upstream river basin development, receiving the leftover drainage flows and pollution loads. Almost 40% of the world's population lives within 100 km of a coastline. These coastal inhabitants live on only 22% of the world's available land mass (WRI, 2001) and population pressure is growing. Coastal zones are areas of intense urban and agricultural water needs, but are also home to valuable aquatic ecosystems, essential to human well-being. Around the globe a total of 25 biodiversity hot spots have been identified, of which 23 are at least partly in the global coastal zone (Shi and Singh, 2003).

Coastal zones are highly susceptible to river basin water stress because changed hydrologic flow regimes and increased pollution loads are felt at the tail end of river systems. A global study of environmental water stress indicates that the problem is severe and growing, posing increasing threats to coastal areas (Smakhtin *et al.*, 2004). Therefore, the health of the coastal zone is a good indicator of river basin health. In spite of calls for integrated water resources management (IWRM), the coastal zone is often looked at in isolation, and its importance underestimated compared to upstream activities in the river basin perspective.

Water for Food: Key Trends and Driving Forces

A starting point in the water–food–livelihood–environment equation is to recognize that food consumers are key drivers of agricultural practices. To produce 1 kg of grain, plants must transform between 400 and 5000 l of water – based on the crop type, climate and management practices – into water vapour through evapotranspiration. Depending on diet and where food is grown, each person is responsible for the conversion of 2000 to 5000 l of water each day into water vapour (Renault and Wallender, 2000). The 2–5 l/day for drinking and 50–200 l/day for household use seem insignificant when compared to the amount of water required for food production.

So, how much water is needed for agriculture to feed the world? Agricultural withdrawals from rivers, wetlands and groundwater are in the order of 2500 km^3/year – in many developing countries, this is more than 90% of all water withdrawn for human uses. From another perspective, of the approximately 100,000 km^3/year

[1] The MDG targets on poverty and hunger aim to halve, between 1990 and 2015, the proportion of people whose income is < US$1/day (Target 1); and halve, between 1990 and 2015, the proportion of people who suffer from hunger (Target 2) (UN, 2002).

[2] The MDG targets on environmental sustainability aim to integrate the principles of sustainable development into country policies and programmes and reverse the losses of environmental resources (Target 9); halve by 2015 the proportion of people without sustainable access to safe drinking water and basic sanitation (Target 10); and have achieved by 2020 a significant improvement in the lives of at least 100 million slum dwellers (Target 11) (UN, 2002).

reaching Earth's surface as rainfall, only 40% or 40,000 km^3 represents the renewable water resource in the form of river run-off and groundwater storage. Of this amount, some 10% or 3800 km^3 is diverted from its natural courses, of which 2500 km^3 is withdrawn for irrigation (based on Shiklomanov, 2000). Presently, approximately 7000 km^3 of water withdrawn from the various water sources is converted to evapotranspiration to produce food (Rockström *et al.*, 1999, 2003).

Provision of water is also central to the livelihood security of many rural poor. The majority of people in developing countries still depend on agriculture for their livelihoods. Approximately 70% of the world's poor live in rural areas with limited livelihood opportunities outside of agriculture (World Bank, 2004). However, the role of fishing and related activities, such as small-scale fish processing and trading, can represent a crucial element in their struggle against poverty and food insecurity, especially for women. In West Africa and in a large part of South-east Asia, for instance, small-scale fish processing and retailing activities are dominated by women. Fish are an important source of animal protein and so can contribute to reducing malnutrition, reaching towards the MDGs. Fishing can greatly improve the food security of millions of both rural and urban poor, with limited strain on water and at the same time provide livelihood security (C. Béné, Egypt 2004, personal communication).

Providing the rural poor with a reliable source of water, whether it is from small-scale water harvesting, large-scale irrigation or access to water bodies for fishing, makes it possible for them to move beyond subsistence farming. A reliable source of water prevents yield losses from short-term drought, which in sub-Saharan Africa may claim one out of every five harvests. It gives farmers the 'water security' they need to risk investing in other productivity-enhancing inputs, such as fertilizers and high-yielding and high-value crops such as vegetables.

It is clear that there is a need to change the way water is developed and managed for agriculture. Breaking old habits, recognizing where things are going well and introducing new ways of doing business are the means to achieving this change. This includes considering key shifts in activities such as fishing, which has relatively low water consumption requirements. Given the high environmental and political uncertainty that characterizes some of these regions facing water scarcity, the capacity of fishing activities to generate instantaneous gains represents an enormous advantage over farming (Béné *et al.*, 2003).

Key water–food–nutrition–livelihood–environment trends (SIWI-IWMI, 2004) are laid out in Box 22.1. These indicate that we have to reverse negative ecological and livelihood trends, mitigate water problems arising from new trends such as urbanization and enhance positive promising trends.

River Basin Drivers of Coastal Zone Change

In response to increased demand, people clear land and withdraw water from rivers and groundwater to grow more food. Water withdrawals and subsequent changes in storage, flows and evapotranspiration influence river hydrology, and consequently affect related ecosystems and the livelihoods they serve. Here we outline the main ways in which agriculture alters coastal zone ecosystems and related biodiversity. Of over 17,000 major sites globally already devoted to the protection of biodiversity, many of which are types of wetland, 45% have at least 30% of their land used for agriculture, and most of the remainder adjoin or are encompassed by agricultural lands. Around the globe, a total of 25 biodiversity hot spots have been identified, of which 23 are at least partly in the global coastal zone (Shi and Singh, 2003).

Notably, agriculture affects more than 50% of the world's more than 1267 wetlands of international importance (Wiseman *et al.*, 2003). Drainage and conversion for agricultural use are the principal and most direct causes of wetland loss globally. Globally, irrigated area continues to expand, particularly in the developing world (FAOSTAT 2000 Database), affecting river systems. By 1985, it

Box 22.1. Key trends in the water–food–livelihood–environment challenge (adapted from SIWI–IWMI, 2004)

Promising trends

- A steady increase in the per capita consumption of food leading to better nutrition for many. Average global calorie intake 1961–2000 increased from 2250 to 2800 kcal.
- A steady increase in land and water productivity – average yields increasing from 1.4 to 2.8 t/ha, and equivalent gains in water productivity from the 1960s to 2000.
- Increases in global trade in food products and consequent virtual water flows offer prospects for better national-level food security and relief of water stress. It has been estimated that global trade 'saved 11% (112 km^3) in 1995' and 19% less water use is forecast for 2025 (Fraiture *et al.*, 2004).
- Aquaculture practised in the coastal zone has been the mode for meeting the increasing demand on fisheries and is also the fastest-growing food sector in the world, with 37.9 million metric tons in 2001 (Kura *et al.*, 2004).

Very disturbing trends

- Average calorie intake in South Asia (2450 kcal) and sub-Saharan Africa (2230 kcal) remains far below norms.
- An increasing number of rivers are reduced to polluted drains because of increased agricultural production and water consumption, and inability to deal with increased pollution.
- An estimated 19% of coastal areas, including mangroves, have been converted to agriculture (including aquaculture) and urbanization, leading to loss and hindrance of natural wetlands, associated biodiversity and ecosystem functions.
- An increase in agriculture upstream and downstream has led to many pressures on the coastal belt, with outbreaks of disease and excessive loads of pollution having spin-off effects on other food chains and on general environmental conditions.
- Groundwater levels are declining rapidly because of overexploitation in densely populated areas of North China, India and Mexico.
- Increasing land and water degradation is resulting from nutrient depletion, soil degradation, salinization and seawater intrusion (de Vries *et al.*, 2003).

Double-edged trends

- Increasing withdrawals for irrigation in developing countries are positive for economic growth and poverty alleviation, but negative for the environment.
- Increasing water demand of cities and industries offers possibilities of income and employment. Developing-country cities are projected to use 150% more water in 2025 than today (IWMI, 2000). This increase often results in a shift of water out of agriculture, putting extra strain on rural communities, and leads to more polluted wastewater.

was estimated that 56–65% of available wetland had been drained for agriculture in Europe and North America, 50% in Australia, 27% in Asia, 6% in South America and 2% in Africa. Though the coastal zone in its own element has a diverse range of functions, an estimated 19% of coastal areas have already been converted for the purposes of agriculture and urbanization (WRI, 2001). Much of this has been at the expense of important coastal ecosystems that are undervalued and underappreciated. Coral reefs and mangroves were proved to be extremely valuable in lessening human loss during the recent Asian tsunami.

Closing River Basins

Especially within the last 50 years, river basins have been significantly altered to serve the needs of growing populations. As increasing amounts of water are withdrawn and depleted for agricultural and urban purposes, river basins change from an open state – where there is still available water to allocate to new uses – to a closed state, where all available water has been developed and allocated across uses (Seckler, 1996; Molden 1997). With increasing hydraulic infrastructure and conversion of land, the discharge at the terminal end of rivers decreases. Changes upstream in a basin ultimately affect the coastal zone – land-use changes, increased withdrawals and depletion of water, hydraulic structures to store and release water change flow patterns and change sediment loads, and increased pollution.

A consequence of closed basins is that a change in use in one part of a basin affects water availability in another part, often in unseen and counter-intuitive ways. The impact of an additional dam for irrigated agriculture can reduce discharge to the coastal zone. The impact of one small water harvesting system may not be felt, but the impact of many may be considerable and reduce the amount of water available to other users in rivers, much like the impact of large dams (Batchelor *et al.*, 2003).

Changing land use is directly related to change in river basin hydrology (Calder, 1998). A net change in evapotranspiration from land cover change affects the volume and temporal pattern of river flows. A change from arid to irrigated land increases evapotranspiration and subsequently decreases river flow. A change from forest to rainfed land in the humid tropics may have the reverse effect of increasing the volume of discharge in a river. Dams and reservoirs alter patterns of flow, often dampening patterns of high and low flows. With extensive development of land and water river flow patterns in terms of volume, the temporal pattern of discharge and pollution loads are very different from natural flow patterns.

To determine whether a basin is closed requires knowledge of flow requirements or allocations for environmental purposes. Some important functions of rivers are to flush out sediments, salts or pollution. Coastal estuaries and their fisheries depend on fresh water and river sediments. In many river deltas, dry-season flows are diminished, leading to increased salt water intrusion. Coastal aquatic ecosystems and fisheries, dependent on sustaining these ecosystems, are susceptible to changing patterns of salinity and flows. The boundaries of river basin management need to be extended to where they influence coastal systems and ecosystems. Basins are considered to be open if water remains in excess of these allocations or requirements. They become closed when only the environmental allocation remains. Some basins have already overcommitted their water to various uses, and are not able to meet environmental consequences. The Yellow and Colorado rivers (USA) and Amu and Syr Darya rivers (Central Asia) are examples of overcommitted rivers.

Scale Issues

One of the biggest difficulties in understanding water resources is the prevalence of cross-scale effects that often happen in a counter-intuitive manner. For example, drip and sprinkle irrigation are widely regarded as water conservation practices because less application to crops is needed, and drainage flows from fields are reduced. This sounds practicable and acceptable, and indeed for individual farmers it may be extremely beneficial, especially if yields and profits are boosted. But the story is not complete until we understand what happened to the 'saved' water. One common occurrence is that farmers use this 'extra' water to irrigate more area, and thus overall evaporation and transpiration increase and downstream flows decrease. Over time, downstream farmers and ecosystems may become dependent on return flows from agriculture. A reduction in drainage flows, without compensating flows, may cause serious downstream impacts as discussed above. The lesson is that actions taken at one scale, such as farm, field or irrigation system, typically have broader basin-wide impacts that must be clearly understood before actions are taken. The Aral Sea basin (Central Asia) is an excellent example of how upstream development caused downstream damage.

Similarly, time is an important but complex dimension. Agricultural water management builds on slow-response variables – changes in groundwater levels, salinity and pollution build-up, all of which influence long-term sustainability of these systems. The sustainability of major food production systems – the Aral Sea basin, the North China plains, northwest India – is in question. Natural ecosystems similarly show lagged responses to agriculture-driven change over many years, such as long-term changes in the geomorphological character of rivers with flow regulation, or progressive declines in inland fisheries with combined pollution, overharvesting and water abstraction effects, leading to their eventual, sometimes sudden, collapse.

Coastal Impacts of River Basin Change

When, how much, how often and in what distributional pattern water flows in rivers or moves through other wetlands affect ecosystem character and aquatic biodiversity (Richter *et al.*, 1997). For instance, major hydraulic structures to store and divert water completely change the character of parts of rivers, in some cases changing a river system to a reservoir cascade system. In other cases, rivers dry out immediately downstream of dams, or in their lower reaches, simply because of additional withdrawals and depletion of water for agriculture.

River depletion leads to several concerns. The deprivation of rich silt that fertilizes flood plains and river deltas (UNEP-WCMC, 2000), widely used for agriculture in countries such as Bangladesh and Vietnam, is one of the results of river depletion. Reduced flows as a result of damming or excessive withdrawals can lead to reduced transport of essential minerals and nutrients, which can create a drop in coastal fisheries. For example, the Aswan High Dam on the Nile caused a very significant reduction in the phosphate (by 96%) and silicate (by 82%) levels that reached the coast. This reduction combined with other factors such as increased salinity, including saline intrusion in monsoon deltas during the dry season. Reduced flows have caused significant reductions in coastal fisheries (FAO, 1995). Because of the trapping of sediments and nutrients upstream, the dense blooms of phytoplankton on the Nile Delta have been eliminated. Since this was a food source for many detritus feeders such as sardines, other pelagic fish and shrimp, their populations have been affected. The average fish catch in 1962 and 1963 was 35,000 tons, which decreased by 75% in 1969 as a result of the dam (UNEP-WCMC, 2000), although others have argued that the changes have enabled other fisheries to develop.

Both excessive sediment run-off and reduced sediment movement, often caused by upstream river basin activities, result in serious degradation of the environment. This can lead to elimination of beaches and backwaters and bring about overall degradation of coastal ecosystems. For example, the Nile Delta coastline has seen erosion rates up to 5–8 m/year (Megeed and Makky, and Stanley and Warne, 1993, cited in UNEP-WCMC, 2000). Degradation of this nature has been reported from other locations too, such as the coastlines of Togo and Benin (downstream of the Volta River) and in the regions of Camargue and Languedoc in the Mediterranean (UNEP-WCMC, 2000).

Aquaculture and fisheries

Of major concern has been the unplanned increase in aquaculture in the coastal belt, which has led to the conversion of vast areas of mangroves, salt pans and rice fields into aquaculture ponds. While affecting directly through the loss of natural habitat, this also affects the productivity of coastal fisheries as most of these ecosystems support nursery stages of fish and economically important invertebrates. Burgeoning aquaculture has brought about major problems of water pollution and disease outbreaks, which could eventually also lead to a loss of productivity and livelihoods.

A wide array of other agriculture-related factors, including overharvesting of food resources, such as those provided by inland live capture fisheries, the impacts of various forms of aquaculture and the enhanced spread of invasive species sometimes associated with agricultural systems, also play roles in altering natural ecosystems.

Shocks and natural hazards

The recent tsunami that hit the coasts of South Asia and parts of eastern Africa brought home another message on the importance of the coastal zone and its role in minimizing the risk posed by natural hazards. Preliminary investigations have already suggested that, where natural coastal ecosystems such as mangroves and sand dunes remained intact, the impacts inland have been less. This further highlights the importance of sustaining coastal ecosystems by ensuring that they get their fair share of the available water resources.

Responses – What Can Be Done?

Several courses of action in the context of river basins could help sustain coastal livelihoods and environments and meet the MDGs. These actions take place at several scales and at different locations, and are the subject of this book.

Integrating coastal zones more effectively into river basin management is surely a key to productive and sustainable coastal zone practices. This requires that coastal zone issues related to river basin management be identified, and brought into river basin management discussions. Coastal zone stakeholders – fishermen, farmers, urban dwellers and others – need to be brought into the discussion.

Within a river basin, identification and allocation of environmental flows – flows required to maintain the character of life-supporting ecosystems – are necessary. The science of environmental flows is rapidly evolving (Arthington *et al.*, 1998; Tharme, 2003; Smakhtin *et al.*, 2004). Implementation requires that these concepts be incorporated into management and allocation procedures, and there is a long way to go in this area for many basins.

Given that coastal zone ecosystems, including agro-ecosystems, will require more water, and that cities will require an increasing share, agriculture has to produce more food with less water. Increasing water productivity (Kijne *et al.*, 2003) combined with allocation procedures are important agricultural actions within a river basin.

In summary, water management in coastal areas has to balance the complex needs and objectives of a set of multiple users, while at the same time responding to pressures brought about by external changes.

Conclusions

Pressure from a growing population for more food also translates into more water and additional stress on production and ecological systems. This increasing pressure will make it more difficult for the poor, who are most vulnerable to water scarcity and environmental degradation. Continuing on our present path will mean more conflict, more environmental degradation and the persistence of poverty and food insecurity.

The deterioration of coastal habitats leads to ecosystem and livelihood loss through direct habitat loss caused by agricultural practices upstream and within coastal areas, including the rearing of crops in river deltas and conversion of inland coastal habitats for the purposes of aquaculture in an unmanaged way. All of these losses will also have adverse effects on the livelihoods of the dependent communities as well as other outside communities that reap benefits from these ecosystems.

Some solutions and responses will have to take place simultaneously and at different scales:

- changes in global policies (subsidies, trade and prices) that influence water management and use,
- changes in consumption patterns – what we eat and how we use water, and
- improved river basin management to take into consideration coastal zone needs, such as: (i) better consideration at the river basin scale of the impact of changes in water management on coastal areas; (ii) environmental flows to determine the existing levels and to determine the maximum levels of abstraction possible while ensuring that the minimum required flows will reach the coastal zone; (iii) an

increase in water productivity so that less water is used to produce more food and to reduce the wastage of water in agriculture; and (iv) managing water for multiple uses is a key issue that cannot be avoided in water management at the present time.

Difficult choices will have to be made to resolve difficult livelihood and environmental concerns. A better understanding of trade-offs and consequences of water use will better reduce controversy and uncertainty, and lead to better-informed decisions.

References

Arthington, A.H., Brizga, S.O. and Kennard, M.J. (1998) *Comparative Evaluation of Environmental Flows Assessment Techniques: Best Practice Framework.* Occasional Paper No. 25/98. Land and Water Resources Research and Development Cooperation, Canberra, Australia.

Batchelor, C.H., Rama Mohan Rao, M.S. and Manohar, S. (2003) Watershed development: a solution to water shortages in semi-arid India or part of the problem? *Land Use and Water Resources Research* 3, 1–10.

Béné, C., Neiland, A., Jolley, T., Ladu, B., Ovie, S., Sule, O., Baba, M., Belal, E., Mindjimba, K., Tiotsop, F., Dara, L., Zakara, A. and Quensiere, J. (2003) Natural resource institutions and property rights in inland African fisheries: the case study of the Lake Chad Basin Region. *International Journal of Social Economics* 30(3), 275–301.

Calder, I.R. (1998) *Water-resource and land-use issues.* SWIM Paper 2. International Water Management Institute, Colombo, Sri Lanka.

de Vries, F.W.T.P., Acquay, H., Molden, D., Scher, S.J., Valentin, C. and Cofie, O. (2003) *Integrated Land and Water Management for Food and Environmental Security.* Comprehensive Assessment of Water Management in Agriculture Research Report No. 1, International Water Management Institute, Colombo, Sri Lanka, 62 pp.

FAO (1995) *Effects of riverine inputs of coastal ecosystems fisheries resources.* FAO Fisheries Technical Paper, No. 349. Rome.

Fraiture, C. De, Cai, X., Amarasinghe, U., Rosegrant, M. and Molden, D. (2004) *Does Cereal Trade Save Water? The Impact of Virtual Water Trade on Global Water Use.* Comprehensive Assessment Research Report No. 4, International Water Management Institute, Colombo, Sri Lanka, 32 pp.

IWMI (2000) World water supply and demand in 2025, document prepared for the World Water Vision. In: Rijsberman, F.R. (ed.) *World Water Scenarios Analyses.* Earthscan Publications, London.

Kijne, J., Barker, R. and Molden, D. (eds) (2003) *Water Productivity in Agriculture: Limits and Opportunities for Improvement.* CAB International, Wallingford, UK.

Kura, Y., Revenga, C., Hoshino, E. and Mock, G. (2004) *Fishing for Answers: Making Sense of the Global Fish Crisis.* World Resources Institute, Washington DC, 70 pp.

Molden, (1997) *Accounting for Water Use and Productivity.* SWIM Report no. 1, International Irrigation Management Institute, Colombo, Sri Lanka.

Renault, D. and Wallender, W. (2000) Nutritional water productivity and diets. *Agricultural Water Management* 45(3), 275–296.

Richter, B.D., Baumgartner, J.V., Winington, R. and Braun, D.P. (1997) How much water does a river need? *Freshwater Biology* 37, 231–249.

Rockström, J., Gordon, L., Falkenmark, M., Folke, C. and Engvall, M. (1999) Linkages among water vapor flows, food production and terrestrial services. *Conservation Ecology* 3(2), 1–28.

Rockström, J., Barron, J. and Fox, P. (2003) Water productivity in rain-fed agriculture: challenges and opportunities for smallholder farmers in drought-prone tropical agroecosystems. In: Kijne, J.W., Barker, R. and Molden, D. (eds) *Water Productivity in Agriculture: Limits and Opportunities for Improvement.* CAB International, Wallingford, UK, pp. 145–162.

Seckler, D. (1996) *The New Era of Water Resources Management: from 'Dry' to 'Wet' Water Savings.* Research Report 1. International Irrigation Management Institute, Colombo, Sri Lanka.

Shi, H. and Singh, A. (2003) Status and interconnections of selected environmental issues in the global coastal zones. *Ambio* 32(2), 145–152.

Shiklomanov, I. (2000) Appraisal and assessment of world water resources. *Water International* 25(1), 11–32.

SIWI–IWMI (2004) Water – more nutrition per drop. In: *12th Session of the Commission on Sustainable Development (CDD).* New York: Stockholm International Water Institute (SIWI) and International Water Management Institute (IWMI). <http://www.siwi.org/downloads/More_Nutrition_Per_Drop.pdf>

Smakhtin, V., Revenga, C. and Doll, P. (2004) *Taking into Account Environmental Water Requirements in Global-scale Water Resources Assessments.* Comprehensive Assessment of Water Management in Agriculture Research Report No. 2. International Water Management Institute, Colombo, Sri Lanka, 24 pp.

Tharme, R.E. (2003) A global perspective on environmental flow assessment: emerging trends in the development and application of environmental flow methodologies for rivers. *River Research and Applications* 19, 397–441.

UN (2002) *UN Millennium Development Goals. Implementation of the United Nations Millennium Declaration.* Report of the Secretary-General, pp. 22–35. <http://www.un.org/millenniumgoals/sgreport2002.pdf?OpenElement>

UNEP-WCMC (2000) *World Commission on Dams and Development: a New Framework for Decision-Making.* Published in association with the World Commission on Dams, 404 pp.

Wiseman, R., Taylor, D. and Zingstra, H. (eds) (2003) Special issue on wetlands and agriculture. *International Journal of Ecology and Environmental Science* 29, 3–8.

World Bank (2004) *World Development Report 2004: Making Services Work for Poor People.* Oxford University Press, UK, 288 pp.

WRI (World Resources Institute) (2001) *World Resources 2000–2001: People and Ecosystems – the Fraying Web of Life.* UNDP, UNEP, WB and WRI. Washington, DC.

Index

Note: page numbers in *italics* refer to tables, figures, maps and boxes

acid sulphate soils 6–7, 51–52
  distribution 115
  flood plain drainage
    cooperation between farmers and researchers 116–117
    drain-water quality 102–103, *104*
    environmental impact 100–101, 110–112, *113*
    institutional and policy developments 119–121
    lake sediment characteristics 103, *105*, *106*
    land management changes 118–119
    soil properties 102, 114–115, 117–118
  knowledge gaps 114–115
  reclamation 57–58
  sugar industry guidelines 119
Africa: mangroves
  ambivalent policies 171
  defective regulatory instruments 172–173
    sustainable alternatives 173–174
  designated protected areas 167, *168*
    Kaw estuary (French Guinea) 167, 169
    Menab mangrove wetland (Madagascar) 169–170
    Northern Rivers 170–171
  destruction of forests *3*, 4
  exploitation in antiquity 164–165
  heterogeneity of mangrove areas 171
  negative attitudes of white colonialists 165
agriculture: water required for production 294–295
aluminium: in water and soil 102–103
  toxic effects on fish 111–112
Asia: mangroves
  destruction 4, 126–128
  exploitation in antiquity 164
Asian Development Bank 88
assimilative capacity *see* environmental capacity

Australia
  coastal flood plain development 109–110
  coastal flood plain drainage
    Acid Sulphate Soil Management Committee 120–121
    attitudes of sugarcane farmers 115–116
    conference on acid sulphate soils 119–120
    cooperation between farmers and researchers 116–117
    drain-water quality 102–103, *104*
    environmental impact 100–101, 110–112, *110–113*
    institutional arrangements 110
    lake sediment characteristics 103, *105*, *106*
    land management changes 118–119
    measurements 101–102
    national strategy 121
    soil properties 102, 114–115, 117–118
    study areas 100, *101*
  coastal stewardship for conflict resolution 109
  guidelines
    Local Environment Plans 119
    wise coastal practice agreements 109
  importance of participatory approaches 108–109
  information lacking on coastal ecosystems 108, 113–115

Bangladesh
  coastal area 61, *63*, *239*
    polders *241*
  Coastal Embankment Project 73–74, 239–240
  employment and livelihoods 7, 242
  land use changes *64*
    conversion from rice to shrimp farming 4, 6, 7, 241

Bangladesh *continued*
loss of agricultural land 6, 238
over the 20th century 238–241
NGOs 283
rice cultivation: new strategies 73
area of study 74
average monthly river salinity *76*, *80*
changes in production and income 82, *83*
conclusions and recommendations 83–84
cropping system and water management 75–76
environmental impact 82–83
experimental set-up 77
farmer participation 76–77
groundwater fluctuation and salinity *79*
monitoring of water and soil salinity 77–78
previous cultivation practice 74–75
rainfall and evaporation pattern *75*
soil salinity in cropped and fallow lands 80, *81*
wet and dry season yields 80–82
yield assessment 79
seasonal variation in land use *240*
shrimp farming
associated problems 73
environmental impact 69–70, 242
gher system 61–62, 62–63, *64*
impact on field crops 63–65, *66*
impact on natural vegetation 65, 67
impact on occupation and income 68–69, 242
survey methodology 62
zoning
achieving consensus 245–246
based on land suitability 242–243
emerging concept 244–245
indicative land zones 246, *247*
programmes and policies 243–244
basins, river *see* river basins
Bayesian networks *see* networks, Bayesian
BMP (best management practice) 8–9
Borneo *see* Indonesia
Brazil: Mangrove Dynamics and Management (MADAM) project
background 154–155
conservation and management 160–161
mangrove products *156*
crab 155, 157–158
fish 158–159
molluscs 159
research area 155, *157*

Cambodia 279, 281
capacity, environmental *see* environmental capacity
capital, social 266–267
cashew 170
children, labour of 281
co-management 160–161, 284
Coastal Embankment Project (Bangladesh) 73–74, 239–240
Coastal Habitats and Resource Management (Thailand) 92
cockles 148–150, 151
Code of Conduct for Responsible Fisheries (FAO) 8, 259–260
codes of conduct
disadvantages 8–9
sugar industry guidelines 119
Thailand 94
wise coastal practice agreements 109
see also regulation
competition 278, 279
conflicts
application of New Institutional Economics in resolution 259, 261–262
case study: shrimp farming in India 262–264
in designated protected areas 167, 169–171
between rice farmers and shrimp producers 49, 93, 242
sources *251*
strategies for resolution
coastal stewardship 109
resource management domains *see under* resource management domains
conservation
Mangrove Dynamics and Management (MADAM) project, Brazil 160–161
mangrove forests 130, 134–135, *136*, 137, 268
designated protected areas 167–171
marine protected areas *285*
and poverty 285–286
regulatory problems 172–173
Convention on Biological Diversity 172
crab 155, 157–158
criminalization 281

decentralization 265, 266, 268–270
decisions: Bayesian networks as support systems *see* networks, Bayesian
devolution 265, 266–267
doi moi (renovation) policy (Vietnam) 30
drainage, coastal flood plain *see under* acid sulphate soils
Drainage Act 1904 (Australia) 110
drains, secondary and tertiary 110, *111*
drought 170

Earth Summit 194
Ecological Mangrove Reserve (REMECAM) *see under* Ecuador: shrimp farming

ecotourism 173
Ecuador: shrimp farming
  impact study: Ecological Mangrove Reserve (REMECAM)
    artisanal fishery 147–148
    cockle gathering 148–150, 151
    community responses 150–151
    demographics 143–144, *145*
    methodology 142–143
    natural resource use and allocation 146
    perceptions of shrimp farming 146
    study area 141–142
    use and perceptions of mangrove ecosystem 144–146
  negative impacts on estuarine resources 10
education and training 137–138 see also *under* integrated coastal zone management (ICZM)
effluent: standards and regulations 8
employment and livelihoods
  changes in livelihood structure 21–23, *25*, 68–69
  child labour 281
  competition and exploitation 278
  data collection techniques 19–20
  diversity 275–276
  economic mobility 44, *45*
  impact of salinity control *see under* Vietnam
  insecurity 279–280
  key trends *296*
  livelihood diversification 286–290
  mangrove dependency case studies *see under* Ecuador: shrimp farming; Thailand
  shrimp farming
    rice–shrimp system 42
    vs. rice cultivation 7
  Sustainable Livelihood Framework 31
  vulnerability 50
  when productivity falls 24–26
environmental capacity 11
environmental impact assessment 11
exclusion 280–281

fish and fisheries 170–171
  Code of Conduct for Responsible Fisheries (FAO) 8, 259–260
  contribution to food and livelihood security 295
  Ecological Mangrove Reserve (Ecuador) 147–148
  effect of acid pollution 57, 59, 100–101, 111–112
  effect of tidal sluices 54, 59
  fisheries management 284
  impacts of river basin changes 298
  insecurity of fishermen 279–280
  Mahakam Delta study of mariculture *vs.* fisheries *see under* Indonesia
  in mangrove areas 130, 158–159
  variables used Bayesian network decision support system 211–213
flood plains
  drainage of acid sulphate flood plains *see under* acid sulphate soils
  historical development 109–110
  institutional arrangements for drainage 110
food
  key trends *296*
  security 295
  water required for production 294–295
Food and Agriculture Organization
  approaches to zoning 194
  Code of Conduct for Responsible Fisheries 8, 259–260
  FAO Framework for Land Evaluation 178
  land-use planning guidelines 180
French Guinea 167, 169
fuel-wood 145

gender: imbalances 281–282
Ghana 279–280, 281, 283
gher shrimp culture system 61–62
Guinea 170
Guinea Bissau 170

heritage, natural: concept 166–167
  legal status 172
Honduras 6

incentives 11
India
  exclusion 280–281
  fishing 279, 280–281, 282, 283
  Integrated Coastal Zone Management and Training Project
    aims and objectives 250–251
    course design 255–256
    training framework 252–254
    training the trainers 251–252
    training tools 254–255
    virtual scenario case study approach *252*
  mangrove protection 268
  shrimp farming
    application of New Institutional Economics 262–264, 267–270
    social impact 7
Indonesia
  Mahakam Delta (Borneo) study of mariculture *vs.* fisheries
    background 219–221
    changes in land use 1984–2004 *224*
    data collected 222–224
    data processing and analysis 224–225
    estimated pond area development *223*

Indonesia *continued*
Mahakam Delta (Borneo) study of mariculture *vs.* fisheries *continued*
fisheries: production trends 227, 230–233
mariculture: production trends 225–227, *228*, *229*, *230*
study area 221–222
study conclusions 233–235
tambak integrated shrimp mangrove system 9
information: deficits in knowledge of coastal ecosystems 108, 113–115
institutions: analysis *see* New Institutional Economics
integrated coastal zone management (ICZM) 9–10
background 250
decentralization and devolution 265–267
implementation 260
limitations with respect to conflict 260–264
in relation to poverty 284–287
training
aims and objectives 250–251
course design 255–256
framework 252–254
inappropriate focus on scientific knowledge 249–250
tools 254–255
of trainers 251–252
uptake at international level 259–260
virtual scenario case study approach *252*
International Convention on Wetlands (1971) 166
iron: in water and soil 102–103

Kaw-Roura Swamplands Natural Reserve (French Guinea) 167, 169

labour, child 281
land-use planning
guidelines 178, 180
Mekong Delta study *see under* Vietnam
conflicts in preference for land use 183, 185, *186*
land suitability classification *181*
methodology: comparison of approaches used 188–191
methodology: FAO-MCE 180–182, *183*, *184*
methodology: LUPAS 185, *187*, 188
methodology: PLUP 182–185, *186*
methodology overview 180
livelihoods *see* employment and livelihoods
LUPAS (land-use planning and analysis system) 178, 180, 185, *187*, 188–191

Madagascar 169–170
management
co-management *see* co-management
integrated coastal zone *see* integrated coastal zone management (ICZM)
mangroves
ambivalent policies 171–172
conservation and replanting 130, 134–135, *136* 137, 160–161, 268
conflict of interests in designated protected areas 167, 169–171
regulatory problems 172–173
sustainable alternatives 173–174
degradation due to forest concessions 129
ecological functions *146*
economic significance 18–19
dependency case studies *see under* Ecuador: shrimp farming; Thailand
estimated losses in selected countries *3*
heterogeneity of mangrove areas 171
history
exploitation and loss 3, 4, 129
negative attitudes of white colonialists 165
places of residence 164–165
rehabilitation 166
impact of shrimp farming
environmental effects 5, 88
forest destruction 2, 3–4, 126–128, 130, 166
importance for shrimp and fish 61, 234–235
importance of consensus between stakeholders 220
integrated shrimp–mangrove systems 9
a 'natural heritage' 163–164, 166–167, 171–172
products *156*
crab 146, 155, 157–158, 170
fish 146, 147, 158–159
geographical indicators 173
molluscs 146, 148–150, 151, 159
wild shrimp and prawns 147–148
wood and charcoal 143, 144–145, 160
social impact of forest destruction 7
stewardship programmes 150–151
tourism 173
use and perceptions of mangrove ecosystem 144–146
Menab mangrove wetland (Madagascar) 169–170
metals: in water and soil 102–103
migrants 280, *281*, 282, 283
Millennium Development Goals 294
molluscs 148–150, 151, 159

networks, Bayesian
advantages 215–216
Bayfish-Bac Lieu model (Vietnam)
consultation process 209–210
variable: definition and weighting 211–213
variable: parameterization 214–215
variable: specification 213, *214*
variables: network *212*
water and food model 210–211
examples of mini-networks *208*
principles 207–209

New Institutional Economics
application to conflict resolution 259, 261–262
case study: shrimp farming in India 262–264, 267–270
decentralization and devolution 265–267
definition and scope 260–261
non-governmental organizations (NGOs) 283–284

open-access resources 22–23
problems of enclosure 26–27
oysters 159

pesticides: as model variable 213, 214
planning *see* land-use planning
PLUP (participatory land-use planning) 178, 180, 182–185, *186*, 188–191
pollution 6, 277
populations 277
poverty 13
in areas of salinity control 38, 40–41, *42*, 56–57
and diversity 275–277
impact of change on the poor 278–282
and integrated coastal zone management 284–287
livelihood enhancement and diversification 286–290
poverty dynamics analysis 31–32, 41–45
responses to coastal poverty 282–284
and vulnerability 50
production: as model variable 214–215

rainfall: as model variable 213, *214*
Ramsar Convention 166
regulation
difficulties 13
effluent 8
see also codes of conduct
Local Environment Plans (Australia) 119
replanting: mangrove forests 130, 134–135, *136*, 137
reserves, extractive 160–161
resource management domains
applied to Mekong Delta, Vietnam
at broad level 196–197
conflicts over land use 195–196
at detailed level 197–202
hamlet clusters *196*, *199*, *200–201*
definition and concept 194–195
role in planning and management process 202–204
rice
abandonment of rice fields 170
acid generation from rice fields 57–58
in areas of controlled salinity 33–34
conflict between rice farmers and shrimp producers 49
conversion of farms to shrimp farming 4, 49, 50–51, 90
environmental impact 6
decreased yield attributed to salinization 63
new cultivation strategies 73
cropping system and water management 75–76
environmental impact 82–83
experimental set-up and parameters 77–79
farmer participation 76–77
variations in water and soil salinity 79–80, *81*
yields and farmers' income 80–82, *83*
productivity
rice-shrimp system 41–42, *43*
short-duration high-yield varieties 73, 76
characteristics *78*
dry season yields 81–82
wet season yields 80, *81*
Rio Earth Summit 163, 259
river basins
closure 296–297
impact of changes on coastal regions 298–299
irrigation: issues of scale 297–298

salinity
addressed by Coastal Embankment Project (Bangladesh) 73–74
average monthly salinity, Kazibachha River, Bangladesh *76*, *80*
cropped *vs.* fallow lands 80
effect of new rice cultivation strategies 82–83
groundwater salinity and water management strategies *79*
impact studies of salinity control *see under* Vietnam
increased by lack of rain 170
monitoring of water and soil 77–78
salinization due to shrimp farms 6, 7, 55
blamed for poor crop yields 63–65
Saloum National Park (Senegal) 171
Senegal 170, 171
sesame 74–75
ship bore worm (*turu*) 159
shrimp farming
application of New Institutional Economics 262–264, 267–270
concerns caused by rapid expansion 2, 127
conflict with rice farmers 49
a critical land-use issue 12
culture systems 2–3
gher system (Vietnam) 61–62
integrated shrimp-mangrove systems 9
intensive systems 88–89
low-salinity systems 89–91
rice-shrimp systems 41–42, *43*
semi-intensive culture 87–88
thammachaat ('natural' shrimp farms, Thailand) 87

shrimp farming *continued*
economic importance 17
farm characteristics *56*
in freshwater areas 91–92
government subsidies 128
growth rates 1970–2000 2
impact of upstream users 6
impact on developing countries 140
investment patterns 127–128
Mahakam Delta study of mariculture *vs.* fisheries *see under* Indonesia
major producers *3*
major source of income and employment 2
negative impacts: environment 5–7, 55–56, 69–70, 88, 242
acid pollution 57
field crops and vegetation 63–67
livestock and poultry 67–68
mangrove destruction 3–4, 126–128, 130, 166
negative impacts: humans *5*, 24–26, 56–57, 68–69, 242
Ecuador study *see under* Ecuador: shrimp farming
perceptions of local communities 146
productivity 20–21
rice–shrimp system 41–42, *43*
regulations and codes of practice 8–9, 94
rice farm conversion 4, 6, 49, 50–51, 90
sources of water pollution 6
sustainability 55–57, 89, 92–95
Sierra Leone 170
silvo-fishery systems 9
'snowball sampling' 19
social capital 266–267
soils, acid sulphate *see* acid sulphate soils
Sri Lanka 10
stewardship, coastal 109, 150–151
strategic environmental assessment 11
sugarcane
attitudes of cane farmers 115–116
best practice guidelines 119
effects of industry on fish 112
land management changes 118–119
survives on shallow groundwater 117
sulphides: in water and soil *see* acid sulphate soils
sustainability
essential requirements 49–50
key challenges in coastal zone management 108
of livelihoods 287–290
shrimp production 55–57, 89, 92–95
problems 127
strategies
aquaculture planning 92
social organization 93–95
water supply infrastructure 92–93
Sustainable Livelihood Framework 31

tambak system 9, 222–223, *224*
production trends 225–227
Taura syndrome virus 92
technology: changes 278
Thailand
conversion from rice to shrimp farming 4, 6
integrated coastal zone management 10
mangrove dependency case studies
conclusions and policy implications 135–138
conservation and replanting 130, 134–135, *136*, 137
labour allocation: impact of mangrove loss 133–134
labour allocation and employment 130–133
mangrove degradation due to forest concessions 129
mangrove loss due to shrimp farming 126–128, 130
methodology 128
study villages 128–129
salinization 6
shrimp farming
codes of conduct 94
in freshwater areas 91–92
government subsidies 128
initiatives to improve sustainability 89
intensive *(phattana)* systems: 1988–1995 88–89
investment patterns 127–128
lack of sustainability 127
low-salinity systems: 1996–2002 89–91
production levels *88*, *90*
recent developments: 2003–2004 91–92
semi-intensive culture: 1972–1987 87–88
strategies for sustainability 92–95
traditional systems: 1930–1971 87
water supply conflicts 93
thammachaat ('natural' shrimp farms, Thailand) 87
tin: mining concessions 128
tourism 173
Traditional Ecological Knowledge 172
training *see* education and training
turu 159
Tweed River Advisory Committee (Australia) 112

United Nations Conference on Environment and Development (UNCED, 1992) 194, 259

Vietnam
changes in livelihood structure 21–23, *25*
conflict resolution using resource management domains *see under* resource management domains
decision support system based on Bayesian networks *see under* networks, Bayesian
doi moi (renovation) policy 30

economic significance of mangrove products 18–19
ethnic minorities *23*
land owners *22, 23*
land use and property systems 18
land-use planning in Mekong Delta *see under* land-use planning
landless households *23*
map of case study sites *19*
Mekong Delta land-use planning study
  background 177–178
  study area 178–180
open-access resources 22–23, 26–27
salinity control: impact studies
  acid sulphate soils 51–52, 57–58
  capital endowment 34–35, *36*
  changes in saline intrusion with time *50*
  conclusions 45–46
  conflict between shrimp producers and rice farmers 49
  effects on fish and fisheries 54, 57, 59
  environmental management strategies *54*
  household incomes 38, *39*, *40*, 44, *45*
  livelihood strategies 35–38, 54–55
  methodology 32, *33*, *34*, 52–54
  poverty 38, 40–41, 42, *42*, 44, 56–57
  production system changes 32–34, 49, 50–51
  short-term impact 54–55
  yields and productivity 41–42
shrimp farming
  conversion from rice cultivation 4, 49, 50–51
  economic importance 17
  farm characteristics *56*
  financial support 18
  integrated shrimp–mangrove systems 9
  negative impacts 24–26, 55–57
  productivity 20–21
  productivity of rice-shrimp system 41–42, *43*
social impact of ecosystem changes 7
viruses 92
vulnerability 50

water
  key trends *296*
  on-farm storage of surface water 75, 77, 83
  pollution by shrimp farms 6
  quality of drain-water from acid sulphate soils 102–103, *104*
  requirements for food production 294–295
  supply infrastructure 92–93
  see also river basins
wetlands: loss through drainage 295–296

zoning 10–11, 12–13, 92
  based on land suitability 242–243
  FAO agro-ecological zones (AEZ) 194
  indicative land zones in Bangladesh 246, *247*
  programmes and policies 243–245
  use of resource management domains *see* resource management domains